SOLAR DYNAMICS AND ITS EFFECTS ON THE HELIOSPHERE AND EARTH

Space Sciences Series of ISSI

Volume 22

The International Space Science Institute is organized as a foundation under Swiss law. It is funded through recurrent contributions from the European Space Agency, the Swiss Confederation, the Swiss National Science Foundation, and the University of Bern. For more information, see the homepage at http://www.issi.unibe.ch/.

SOLAR DYNAMICS AND ITS EFFECTS ON THE HELIOSPHERE AND EARTH

Edited by

D. N. BAKER
Laboratory for Atmospheric and Space Physics, Boulder CO, USA

B. KLECKER
Max-Planck-Institute für extraterrestriche Physik, Garching Germany

S. J. SCHWARTZ
Blackett Laboratory, Imperial College London, London, UK

R. SCHWENN
Max-Planck Institut für Sonnensystemforschung, Katlenburg-Lindau, Germany

R. VON STEIGER
International Space Science Institute (ISSI), Bern, Switzerland

Reprinted from *Space Science Reviews*, Volume 124, Nos. 1–4, 2006

Springer

INTERNATIONAL SPACE SCIENCE INSTITUTE
Space Sciences Series of ISSI

A.C.I.P. Catalogue record for this book is available from the Library of Congress

ISBN: 978-0-387-69531-0

Published by Springer
P.O. Box 990, 3300 AZ Dordrecht, The Netherlands

Sold and distributed in North, Central and South America
by Springer,
101 Philip Drive, Norwell, MA 02061, U.S.A.

In all other countries, sold and distributed
by Springer,
P.O. Box 322, 3300 AH Dordrecht, The Netherlands

Printed on acid-free paper

Printed in the Netherlands

TABLE OF CONTENTS

Foreword vii–viii

SECTION I: SOLAR DYNAMICS

A. G. KOSOVICHEV and T. L. DUVALL, JR / Active Region Dynamics. *Recent Helioseismology Results* 1–12

N. O. WEISS / Sunspot Structure and Dynamics 13–22

L. GOLUB / Coronal Dynamics and the AIA on SDO 23–33

ESTER ANTONUCCI / Wind in the Solar Corona: Dynamics and Composition 35–50

SECTION II: SOLAR WIND CHAIN

R. SCHWENN / Solar Wind Sources and Their Variations Over the Solar Cycle 51–76

THOMAS H. ZURBUCHEN / Heliospheric Physics: Linking the Sun to the Magnetosphere 77–90

RUMI NAKAMURA / Substorms and Their Solar Wind Causes 91–101

J. BIRN, M. HESSE and K. SCHINDLER / Modeling of the Magnetospheric Response to the Dynamic Solar Wind 103–116

M. LESTER, S. E. MILAN, G. PROVAN and J. A. WILD / Review of Ionospheric Effects of Solar Wind Magnetosphere Coupling in the Context of the Expanding Contracting Polar Cap Boundary Model 117–130

SECTION III: CME CHAIN

S. E. GIBSON, Y. FAN, T. TÖRÖK and B. KLIEM / The Evolving Sigmoid: Evidence for Magnetic Flux Ropes in the Corona Before, During, and After CMEs 131–144

NAT GOPALSWAMY / Properties of Interplanetary Coronal Mass Ejections 145–168

H. E. J. KOSKINEN and K. E. J. HUTTUNEN / Geoeffectivity of Coronal Mass Ejections 169–181

IOANNIS A. DAGLIS / Ring Current Dynamics 183–202

J. GOLDSTEIN / Plasmasphere Response: Tutorial and Review of Recent Imaging Results 203–216

M. WILTBERGER and D. BAKER / End-to-End Modeling of the Solar Terrestrial System 217–231

SECTION IV: SEP CHAIN

R. P. LIN / Particle Acceleration by the Sun: Electrons, Hard X-rays/ Gamma-rays 233–248

PETER J. CARGILL, LOUKAS VLAHOS, RIM TURKMANI, KLAUS GALSGAARD and HEINZ ISLIKER / Particle Acceleration in a Three-Dimensional Model of Reconnecting Coronal Magnetic Fields 249–259

M. I. DESAI, G. M. MASON, J. E. MAZUR and J. R. DWYER / The Seed Population for Energetic Particles Accelerated by CME-driven Shocks 261–275

JOE GIACALONE and JÓZSEF KÓTA / Acceleration of Solar-Energetic Particles by Shocks 277–288

B. KLECKER, E. MÖBIUS and M. A. POPECKI / Solar Energetic Particle Charge States: An Overview 289–301

R. A. MEWALDT / Solar Energetic Particle Composition, Energy Spectra, and Space Weather 303–316

SECTION V: COMMONALITIES

GERHARD HAERENDEL / Commonalities Between Ionosphere and Chromosphere 317–331

STEVEN J. SCHWARTZ / Shocks: Commonalities in Solar-Terrestrial Chains 333–344

J. BÜCHNER / Theory and Simulation of Reconnection. *In memoriam Harry Petschek* 345–360

MARKUS J. ASCHWANDEN / The Localization of Particle Acceleration Sites in Solar Flares and CMEs 361–372

FOREWORD

The SOHO and Cluster missions form a single ESA cornerstone. Yet they observe very different regions in our solar system: the solar atmosphere on one hand and the Earth's magnetosphere on the other. At the same time the Ulysses mission provides observations in the third dimension of the heliosphere, and many others add to the picture from the Lagrangian point L1 to the edge of the heliosphere. It was our aim to tie these observations together in addressing the topic of Solar Dynamics and its Effects on the Heliosphere and Earth with a workshop at the International Space Science Institute (ISSI), under the auspices of the International Living With a Star (ILWS) program. It started out with an assessment and description of the reasons for solar dynamics and how it couples into the heliosphere. The three subsequent sections were each devoted to following one chain of events from the Sun all the way to the Earth's magnetosphere and ionosphere: The normal solar wind chain, the chain associated with coronal mass ejections, and the solar energetic particles chain. The final section was devoted to common physical processes occurring both at the Sun and in the magnetosphere such as reconnection, shock acceleration, dipolarisation of magnetic field, and others.

This volume is the result of an ISSI Workshop held in April 2005. An international group of about forty experimenters, ground-based observers, and theoreticians was invited to present and debate their data, models, and theories in an informal setting. The group was convened by Madhulika Guhathakurta (NASA HQ), Gerhard Haerendel (then at IU Bremen), Hermann Opgenoorth (ESA-ESTEC), Roger M. Bonnet, Götz Paschmann, and Rudolf von Steiger (all ISSI).

It is a pleasure to thank all those who have contributed to this volume and to the workshops in general. First of all, we thank the authors for writing up their contributions. All papers were peer-reviewed by referees, and we thank the reviewers for their critical reports. We also thank the directorate and staff of ISSI for selecting this topic for a workshop and for their support in making it happen, in particular Roger M. Bonnet, Brigitte Fasler, Vittorio Manno, Saliba F. Saliba, Irmela Schweizer, and Silvia Wenger.

December 2006

D. N. Baker, B. Klecker, S. J. Schwartz, R. Schwenn and R. von Steiger

Group photograph; from left to right (nose tip counts): Markus Aschwanden, Rainer Schwenn, Hannu Koskinen, Hermann Opgenoorth, Alexander Kosovichev, Peter Cargill, Mark Lester, Ester Antonucci, John Leibacher, Rudolf von Steiger, Steve Schwartz, Joachim Birn, Rumi Nakamura, Dan Baker, Mihir Desai, Roger-Maurice Bonnet, Lika Guhathakurta, Sarah Gibson, Thomas Zurbuchen, Silvia Wenger, Berndt Klecker, Jerry Goldstein, Brigitte Fasler, Yannis Daglis, Richard Mewaldt, Jon Linker, Götz Paschmann, Ruth Esser, Jörg Büchner, Bob Lin, Dave Sibeck, Joe Giacalone, Nat Gopalswamy, Bernhard Fleck, Mike Wiltberger, Gerhard Haerendel (picture taken by Stein Haaland).

ACTIVE REGION DYNAMICS
Recent Helioseismology Results

A. G. KOSOVICHEV[1,*] and T. L. DUVALL, JR[2]
[1]*Stanford University, Stanford, CA 94305, USA*
[2]*Laboratory for Astronomy and Solar Physics, NASA/Goddard Space Flight Center, Greenbelt, MD 20771, USA*
*(*Author for correspondence, E-mail: akosovichev@solar.stanford.edu)*

(Received 20 October 2005; Accepted in final form 7 March 2006)

Abstract. New methods of local helioseismology and uninterrupted time series of solar oscillation data from the Solar and Heliospheric Observatory (SOHO) have led to a major advance in our understanding of the structure and dynamics of active regions in the subsurface layers. The initial results show that large active regions are formed by repeated magnetic flux emergence from the deep interior, and that their roots are at least 50 Mm deep. The active regions change the temperature structure and flow dynamics of the upper convection zone, forming large circulation cells of converging flows. The helioseismic observations also indicate that the processes of magnetic energy release, flares and coronal mass ejections, might be associated with strong (1–2 km/s) shearing flows, 4–6 Mm below the surface.

Keywords: Sun: activity, Sun: heliseismology, Sun: interior, Sun: magnetic field, sunspots

1. Introduction

Active regions are the most important source of heliospheric disturbances. They are formed by magnetic fields generated by dynamos in the convection zone and emerging from the Sun's interior. Magnetic field topology and magnetic stresses in the solar atmosphere are likely be controlled by motions of magnetic flux footpoints in the sub-photosphere. However, the depth of these motions is unknown. Twisting and shearing of the magnetic field of active regions by subphotospheric motions as well as its interaction with new emerging magnetic field result in flares and CMEs. Helioseismology provides tools for diagnosing the subsurface structures and dynamics, and allows us to investigate the origin of solar magnetic fields, formation and evolution of active regions, the relationship between the internal dynamics and activity, and to develop methods for predicting the emergence and evolution of active regions and their activity. The helioseismic investigation of the dynamics of active regions has only just begun, and the results are still very preliminary. However, we are beginning to develop a new understanding of the lifecycle of active regions, their emergence, evolution and decay, as well as the relationship between their activity and internal dynamics. Specifically, some of the questions that are studied by local helioseismology are:

Space Science Reviews (2006) 124: 1–12
DOI: 10.1007/s11214-006-9112-z

- How deep are the roots of sunspots and active regions?
- How fast do active regions emerge?
- What is the basic mechanism of formation of active regions: are they formed by a large magnetic Ω-loop breaking into smaller parts near the surface, or by merging together fragmented small-scale magnetic structures in the subphotospheric layers?
- Why do active regions tend to appear in the same place forming long-living complexes of activity ('active longitudes')?
- How are the twisted ('δ-type') magnetic configurations, which produce the most energetic flares and CMEs, formed?
- How can surface and subsurface plasma flows affect stability and magnetic energy release of active regions?
- How do sunspots and active regions decay?
- What determines the dissipation time scale, and is there submergence of magnetic flux when active regions decay?

In this article, we present some recent results on the dynamics of active regions obtained by time-distance helioseismology (Duvall *et al.*, 1993), addressing some of these questions.

2. New Methods of Investigating Solar Dynamics

Time-distance helioseismology measures travel times of acoustic waves propagating to different distances, and uses these measurements to infer variations of the wave speed along the wave paths. Turbulent convection excites acoustic waves which propagate deep into the solar interior. Because the sound speed increases with depth these waves are refracted and come back to the solar surface. The wave speed depends on temperature, magnetic field strength and flow velocity field in the region of the wave propagation. By measuring reciprocal travel times of acoustic waves propagating along the same ray paths in opposite directions, and then taking the mean and the difference of these travel times it is possible to separate the flow velocity (advection) effect from temperature and magnetic field perturbations (Kosovichev and Duvall, 1997). However, in order to disentangle contributions of temperature and magnetic field to the mean travel times it is necessary to measure the travel-time anisotropy, and this has not been accomplished. Therefore, the current helioseismic results represent maps of sub-photospheric variations of the sound (magneto-acoustic) speed and flow velocities.

The travel times are typically measured from a cross-covariance function of solar oscillation signals for various distances and time lags. When for a given distance the time lag corresponds to the propagation time of acoustic waves for this distance, a wavepacket-like signal appears in the cross-covariance function. The

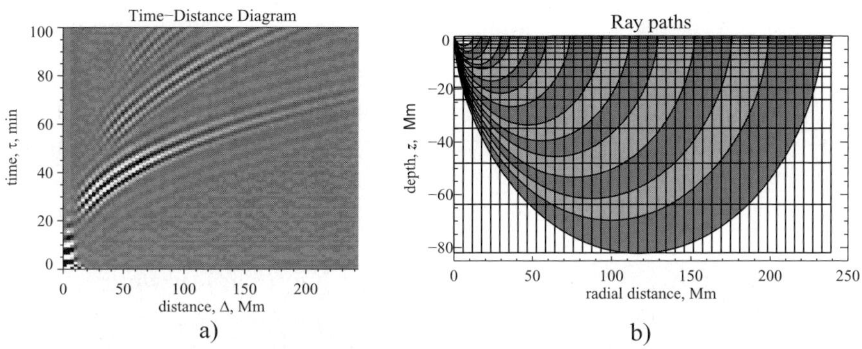

Figure 1. (a) Cross-covariance functions of solar oscillations as a function of distance between measurement points on the solar surface and time lag. The lowest ridge is formed by the acoustic wave packets propagating between these points through the solar interior. The solid curve shows the time-distance relation in the ray approximation. The higher ridges are formed by the wave packets with additional bounces at the surface. (b) a sample of acoustic ray paths used for time-distance helioseismology, shown in a vertical plane. The shadowed regions illustrate ranges of averaging. The vertical and horizontal lines show a grid used for inversion of acoustic travel time data.

cross-covariance plotted as a function of the distance and time lag displays a set of ridges formed by the wave-packet signals (Figure 1a), representing an analog of a solar 'seismogram'. Since the solar oscillations are stochastic it is necessary to use the oscillation signals at least 4–8 hour long and also average them over some surface (typically, circular) areas in order to obtain a sufficient signal-to-noise ratio. Then, the travel times are determined by fitting a wavelet to this function (e.g. Kosovichev and Duvall, 1997), or by measuring displacement of the ridges (Gizon and Birch, 2002).

The relationship between the observed travel-time variations and the internal properties of the Sun is given by so-called sensitivity kernels through integral equations. These integral equations are solved by standard mathematical inversion techniques such as LSQR and Multi-Channel Deconvolution (MCD) (Couvidat *et al.*, 2004). The sensitivity functions are calculated using a ray theory (Figure 1b) or more complicated wave perturbation theories, e.g. Born approximation, which takes into account the finite wave-length effects (Birch and Kosovichev, 2000). These theories can also take into account stochastic properties of acoustic sources distributed over the solar surface (Gizon and Birch, 2002; Birch *et al.*, 2004).

3. Lifecycle of Active Regions

Helioseismic observations show that the flow dynamics changes during the evolution of active regions. One of the important tasks is to develop diagnostics of emerging active regions in the interior. For space weather predictions it would be

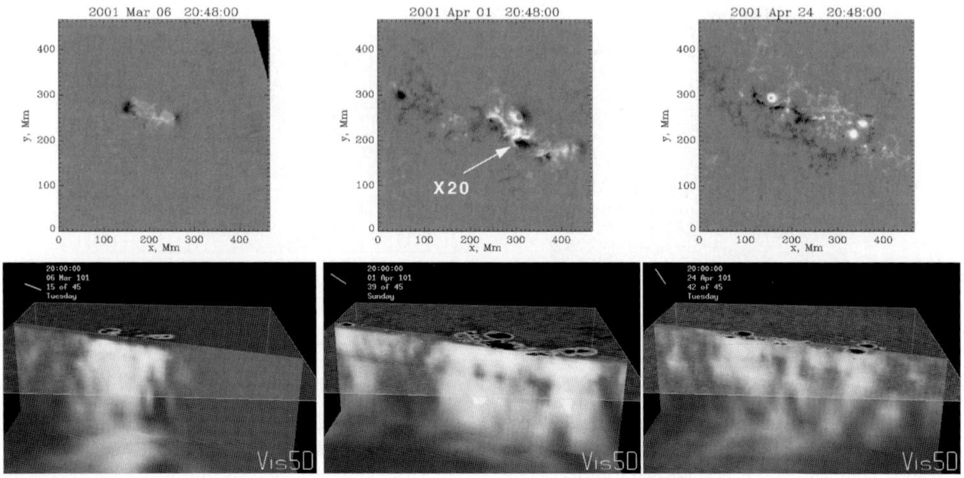

Figure 2. The sound-speed perturbation below the surface and photospheric magnetograms in the emerging active region NOAA 9393. The upper panels are MDI magnetograms showing the surface magnetic field of positive (light) and negative (dark) polarities. The perturbations of the sound speed shown in the vertical cut and the bottom horizontal panel, are approximately in the range from –1 to +1 km/s. The positive variations are shown in light color, and the negative ones in dark. The top (semitransparent) panels are white-light images, the bottom panels show the sound-speed maps 57 Mm deep. The arrow shows the location of the powerful X20 flare on April 2, 2001.

very important to detect active regions before they emerge. However, this task has proven to be very difficult because the emerging magnetic flux propagates very rapidly in the upper convection zone with a speed exceeding 1 km/s (Kosovichev *et al.*, 2000).

Here we present as an example the evolution of active region NOAA9393, which was observed in March to May 2001 during the Dynamics Program for the MDI instrument on SOHO. Almost uninterrupted series of full-disk Dopplergrams with the resolution 2 arc sec per pixel and 1-min cadence were obtained. For the time-distance analysis, travel distances from 0.3 to 24 degrees were used (Kosovichev and Duvall, 2003). The inversion results (Figures 2–6) produced 3D maps of the sound-speed variations and mass flows in a cube of $400 \times 400 \times 80$ Mm for 3 periods, when the active region was on the front side of the Sun, during Carrington rotations 1973, 1974 an 1975. The total number of interior maps included in this analysis is 45. The integration time for a single map was 8 hours. Therefore, typically three maps per day were obtained. The analyzed dates are: March 2–6, March 25–April 1, and April 24–25, 2001. These include the periods of emergence, maximum activity and decay of this active region. The initial results show complicated patterns of rapidly evolving sound-speed perturbations most likely associated with multiple interacting magnetic flux tubes.

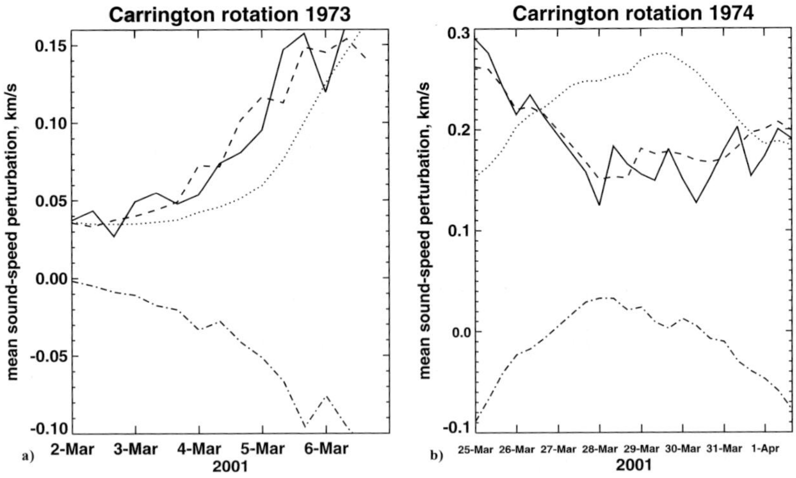

Figure 3. The evolution of the total unsigned magnetic flux (dotted curves) and the mean sound-speed perturbation at 0–3 Mm (dot-dashed curves), 4–12 Mm (dashed curves) and 15–34 Mm (solid curves), during Carrington rotations 1973 and 1974, periods of emergence and maximum development of the active region, NOAA 9393.

The evolution of the total photospheric magnetic flux and mean sound-speed perturbations at various depths in this active region is shown in Figure 3. It appears that during the emergence and development phase (Figure 3a) the sound-speed perturbations in the deeper layers, 4–34 Mm, grow somewhat faster than the magnetic flux, and in the subsurface layer (0–3 Mm) the sound-speed rapidly decreases. During the maximum phase (Figure 3b) the sound-speed behavior is opposite. It decreases in the deep interior in antiphase with the magnetic flux, and in the near-surface it changes almost in phase with the magnetic flux which, however, lags the sound-speed variations.

Time-distance helioseismology also provided maps of plasma flows beneath this active region. Figure 4 shows the distribution of the photospheric magnetic field and horizontal and vertical flow maps in the subsurface layers 2 and 6.4 Mms deep, shown in Figure 4b and c. Beside the usual supergranular flows these maps do not reveal any specific flow pattern that could be associated with emergence of a large-scale structure, e.g. a large-scale outflow or upflow. However, a localized shearing flow appears at the place of emergence. Soon after the emergence, the dominant flow pattern consists of converging downflows around the active regions (Kosovichev, 1996; Zhao *et al.*, 2001). Figure 5 shows the active region dynamics during the maximum activity phase. The flow structure is quite complicated. In addition to the converging downflows surrounded by upflows we see a diverging flow around a rapidly evolving leading spot. Also, there is evidence for strong shear flows in the central part of this region where a very strong flare occurred 3 days later, on April 2. The decaying phase shown in Figure 6 is characterized by predominant outflows.

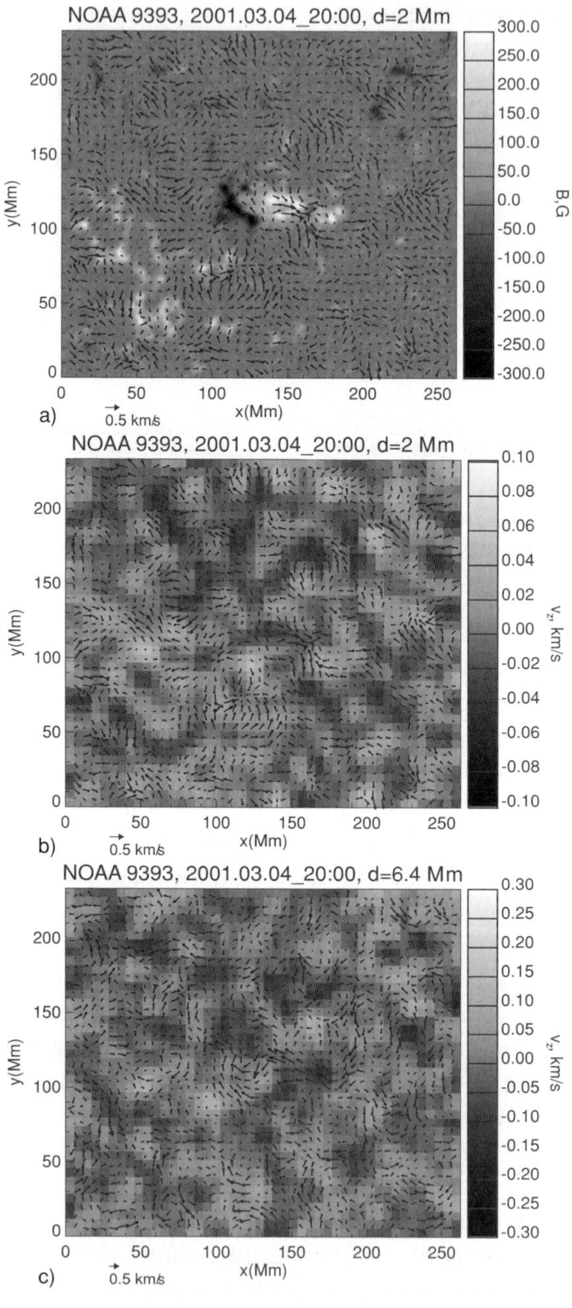

Figure 4. Emergence of active region NOAA 9393 (March 4, 2001): (a) the photospheric magnetic field and the horizontal velocity field at a depth of 2 Mm; (b) the vertical (the grayscale map; positive – upflows, negative – downflows) and horizontal velocity fields at the depth of 2 Mm; (c) the vertical and horizontal velocities at the depth of 6.4 Mm.

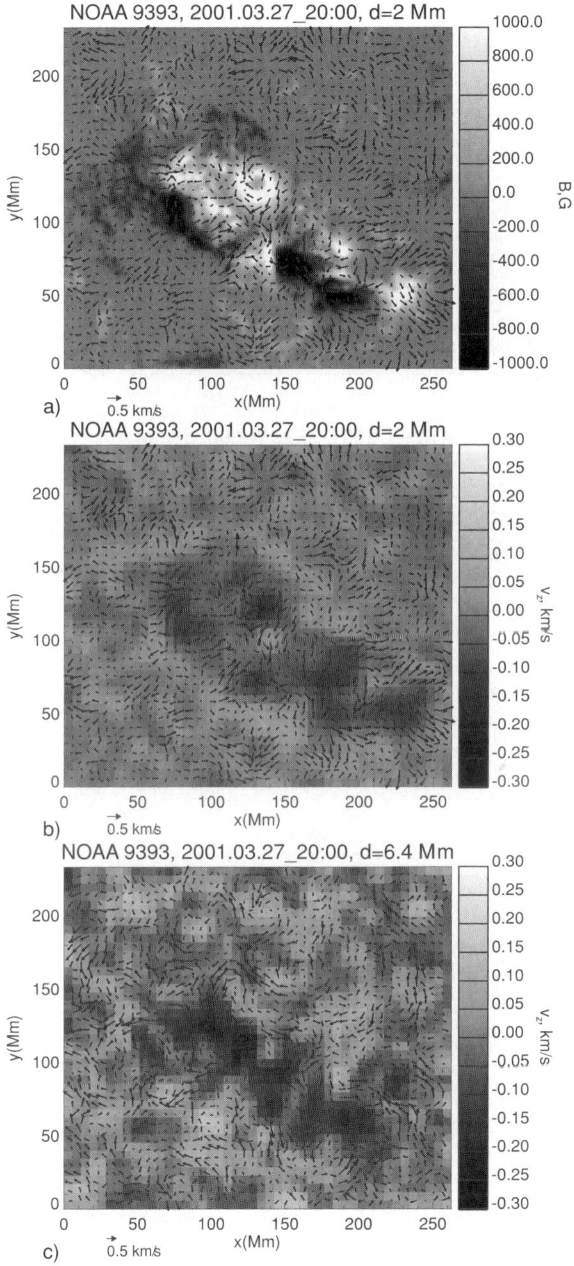

Figure 5. The maximum activity phase of AR 9393 (March 27, 2001): (a) the photospheric magnetic field and the horizontal velocity field at a depth of 2 Mm; (b) the vertical (the grayscale map; positive – upflows, negative – downflows) and horizontal velocity fields at the depth of 2 Mm; (c) the vertical and horizontal velocities at the depth of 6.4 Mm.

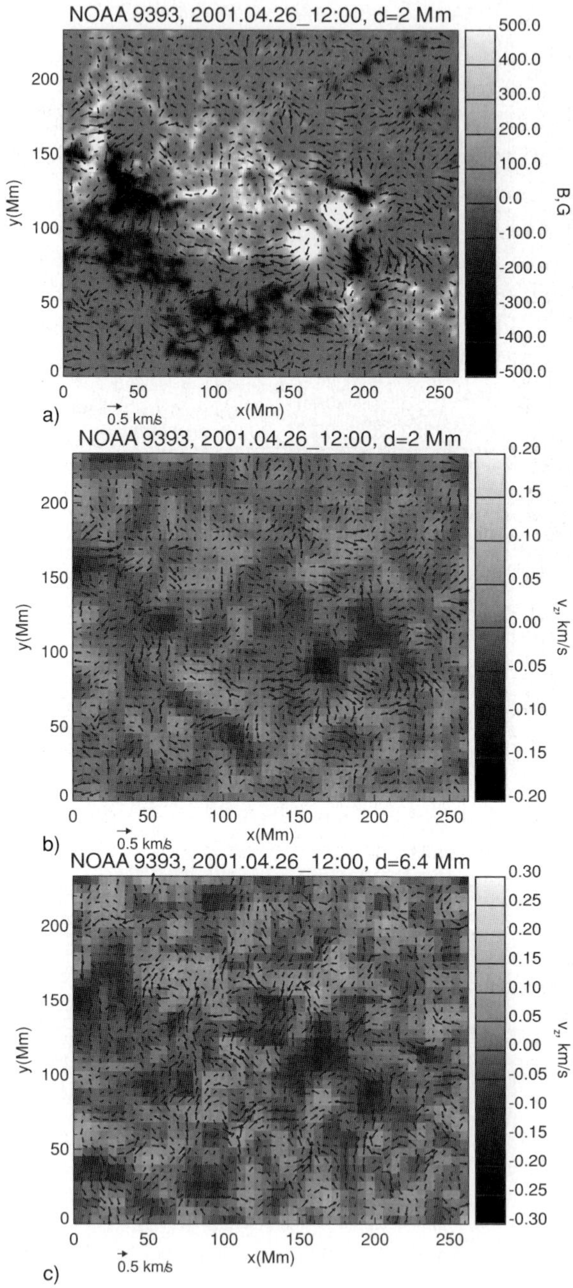

Figure 6. The decay phase of AR 9393 (April 26, 2001): (a) the photospheric magnetic field and the horizontal velocity field at a depth of 2 Mm; (b) the vertical (the grayscale map; positive – upflows, negative – downflows) and horizontal velocity fields at the depth of 2 Mm; (c) the vertical and horizontal velocities at the depth of 6.4 Mm.

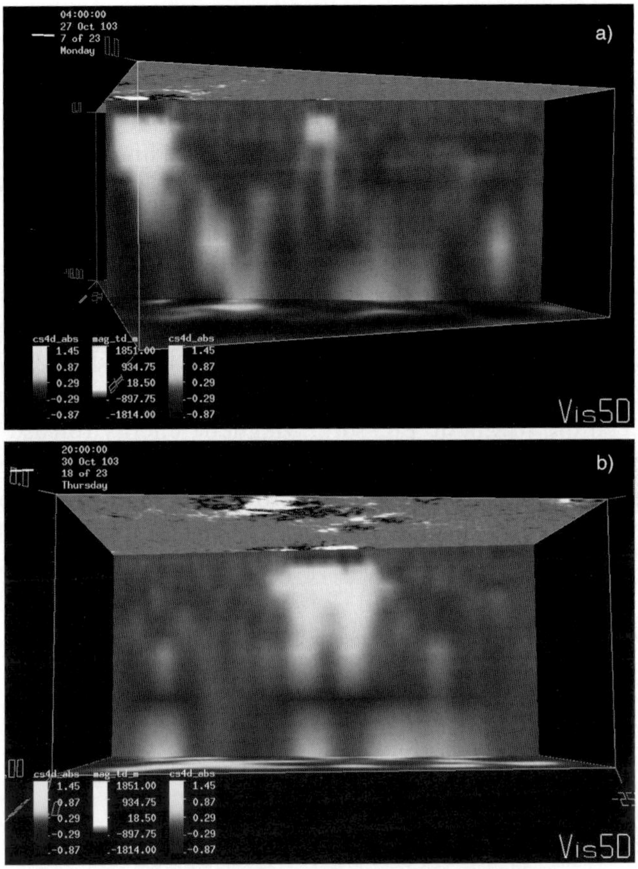

Figure 7. Sound-speed variations (vertical cuts) associated with developing active region NOAA 10488 (October 2003): (a) emergence of the active region in the middle of the domain (the structure near the right boundary is AR10486; (b) fully developed AR 10488. The depth of the box is 48 Mm, the horizontal size is about 540 Mm. The sound-speed scale is from –1 to 1.5 km/s, the scale of the photospheric magnetic field shown in the upper panel (view from below the surface) is from –1800 to 1800 Gauss.

The sound-speed and flow maps reconstructed up to a depth of 60 Mm reveal that the subsurface structure of the active region is as complicated as its surface structure, and also rapidly evolving. From these observations, we find no evidence for a large magnetic Ω-loop emerging from the interior and forming this active region. The active region was rather formed by fragmented magnetic flux emerging during an extended period of time. However, the sound-speed image of another large active region NOAA 10488 reveals a large-scale loop-like structure below the surface (Figure 7). Obviously, more observations are needed for understanding the structure and evolution of active regions.

4. Dynamics of Active Regions and Sources of Heliospheric Disturbances

During the maximum of activity the helioseismic observations show large-scale converging downflows accompanied by complicated shearing motions which may be related to flaring activity (Dzifcakova *et al.*, 2003; Kulinova *et al.*, 2003). During the decay phase the downflows become significantly weaker, and diverging flows around decaying sunspots are observed.

A series of 9 X-class flares produced during Oct. 23–Nov. 4, 2003, by the two active region 10486 and 10488 was one of the most powerful in the history of solar observations. It is well-known that flares usually occur in complex sheared and twisted magnetic configurations which are presumably produced by shearing and twisting plasma flows below the surface where the dynamic pressure of plasma flows may exceed the magnetic pressure. Magnetic energy release in solar flares typically happens around neutral lines of the line-of-sight (vertical) component of magnetic field. These places can be identified by rapid permanent changes of the photospheric magnetic flux on both sides of the neutral line. The true height

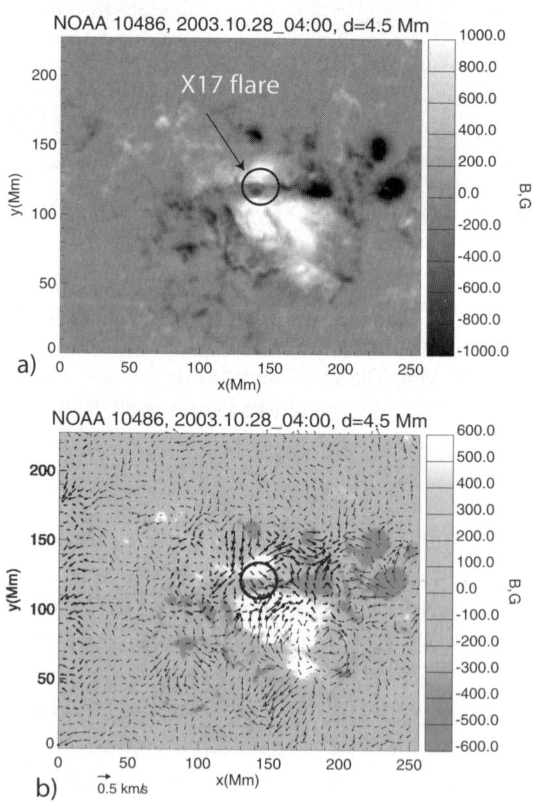

Figure 8. Surface magnetograms and subsurface flows during the X17 flare of October 28, 2003.

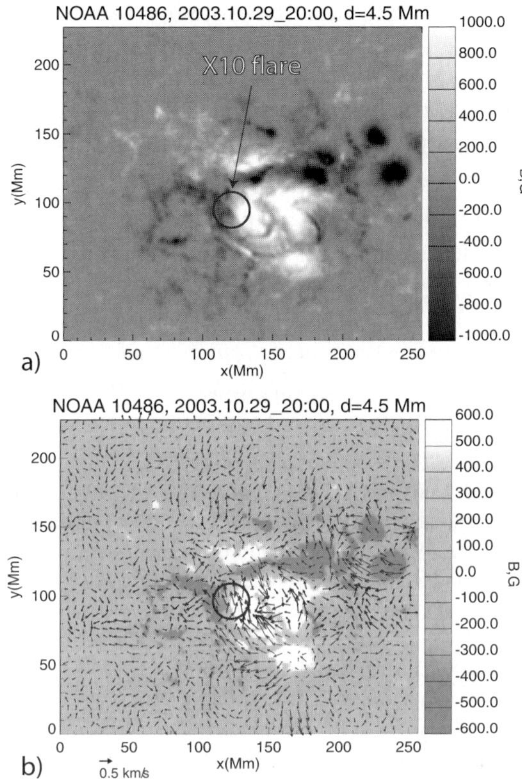

Figure 9. Surface magnetograms and subsurface flows during the X10 flare of October 29, 2003.

of the magnetic energy release is still not established. Presumably the energy is released mostly in magnetic structures in the upper atmosphere, covering some range of heights, but evidently these structures are connected to the places in the photosphere where we see significant permanent magnetic flux changes during the impulsive phase of solar flares.

The black circles in Figures 8 and 9 indicate the sites of the magnetic energy release for two strong flares, X17 started at 9:51 UT on October 28, and X10 started at 20:37 UT on October 29. It is intriguing that the flow maps inferred by time-distance helioseismology at the depth of 4–6 Mm reveal strong (with speed about 1–2 km) shearing flows directed to the sites of the magnetic energy release during these flares. This is particularly evident from the flow map (Figure 8b, taken for the 8-hour periods: 0–8 UT on October 28 (labeled as 2003.10.28_04: 00, just before the X17 flare), and from the flow map (Figure 8b, obtained for 16–24 UT on October 29 (2003.10.29_20: 00), before and during the X10 flare.

Obviously, the 8-hour resolution of our time-distance measurement does not allow us to follow the plasma dynamics during the flares which happen on a much shorter time scale. Nevertheless, these results indicate that some interesting dynamics associated with flaring activity probably occurs in subsurface regions, 4–6 Mm deep, just below the zone where the strong magnetic field of sunspots inhibits convection (Zhao *et al.*, 2001).

In conclusion, the new methods of local helioseismology provide powerful diagnostics of sub-photospheric dynamics of active regions, which allow us to investigate the birth and evolution of active regions, and origins of solar activity. Further analysis should include more accurate flow maps with high temporal and spatial resolutions, and determine links between the interior dynamics and coronal magnetic fields of active regions.

References

Birch, A. C., and Kosovichev, A. G.: 2000, *Solar Phys.* **192**, 193.

Birch, A.C., Kosovichev, A. G., and Duvall, T. L., Jr.: 2004, *Astrophys. J.* **608**, 580.

Couvidat, S., Birch, A. C., Kosovichev, A. G., and Zhao, J.: 2004, *Astrophys. J.* **607**, 554.

Duvall, T. L., Jr., Jefferies, S. M., Harvey, J. W., and Pomerantz, M. A.: 1993, *Nature* **362**, 430.

Dzifcakova, E., Kulinova, A., and Kosovichev, A. G.: 2003, in Sawaya-Lacoste, H. (ed.), *Proceedings of SOHO 12/GONG+ 2002. Local and Global Helioseismology: The Present and Future*, ESA SP-517, Noordwijk, p. 263.

Gizon, L., and Birch, A. C.: 2002, *Astrophys. J.* **571**, 966.

Kosovichev, A. G.: 1996, *Astrophys. J.* **461**, L55.

Kosovichev, A. G., and Duvall, T. L., Jr.: 1997, in Pijpers, F. P. *et al.* (ed.), *SCORe'96 : Solar Convection and Oscillations and their Relationship*, Ast. Space Sci. Lib. vol. 225, p. 241.

Kosovichev, A. G., Duvall, T. L., Jr., and Scherrer, P. H.: 2000, *Solar Phys.* **192**, 159.

Kosovichev, A. G., and Duvall, T. L., Jr.: 2003, Stephen L. Keil, and Sergey V. Avakyan (eds.), *Innovative Telescopes and Instrumentation for Solar Astrophysics, Proceedings of the SPIE*, vol. 4853, p. 327.

Kulinova, A., Dzifcakova, E., Duvall, T. L., Jr., and Kosovichev, A. G.: 2003, in Wilson, A. (ed.), *Solar Variability as an Input to the Earth's Environment*, ESA SP-535, Noordwijk, p. 125.

Zhao, J., Kosovichev, A. G., and Duvall, T. L., Jr.: 2001, *Astrophys. J.* **557**, 384.

SUNSPOT STRUCTURE AND DYNAMICS

N. O. WEISS
DAMTP, Centre for Mathematical Sciences, Wilberforce Road, Cambridge CB3 0WA, UK
(E-mail: now@damtp.cam.ac.uk)

(Received 8 July 2005; Accepted in final form 12 May 2006)

Abstract. Sunspots are the most prominent magnetic features on the Sun but it is only within the last few years that the intricate structure of their magnetic fields has been resolved. In the penumbra the fields in bright and dark filaments differ in inclination by 30°. The field in the bright filaments is less inclined to the vertical, while the field in dark filaments becomes almost horizontal at the edge of the spot. Recent models suggest that this interlocking-comb structure is maintained through downward pumping of magnetic flux by small-scale granular convection, and that filamentation originates as a convective instability. Within the bright filaments convection patterns travel radially owing to the inclination of the field. A proper understanding of these processes requires new observations, from space and from the ground, coupled with large-scale numerical modelling.

Keywords: sunspots – Sun: magnetic fields

1. Introduction

Sunspots have been observed through telescopes for almost 400 years and early observers, such as Galileo, Scheiner and Hevelius, were already able to distinguish the dark central core of a spot (the *umbra*) from the fuzzier annulus (the *penumbra*) that surrounds it. The filamentary structure of the penumbra was not recognized till two centuries later, when achromatic lenses were available, and it was only in 1908 that Hale used the Zeeman effect to show that sunspots were the sites of kilogauss magnetic fields. Forty years later, magnetohydrodynamics had been established and it was realised that sunspots were dark because normal convective transport was inhibited by their strong magnetic fields. However, it is only within the last ten years that it has become possible to observe the fine structure of the penumbral magnetic field, with telescopes that are capable of arc-second or sub-arc-second resolution – and these measurements have posed questions that theorists are still struggling to answer. So the structure of sunspots may be an old problem but it raises issues that are very much alive today.

There are several recent reviews of this subject, by Solanki (2003), by Thomas and Weiss (2004) and by Tobias and Weiss (2004), in order of decreasing detail. In this brief survey I shall first summarize the observational results and outline the physical picture that arises from them. Next, in Section 3, I shall discuss the crucial mechanism of flux pumping, which appears to be responsible for maintaining the

Space Science Reviews (2006) 124: 13–22
DOI: 10.1007/s11214-006-9096-8

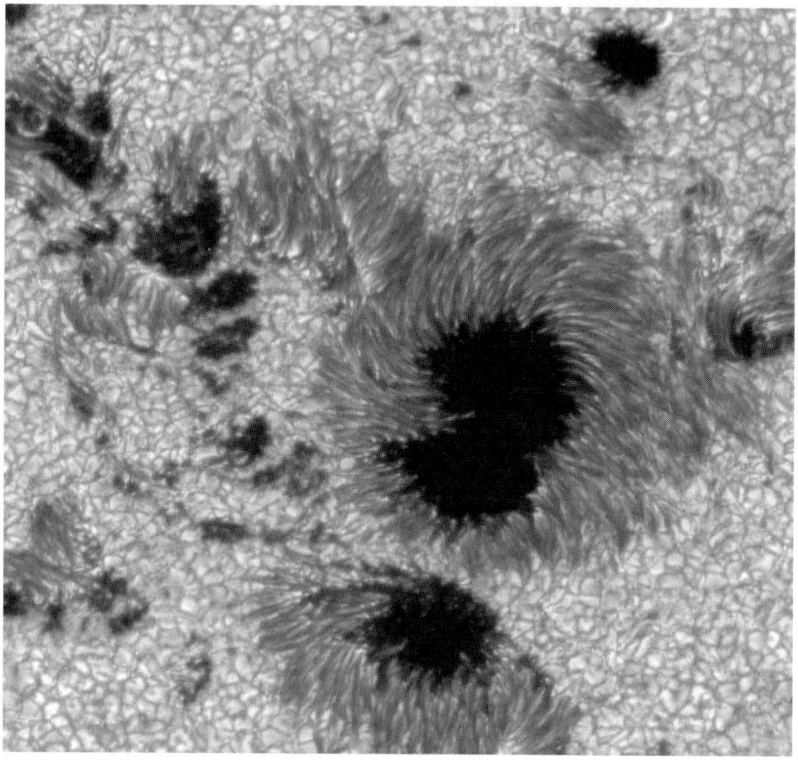

Figure 1. Sunspots observed with the 1-m Swedish Solar Telescope on La Palma, at 0.1 arc-sec resolution. The dark umbra of the central spot is surrounded by a filamentary penumbra with a total diameter of about 25 Mm. There are several dark pores (e.g. at top right) without penumbrae as well as smaller micropores. The background pattern of convection cells, with diameters of order 1 Mm, is the solar granulation. This image, obtained in the CH G-band, also shows tiny bright features, which correspond to small magnetic flux elements nestling between the granules. (Courtesy of the Royal Swedish Academy of Sciences.)

penumbra's filamentary structure. Then, in the final section, I shall comment on some outstanding problems and point to future progress.

2. The Magnetic Structure of a Sunspot

The remarkable high-resolution image in Figure 1 shows two sunspots with filamentary penumbrae, as well as several pores (without penumbrae) and various smaller magnetic features (Scharmer *et al.*, 2002; Rouppe van der Voort *et al.*, 2004). The strong magnetic fields in the spots suppresses the normal pattern of small-scale convection – the solar granulation – in the photosphere surrounding them, where bright hot plumes are enclosed by a network of cooler sinking gas.

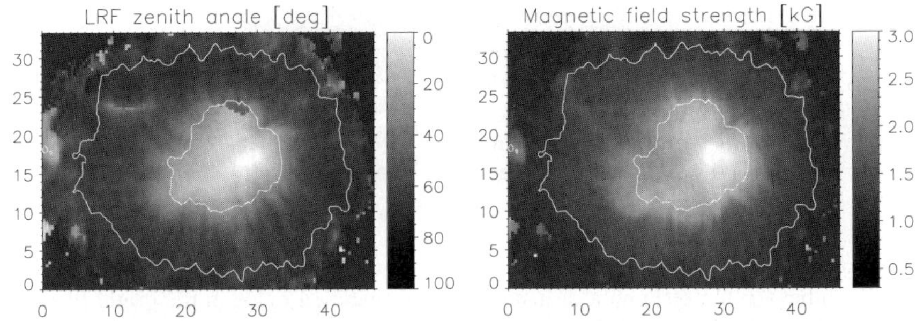

Figure 2. Fine structure of the penumbral magnetic field. Right panel: Field strength. Left panel: inclination of the field to the local vertical. The average inclination increases towards the edge of the spot but azimuthal variations in inclination are clearly visible. (After Bellot Rubio, 2003.)

The azimuthally averaged magnetic field is vertical at the centre of an isolated sunspot and its inclination to the vertical increases with increasing radius, reaching a value of 70° at the edge of the penumbra, as shown in Figure 2. It has long been known, however, that there is a persistent horizontal outflow (the Evershed flow) in the outer part of the penumbra. Since the velocity should be parallel to the field in such a highly conducting plasma, this raises an apparent contradiction (Adam and Petford, 1990), which can only be resolved by assuming an inhomogeneous magnetic structure (Beckers and Schröter, 1969).

High-resolution observations (e.g. Title *et al.*, 1993; Lites *et al.*, 1993; Solanki and Montavon, 1993; Stanchfield *et al.*, 1997; Bellot Rubio, 2003; Bellot Rubio *et al.*, 2003; Borrero *et al.*, 2004; Bellot Rubio, *et al.*, 2004; Bello González *et al.*, 2005) have subsequently confirmed that the inclinations of the fields in bright and dark filaments do indeed differ by 30–40°, as can be seen in Figure 2. The fields in the dark filaments (which carry the Evershed flow) are more inclined to the vertical, becoming almost horizontal at the outer edge of the spot. The most recent measurements, obtained at exceptionally high resolution with the Swedish Solar Telescope on La Palma (Langhans *et al.*, 2005), clearly distinguish between a darker component, with a weaker field that is more inclined, and a brighter component with a stronger field that is more nearly vertical (though the anticorrelation between field strength and inclination is more marked than that between brightness and inclination). Thus the penumbral magnetic field has the improbable interlocking-comb structure that is shown schematically in Figure 3. Moreover, the two families of field lines are apparently distinct, for the loops that follow field lines emerging from bright filaments extend across vast distances, as is apparent from the TRACE image in Figure 4, while the fields associated with dark filaments either hug the surface (forming a superpenumbra in Hα) or actually plunge beneath it.

The observed Evershed flow is confined to thin channels, which are indeed aligned with the most nearly horizontal fields (Bellot Rubio *et al.*, 2003, 2004; Tritschler *et al.*, 2004; Schlichenmaier *et al.*, 2004), though the correlation with

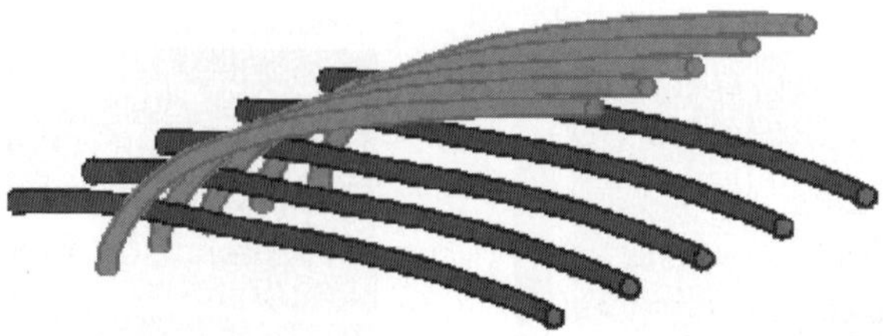

Figure 3. Sketch showing the interlocking-comb structure of the magnetic field (represented schematically by flux tubes) in the penumbra of a sunspot, with inclined fields in the bright filaments and almost horizontal fields in the dark filaments. (Courtesy of N. H. Brummell.)

Figure 4. TRACE image of a sunspot pair, showing coronal loops that follow magnetic field lines emerging the penumbrae of the spots and extending far across the surface of the Sun. (Courtesy of the Lockheed-Martin Solar and Astrophysics Laboratory.)

dark filaments is somewhat weaker (Rimmele, 1995a; Stanchfield *et al.*, 1997; Schlichenmaier *et al.*, 2005). Furthermore, many of the flow channels that emerge in the penumbra actually turn over and dive down either just outside it or even within it, carrying both the flow and its associated magnetic field with them (e.g. Rimmele, 1995b; Stanchfield *et al.*, 1997; Westendorp Plaza *et al.*, 1997; del Toro Iniesta *et al.*, 2001; Bellot Rubio *et al.*, 2003, 2004; Tritschler et al., 2004; Schlichenmaier *et al.*, 2005; Borrero *et al.*, 2005; Langhans *et al.*, 2005). It is generally supposed that the Evershed flow is in fact a siphon flow along these flux tubes, driven by pressure differences between their footpoints (Meyer and Schmidt, 1968; Montesinos and Thomas, 1997).

3. Flux Pumping by Convection

These observations raise serious theoretical problems. First of all, we need to explain the interlocking-comb magnetic structure illustrated in Figure 3, which is really a structure of interlocking sheets, for continuity of magnetic flux requires that the two families of field lines must have a finite vertical extent. (This structure is often referred to as 'uncombed', following Solanki and Montavon, 1993.) In addition, the unexpected reversal of the vertical component of the magnetic field in the outer penumbra demands an explanation. In fact, it is this reversal that offers a key to understanding how this strange coherent structure can be maintained.

There are two effects that resist downward bending of magnetic flux tubes: the magnetic curvature force tends to straighten field lines, while magnetic buoyancy makes an isolated flux tube rise. Hence there has to be some other effect that drags them down below the surface, either inside the penumbra itself or just outside it. The obvious candidate is downward pumping of magnetic flux by the small-scale, turbulent granular convection within the large annular 'moat' cell that surrounds a well-developed spot. This process leads to the overall picture of a sunspot that is shown schematically in Figure 5 (Thomas *et al.*, 2002; Weiss *et al.*, 2004). (Note that the magnetic field actually fills the space above the sunspot and has an interlocking-sheet structure in the penumbra; it is nevertheless convenient to represent this field

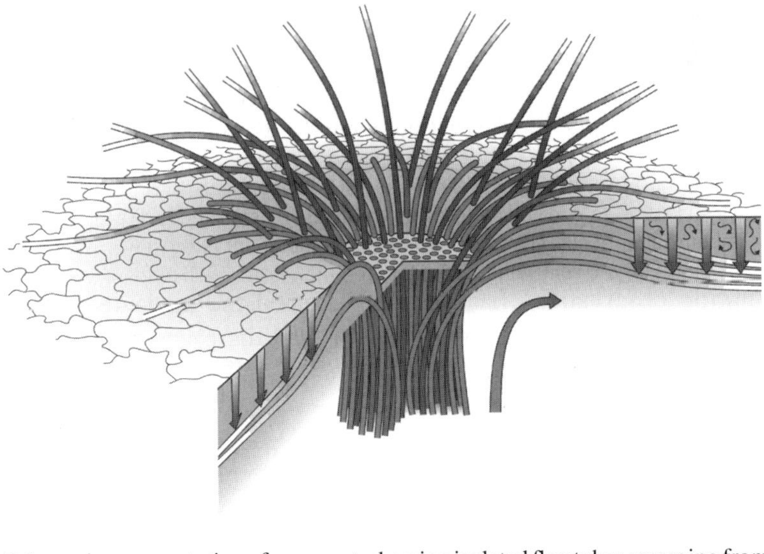

Figure 5. Schematic representation of a sunspot, showing isolated flux tubes emerging from the umbra and penumbra. Flux tubes that emerge from the penumbra either form a canopy over the photosphere or are pumped downwards by granular convection outside the sunspot and held below the surface. There is also a large-scale radial outflow in the annular moat cell that surrounds the sunspot. (From Weiss et al., 2004.)

by depicting isolated flux tubes.) In this picture it is supposed that the Evershed flow is carried by flux tubes that arch above the penumbral photosphere before returning below it, and are then kept submerged by the downdrafts at the boundaries of granules, This picture is supported by the behaviour of moving magnetic features in the moat, which correspond to a stitch of field emerging as a bipolar feature and travelling radially outwards (see Thomas and Weiss, 2004 for further details).

Flux pumping has been studied numerically in some highly idealized configurations, first in relation to the solar tachocline (e.g. Tobias et al., 2002; Dorch and Nordlund, 2002), and then in the present context. Two processes are involved: one is the expulsion of magnetic flux down the gradient of turbulent intensity (e.g. Tao *et al.*, 1998) and the other is the tendency of convection in a stratified layer to pump magnetic flux preferentially downwards. Numerical simulations show a distinction between broad, gently rising plumes that expand as they move upwards and narrow, vigorously sinking plumes that entrain material as they descend. As a result, an initially horizontal field is pumped downwards out of a vigorously convecting region and can accumulate in an adiabatically (or mildly superadiabatically) stratified layer beneath it (Weiss *et al.*, 2004). Figure 6 shows some results with a somewhat more realistic configuration (Brummell et al., 2006). The strongly unstable region has an aspect ratio of $6 \times 6 \times 1$ but the full computational box extends further downwards, with aspect ratio $6 \times 6 \times 3$, and the lower part is mildly subadiabatically stratified. Once convection is fully established, a strong magnetic field is added, with the double-arched structure shown in the upper panel (and periodic lateral boundary conditions). After the calculation has reached a statistically steady state, the horizontal fields are pumped downwards, excluded from the vigorously convecting region, and stored in the stably stratified layer below. More elaborate calculations are clearly needed but it appears already that this process is robust and able to explain the observed behaviour of the fields that carry the Evershed flow in sunspots.

4. Outstanding Problems

High-resolution observations have finally revealed the intricate structure of the magnetic field in the penumbra of a sunspot, and there is a plausible theoretical picture of how this structure is maintained. There is, however, a range of associated problems where theory is in a weaker state.

It is natural to ask how the interlocking-comb structure originates as a sunspot is formed. Sunspots are formed by the amalgamation of smaller pores, which resemble isolated umbrae. (On close inspection, the small pores in Figure 1 do themselves have a very fine-scale fluted structure at their edges.) Model calculations have confirmed that the average inclination of the field at the edge of a pore increases as the magnetic flux in the pore itself increases, and it has been conjectured that a subcritical fluting instability sets in when the inclination reaches a critical value (Rucklidge *et al.*, 1995). Simplified model calculations in Cartesian geometry have

CORONAL DYNAMICS AND THE AIA ON SDO

L. GOLUB
Harvard-Smithsonian Center for Astrophysics, 60 Garden Street, Cambridge MA, 02138, USA
(E-mail: lgolub@cfa.harvard.edu)

(Received 25 July 2005; Accepted in final form 14 February 2006)

Abstract. We provide a brief overview of present-day studies of inner corona dynamics, with examples of mass ejections (CME), flares and active region dynamics. While the names of the topics have not changed in several decades, the internal details and the language used to express the nature of the problem have changed considerably. We conclude with a short discussion of the contribution to studies of coronal dynamics to be expected from the Atmospheric Imager Assembly (AIA) on the Solar Dynamics Observatory.

Keywords: solar corona, coronal activity, solar dynamics, SDO

1. Introduction

Studies of coronal dynamics have been greatly advanced in recent years by observations from space. Satellites such as SoHO (1995) and TRACE (1998) are providing new views of the inner corona with unprecedented spectral and spatial coverage. Yet observation of the solar atmosphere continues to be complicated by at least four factors:

1. The temperature range is $<10^4$ to $>10^7$K, with a comparable spread in the wavelength of the emitted radiation (viz. Figure 1a), effectively preventing a single instrument from observing a complete atmospheric structure.
2. Because of the high temperature (and perhaps also the energy deposition processes) the atmosphere is greatly extended, requiring a large field of view.
3. Because of the magnetic field, the atmosphere is in a highly dynamic state, with fine structure transverse to the field (viz. Figure 1b). Combined with the large field of view, this means that the instrument used must have a very large number of pixels, which places an enormous burden on the resources of the spacecraft.
4. The atmosphere is, to a great extent, optically thin so that many structures add up along any line of sight. It is therefore usually quite difficult to isolate the feature under study and separate it from other structures in the atmosphere.

Further progress is thus hampered by the lack of spectral coverage, combined with limitations on field of view and temporal resolution, the latter due to limitations on data storage and data downlink rates. In this paper we provide a sampling of

Space Science Reviews (2006) 124: 23–33
DOI: 10.1007/s11214-006-9097-7

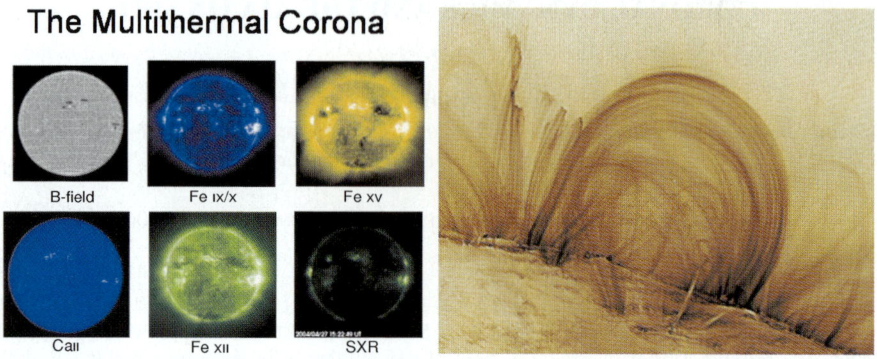

Figure 1. (a) The inner corona is multithermal, presenting markedly different appearance is slightly different temperatures. (b) Coronal fine structure due to the interplay between the hot plasma and the magnetic field.

the present state of solar dynamics studies, and then we discuss aspects of the contribution that will be made by the Atmospheric Imager Assembly (AIA).

2. Dynamics of the Inner Corona

X-ray and EUV observations of the inner corona with high spatial and temporal resolution have shown that the hot magnetized plasma is highly dynamic and variable on nearly all scales viz. (Golub *et al.*, 1999). The list of topics that form the core of present-day research in understanding coronal dynamics is a familiar one: coronal heating, coronal mass ejections (CMEs), reconnection and jets, flare energy release, and photosphere-corona coupling. However, the internal details of the discussion have evolved dramatically, especially with regard to the theoretical framework (Priest and Forbes, 2000). The present view is that the coronal volume is divided into magnetic field regions which are active at the boundaries and at regions of singularity (separatrices, spines, fans) that divide the space into volumes of field having different topological connectivities. These topological structures may even be responsible for loop heating within an active region, in the form of tangential discontinuities on a micro-structure scale, or in more recent terminology, as the formation of current sheets along separatrices and separators (Priest *et al.*, 2005). For the observer, this new view means that it may be the skeletal structure of the field – which is not directly visible – that matters in explaining coronal activity (Longcope, 2005).

3. The Observational Problem

From an instrumental viewpoint, the main problem we encounter in attempting to study the dynamics of the solar atmosphere is that in a hot magnetized plasma the physical processes involved in the dynamics occur on very small spatial scales, while the consequences of these processes are usually observed on very large spatial scales (Karpen, 2003). One must therefore construct an instrument with very high spatial resolution and also with a large field of view. This translates into a requirement for an extremely large number of pixels – of order 10^{12} if we are to observe both regions directly. When we also take into account the temporal scales involved in observing changes at these spatial scales at either the sound speed or the Alfvén speed ($\sim$1 sec), plus the requirement that we have enough photons per pixel to obtain a good measurement of the intensities ($\sim 10^3$ for photometric accuracy), we find that the imaging device must collect more than 10^{15} photons per second.

Because a typical strong coronal emission line in the EUV or soft X-ray emits $\sim 10^9$ photons/sec/cm^2 at 1 AU from a bright coronal active region, we need a collecting area of $> 10^6$ cm^2 to gather enough photons. This means a mirror diameter of at least 13 meters, which has never been done. Every instrument ever flown for coronal studies has therefore compromised either spatial, spectral or temporal resolution.

3.1. The 4-D Data Cube

The observational problem is thus seen to be four-dimensional: when viewing a source we see two spatial dimensions across our line of sight, we can analyze the light into a spectral dimension, and the data are variable in time. Ideally, an array detector with spectral resolution could be used to obtain the entire 4D data cube. However, while suitable devices, such as X-ray calorimater arrays, may someday exist, they have not yet achieved the numbers of elements needed for solar studies.

We are therefore left with more standard techniques. There are four basic possibilities, each requiring a compromise of some sort: imaging, spectroscopy, spectroscopic imaging and imaging spectroscopy. Imaging *per se* is necessary for establishing overall knowledge of the phenomenon under consideration, but reveals very little about the detailed physics responsible for the dynamics. In contrast, spectroscopy alone can provide information about the physical processes, but without imaging it is often impossible to know where the photons originate (Figure 2).

We therefore need *both* imaging and spectroscopy. There are two main methods of carrying this out. We can construct a spectrograph which has imaging capability, or we can construct an imager with good spectrographic capability. An example of the former is the EIS instrument onboard the Solar-B spacecraft (EIS, 2005), while the latter is represented by the AIA instrument, discussed in detail below.

The main issue is that the data "cube" is four-dimensional: the image as presented to the eye has two spatial dimensions, there is a time dimension because of changes in the target under observation, and there is a spectral dimension which must also be recorded. Detector arrays with spectral capability exist – e.g. the imaging proportional counter produces a pulse whose magnitude depends on the incoming

Figure 2. TRACE 173 Å observation of an active region near the limb. The inner corona is complex, filled with structures which often pile up along the line of sight.

photon energy and the bolometric array can measure the photon energy with great precision – but no detector exists at the present time which has the combination of spatial, temporal and spectral resolution that solar observations require.

3.2. Dynamics of the Hot Magnetized Plasma

A short review article cannot do justice to the full range of dynamics observed in the solar corona. Here we will concentrate on one aspect of one major phenomenon: the large solar flare. Analysis of TRACE data indicates that the impulsive phase of a flare – the phase in which the energy release rate increases dramatically – is often characterized by a transition from a strong magnetic shear to a weak shear. This appears to be due to the structure of the magnetic field which supports prominences, and to the change in that structure during the eruption. This subject is reviewed in a recent article (Zhang and Low, 2005).

3.3. Shear Decrease During a Flare

In recent work, analysis of the well-known 28 October 2003 X17 flare (Su, *et al.*, 2005) shows a clear example of the change in magnetic shear of the reconnected loops during the rise phase of the flare. High time resolution hard X-ray data were obtained by the anticoincidence system (ACS) of the SPI spectrometer onboard the ESA INTEGRAL spacecraft and 8 peaks can be seen during the impulsive phase of the solar flare. Individual peaks in thehard X-ray light curve are found to be associated with specific EUV bright kernels in the flare ribbons. The initial EUV footpoint brightenings are on opposite sides of the neutral line (Figure 3) but are very widely separated along the neutral line, indicative of strong shear. As

Figure 3. Decrease in shear during the impulsive phase of the 28 October 2003 X17 flare. Images taken at 11:01, 11:04 and 11:06 UT; the impulsive phase occurred at 11:04 UT.

the event progresses, the shear decreases, ending up with ribbons separating nearly perpendicular to the neutral (polarity inversion) line, ie, with essentially no shear.

It is of particular interest to note that the change from strong shear to no shear coincides in time with the impulsive phase of the flare (Figure 4), apparently coinciding with the progressive motion of the reconnection site to higher altitudes within the region. These topological factors have long been expected on theoretical grounds (e.g. DeVore *et al.*, 2000), but observational support has been difficult to obtain.

4. The Atmospheric Imager Assembly (AIA)

As discussed above, the TRACE Observatory has shown the tremendous benefits to be obtained by observing the EUV corona with high spatial and temporal resolution. Because TRACE was a low-cost SMEX mission, it was built according to the "faster, better, cheaper" approach used by NASA. This resulted in some limitations, particularly in small field of view and in the use of a limited number of wavelengths. It is therefore especially fortunate that the Solar Dynamics Observatory will have onboard the Atmospheric Imaging Assembly (AIA), which consists of several TRACE-like telescopes which cover the full Sun with high spatial resolution, while providing increased spectral coverage (Figure 5).

The AIA will have seven EUV passbands plus one multi-channel UV passband. One EUV channel will image He II 304Å , while the remaining six channels will provide spectroscopic imaging in a broad range of ionization stages of iron, from Fe VIII through Fe XXIV. The equivalent temperature coverage is therefore from the Transition Region through flare temperatures.

4.1. Spatial and Temporal Resolution

One of the most important contributions made by the TRACE instrument is the dramatic increase in spatial resolution over previous satellite observations. Figure 6

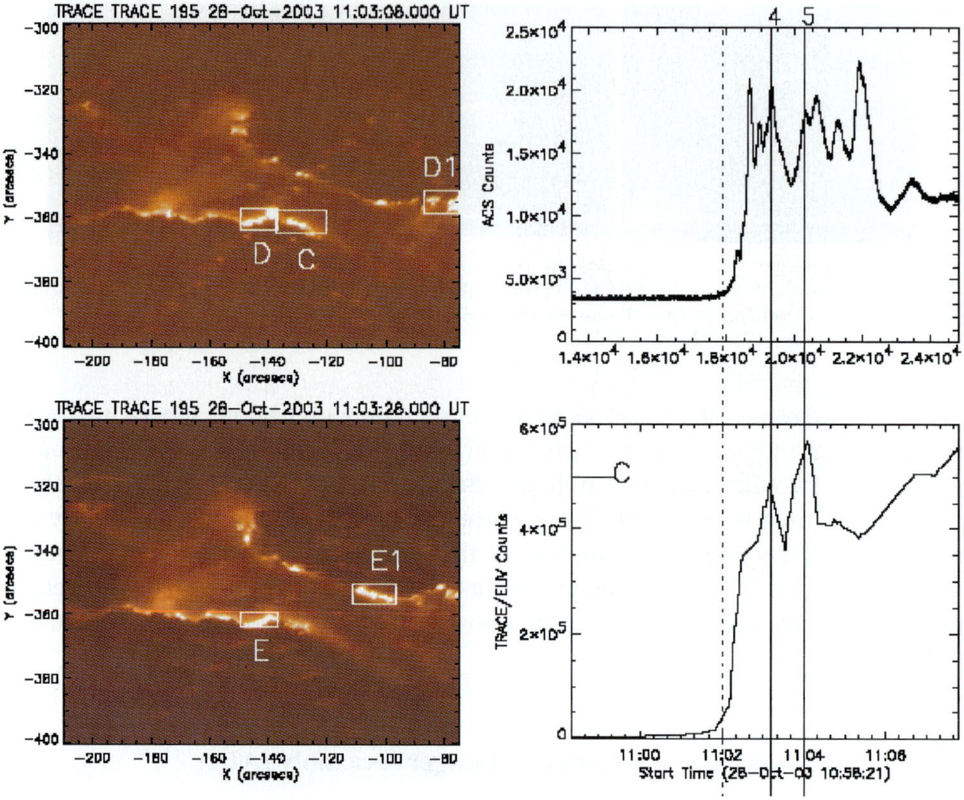

Figure 4. Comparison between hard X-ray bursts seen by the ACS (top right panel) with light curves (bottom right panel) of the EUV bright kernel C labelled in the left-hand panels.

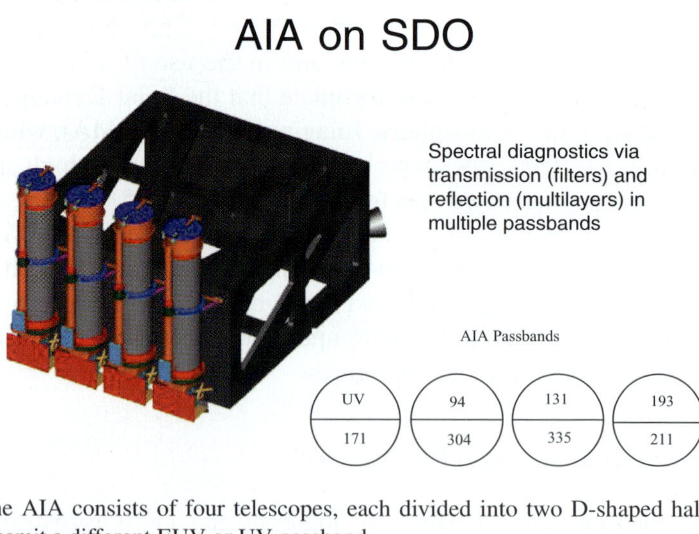

Figure 5. The AIA consists of four telescopes, each divided into two D-shaped halves, each half coated to transmit a different EUV or UV passband.

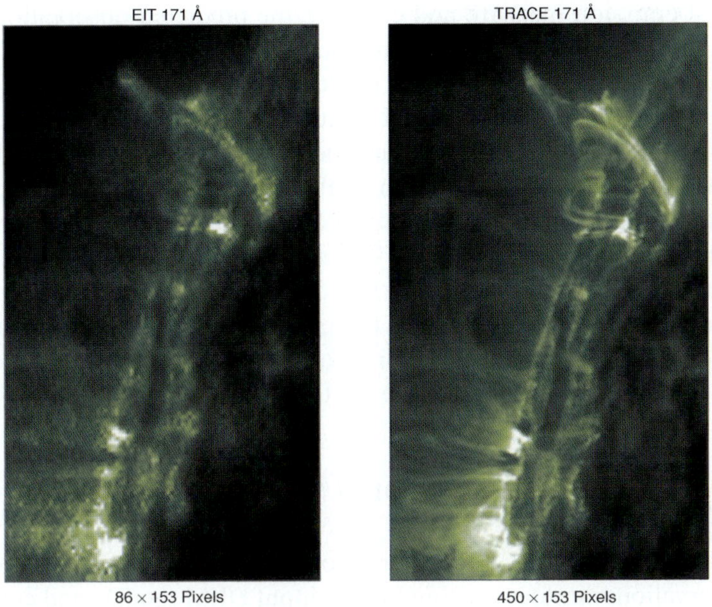

Figure 6. The benefits of resolution: comparison of simultaneous TRACE *vs.* EIT observation of the same coronal region, same passband, same image scale.

illustrates clearly the simple fact that in order to study the coronal dynamics one must first be able to see the structures in the corona. While the AIA will not improve on TRACE's spatial resolution (nor will any other presently planned instrument) it will provide full Sun coverage and greatly increased data rates for increased temporal resolution and increased wavelength (i.e. temperature) coverage.

The next step after one is able to see the coronal structures is to determine their physical properties. Because the AIA on SDO will have unprecedented capability for determining the quantity of material as a function of temperature (the "differential emission measure" or DEM) in the corona, we devote the remainder of this paper to a close look at the methods used for such spectroscopic analysis.

4.2. Iterative Forward Modelling of DEMs

The procedure we use for finding the "best" fitting DEM for a given set of observations in several spectral channels has been described in Weber *et al.* (2005). We consider a set of AIA images taken of an active region (AR) and we discuss how we estimate the DEM in a given pixel. Our procedure produces an iterative least-squares fit to the observations using a DEM represented by a spline with evenly spaced knots in $\log(T_e)$ space. With the forward modeling approach, we assume a differential emission measure and compare the predicted observations for each filter with the real observations, iterating the DEM until an acceptable fit is found.

A basic problem with the forward modeling approach is determining the relevance of the best-fit solution. The AIA multilayer wavelength channels are distinguished by the emission lines that are included in each passband. These filters approximate but do not strictly form an "orthogonal" basis set for temperature and emission measure determinations. In practice there may be multiple nearby minima that have substantially different DEM(T) curves. To address this issue we expand our nominal set of observations into 100 different Monte Carlo realizations by adding random noise, consistent with the photon noise, to the observations. The best least-squares fit to each of the realizations is then determined. Our confidence in the fit is measured by the fluctuations in the fits about the median best fit.

5. Results of DEM Modeling Tests

To assess the efficacy of DEM reconstruction, we have applied the DEM software to AIA observations simulated using known input DEM models and compared the results for the derived DEMs with the models. In this paper, we discuss a selection of three analyses that illustrate the quality of DEM reconstruction that can be achieved with the AIA.

In each of the following cases, a nominal observation is calculated for a given DEM model, and then the procedure samples the observation 100 times by including random photon noise at the 3% level. The distribution of calculated DEM curves (relative to the known DEM model) indicates the accuracy and robustness of the analysis method, as discussed in the previous section. We have indicated the median member of each set of realizations as an estimate of the model DEM.

5.1. "Real" AR DEM & the Value of Many Channels

The corona is known to be highly inhomogeneous in temperature, density, and magnetic field – the isothermal approximation is often inadequate for describing the optically thin solar atmosphere across length scales comparable to the span of an AIA pixel. The actual DEM distribution in an active region is thus expected to include material across a wide temperature range. We analyzed our DEM procedure using a realistic DEM model that is included in the CHIANTI database (Young *et al.*, 2003, and references therein). This model of an active region DEM (CHIANTI file "active_region_oso6.dem") was derived by K.P. Dere from observations of the Sun (Dupree *et al.*, 1973) by the scanning spectrometer on OSO-6 (Huber *et al.*, 1973).

There is, in principle, no limit to the complexity of the model DEM which could be chosen. However, the physics of the situation – primarily the Boltzmann width of the spectral lines, causing them to be formed over a fairly wide temperature

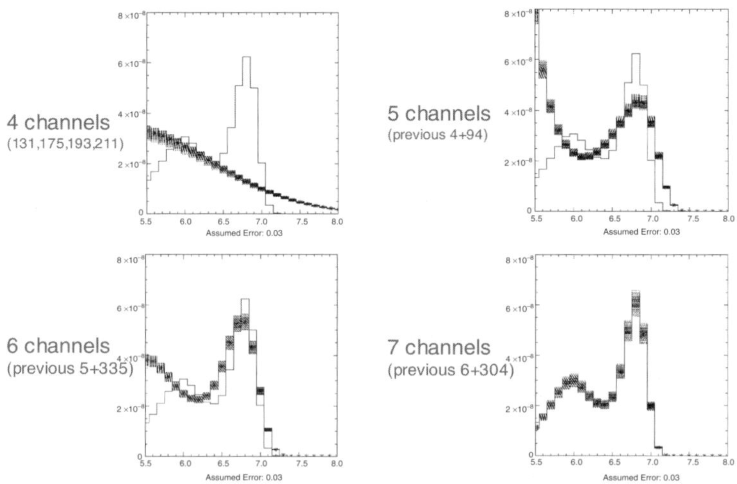

Figure 7. Fitting a "real" AR DEM. Increasing the number of channels dramatically increases the quality of the reconstruction.

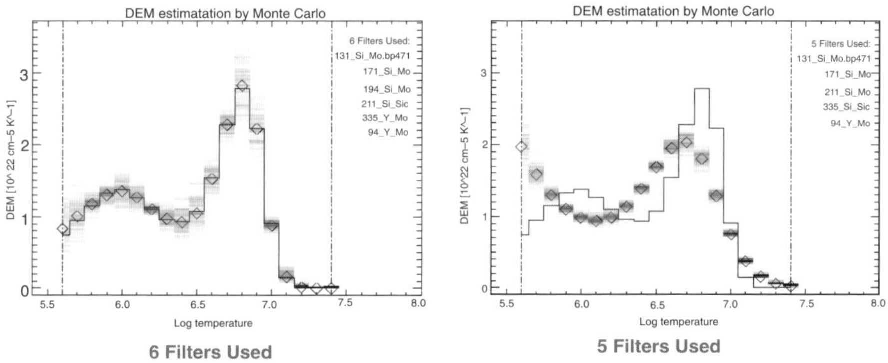

Figure 8. A minimum of six independent channels is needed to provide adequate resolution and coverage in the corona.

range in the corona – provides a fundamental limitation to the resolving power of any spectroscopic analysis (Craig and Brown, 1976).

Figures 7 and 8 suggest how well the input DEM can be reconstructed as a function of the number of observing channels used. In Figure 7, four AIA channels have been used to perform fits. The figure shows the model AR DEM (solid line with two humps), the distribution of fitted DEMs (grayscale), and the median values of the 100 DEM runs (diamonds). The DEM is fitted over the log temperature range 5.5–8.0 and 3% noise is assumed. These relatively high-T AIA filter channels determine the presence of the hotter peak of material, as indicated by the convergence of the median fit to the model DEM curve, but fail to detect the cooler material. The narrow uncertainty bands indicate that the fits are robust or, in the words of one author, "reliably bad".

In Figure 8, the same model is fitted with seven AIA channels; that is, we have included thinner filters in the analysis. It is obvious that the fitted DEMs more accurately reproduce the model DEM curve across the entire temperature range. Even though the uncertainty bands are not as constrained as in Figure 7, they adequately indicate the presence and temperature of the cool component. To achieve good results in DEM reconstruction with AIA data, it is thus important to have observations in many (independent) channels.

5.2. The Number of Channels

The physics of ionization fraction formation under coronal conditions combined with the range of temperatures found in the corona leads to a definite requirement on the number of independent channels that need to be recorded in order to reconstruct the emission measure distribution. The narrowness of the temperature range over which a typical ionization state is formed in the corona, combined with the requirement to have complete but non-reduntant coverage, means that at least six channels must be used. This is shown explicitly in Figure 8, where we examine how the removal of even a single channel affects the quality of the DEM reconstruction for a typical active region. This is the major reason for the choice of six Fe channels in the AIA instrument.

Acknowledgements

The AIA work at SAO is supported by a contract from Lockheed Martin. I wish to thank Yingna Su for permission to show results of her work prior to publication.

References

Craig, I.J.D., and Brown, J. C.: 1976, *Astron. and Astrophys.* **49**(2), 239.
DeVore, C., R., and Antiochos, S.: 2000, *Astrophys. J.* **539**, 954.
EIS URL http://www.mssl.ucl.ac.uk/www_solar/solarB/outreach.html
Golub, L. *et al.*: 1999, *Phys. Plasmas* **6**(5), 2205.
Karpen, J. T.: 2003, *Proc. SPIE* **4853**, 453.
Krieger, A. S., Timothy, A. F., and Roelof, E. C.: 1973, *Solar Phys.* **29**, 505.
Longcope, D.: 2005, *Living Reviews in Solar Physics* **2**(7).
Priest, E., and Forbes, T.: 2000, *Magnetic Reconnection* Cambridge, University Press, UK.
Priest, E., Longcope, D., and Heyvaerts, J.: 2005, *Astrophys. J.* **624**, 1057.
SoHO URL – http://sohowww.nascom.nasa.gov.
Su, Y., Golub, L., Gros, M., and Van Ballegooijen, A.: 2006, *Sol. Phys.* **236**, 325.
TRACE URL – http://vestige.lmsal.com/TRACE/.

Weber, M. *et al.*: 2005, *Proc. IAU Symp. 223*, Vol. 2004, *Multi-Wavelength Investigations of Solar Activity.*
Zhang, M., and Low, B. C.: 2005, *Ann. Rev. Astron. Astrophys.* **43**, 103.

WIND IN THE SOLAR CORONA: DYNAMICS AND COMPOSITION

ESTER ANTONUCCI
INAF Osservatorio Astronomico di Torino, Italy
(E-mail: eantonucci@to.inaf.it)

(Received 6 July 2006; Accepted in final form 28 July 2006)

Abstract. The dynamics of the solar corona as observed during solar minimum with the Ultraviolet Coronagraph Spectrometer, UVCS, on SOHO is discussed. The large quiescent coronal streamers existing during this phase of the solar cycle are very likely composed by sub-streamers, formed by closed loops and separated by open field lines that are channelling a slow plasma that flows close to the heliospheric current sheet. The polar coronal holes, with magnetic topology significantly varying from their core to their edges, emit fast wind in their central region and slow wind close to the streamer boundary. The transition from fast to slow wind then appears to be gradual in the corona, in contrast with the sharp transition between the two wind regimes observed in the heliosphere. It is suggested that speed, abundance and kinetic energy of the wind are modulated by the topology of the coronal magnetic field. Energy deposition occurs both in the slow and fast wind but its effect on the kinetic temperature and expansion rate is different for the slow and fast wind.

Keywords: solar wind, solar corona, composition

1. Introduction

The observations of the outer corona, beyond 1.5 solar radii, obtained with the Ultraviolet Coronagraph Spectrometer (UVCS) on the Solar and Heliospheric Observatory (SOHO), have allowed us to observe for the first time the solar wind in the corona, thus leading to new, surprising results on the physical properties of the regions where the solar wind is accelerated and on the signatures of the plasma acceleration processes. With respect to previous coronagraphs, the novelty of UVCS resides on the capability to investigate spectroscopically the coronal sources of the solar wind and thus to diagnose the composition of the coronal plasma and the macroscopic and microscopic velocity fields of several species. During low solar activity, the corona is shaped by a magnetic dipole or quadrupole, and its emission is extremely dim out of the equatorial belt. In this period UVCS can explore at best these rarified regions and the initial propagation of the fast and slow wind. The study of the most intense ultraviolet lines emitted in the extended corona, primarily, the HI Lyman α line at 1216 Å from neutral hydrogen and the OVI doublet at 1032 Å and 1037 Å from the five time ionized oxygen ions, shows that in the fast wind regions the hydrogen and oxygen wind components exhibit increasingly large kinetic temperature with height, and that this quantity progressively differentiates for the two species. In particular the oxygen component achieves extremely high

Space Science Reviews (2006) 124: 35–50
DOI: 10.1007/s11214-006-9098-6

kinetic temperatures, about 10^8 K above 2 solar radii, a region where it is possible to establish without doubt that the oxygen ion velocity distribution is anisotropic across the coronal magnetic field. This turns out to be the clearest signature of the physical process responsible for the acceleration of the oxygen component of the fast wind.

The main results obtained in a decade by analyzing spectroscopically the outer corona with UVCS, including the discoveries related to the acceleration and the evolution over a cycle of the solar wind, have been recently reviewed in an extensive paper by Kohl *et al.* (2006). The reader is referred to this paper for acquiring an exhaustive view of the physics of the solar outer corona. Here we intend to discuss the physics of the fast and slow wind in the outer corona in the light of the most recent results obtained with the UVCS data. They refer to the solar minimum period, when the simplified coronal configuration allows an easier interpretation of the observations of coronal expansion.

2. Fast Expansion in the Core of Large Polar Coronal Holes

In the tenuous expanding polar corona, protons and electrons depart from thermal equilibrium in the inner layers, at about 1.2–1.3 solar radii, whilst protons and heavy ions do so farther out, at about 1.5 solar radii (Withbroe *et al.*, 1982; Esser *et al.*, 1999). Different species therefore have their own temperature as observed with UVCS and the SUMER spectrometer on SOHO. The broadenings of the coronal spectral lines provide information on the temperature by probing the atom/ion velocity distribution along the line of sight, which includes both thermal and non-thermal velocities. The kinetic temperature, T_k, of the species is related to the line width by $T_k = \frac{mA}{k}(\frac{\sigma(\lambda)c}{\lambda})^2$, where m is the proton mass, A the atomic mass number, k the Boltzmann constant, $\sigma(\lambda)$ and λ, the standard deviation and wavelength of the spectral line. The most surprising feature in the temperature structure of a polar corona hole is the extremely high oxygen kinetic temperature, almost two orders of magnitude higher than that of neutral hydrogen, which never exceeds 3×10^6 K in the range of the UVCS observation (Kohl *et al.*, 1997; Antonucci, 1999; Antonucci *et al.*, 2000a). Recently the influence on line broadening of the coronal expansion, which Doppler shifts the emission in a symmetric way relative to the plane of the sky, has been assessed by Telloni *et al.* (2006) on the basis of a set of polar holes observed in the years 1996–1997 (Figure 1). The oxygen kinetic temperature, when subtracting the effect of the Doppler shifts due to coronal expansion, remains of the order of 10^8 K and peaks around 2.8 solar radii, thus confirming the estimate of the bulk motions effect on the coronal oxygen temperature given by Antonucci *et al.* (2000a). The analysis of the UVCS data by Telloni *et al.* (2006) extends the measurements out to 5 solar radii, thus finding that the kinetic temperature curve is flattening beyond 2.8 $R_\odot$. The large increase of the oxygen kinetic temperature with heliocentric distance clearly points to the existence of a physical process that

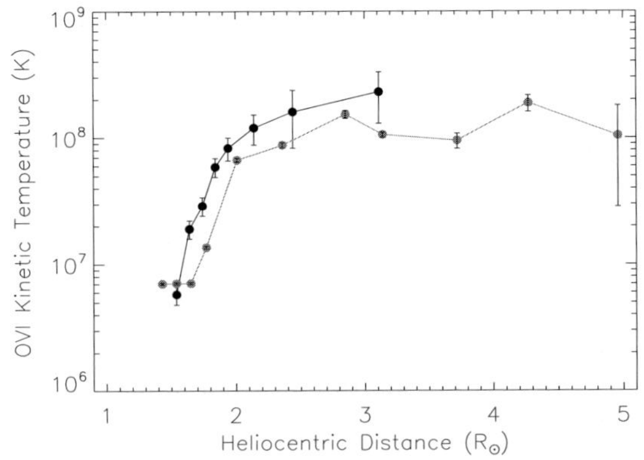

Figure 1. O VI kinetic temperature, T_{OVI}, k, across the magnetic field in the core of polar coronal holes during solar minimum: the black line shows the results without correction for the effect of coronal expansion (Antonucci *et al.*, 2004), and the grey line, accounts for this effect (Telloni *et al.*, 2006). (Courtesy of Daniele Telloni.)

acts in the outer corona to accelerate the ions. Energy is already deposited in the inner corona in order to increase the oxygen temperature to about 10^7 K as observed at 1.5 $R_\odot$. Farther out the abrupt jump in kinetic temperature observed at 1.7 $R_\odot$ is suggesting a discontinuity either in the energy deposition process or in its efficiency. An almost steady regime is then established above 2.8 $R_\odot$.

Line broadening can only probe the ion velocity distribution along the line of sight. This direction however is crucial being the direction perpendicular to the magnetic field, which is oriented almost radially in the core of polar holes. Fortunately the spectroscopic diagnostics proposed by Noci *et al.* (1987) to derive the outflow velocity, based on the ratio of the OVI 1037 to the 1032 line, has provided an important tool to investigate the 3D ion distribution. By applying this technique it is indeed possible to prove the existence of an anisotropy in the velocity distribution and to constrain the value of the anisotropy degree, at least in the region above 1.8 solar radii, where the oxygen kinetic temperature across the field is very large.

In the tenuous outer corona radiative excitation of the OVI lines, linearly dependent on the electron density, becomes dominant relative to collisional excitation, which depends on the density squared. In an expanding corona, the radiative emission is Doppler dimmed. The ratio of the intensities of the OVI lines is an excellent diagnostic technique since the two lines are differently affected by Doppler dimming. The disk spectrum at the blue side of the OVI 1037.61 line center is populated by lines that, red-shifted proportionally to the coronal outflow velocity can be absorbed by the OVI 1037.61 coronal ions. The largest pumping effect on oxygen of the most significant lines at its blue side, namely CII 1037.02 Å and 1036.34 Å, occurs for an expansion speed of 170 $\mathrm{km\,s^{-1}}$ and 370 $\mathrm{km\,s^{-1}}$ respectively. (The effect of pumping of the C II lines, the extreme broadening of the coronal absorbing profiles, and the degree of anisotropy of the ion velocity distribution, have been fully included in the diagnostic codes developed in 1998 by Dodero *et al.* (1998) and Li *et al.* (1998)). For models of coronal hole density and temperature compatible with the observations, the ratios of the Doppler dimmed OVI lines as a function of the coronal outflow velocity are not consistent with the observed ratios in an extended region of the outer corona, unless a significant degree of anisotropy is introduced in the ion velocity distribution. However, only the lowest degree of anisotropy compatible with the data can be established. The Doppler dimming analysis of the OVI doublet thus does provide partial information of the velocity distribution of the oxygen component of the solar wind along the magnetic field lines. The oxygen distribution is then bound to be anisotropic above 1.8 $R_\odot$, just where an abrupt increase of the oxygen kinetic temperature across the magnetic field occurs. The temperature along the field is not increasing at the same rate as the temperature across the field, thus the anisotropy proves the existence of a strong preferential acceleration of the oxygen ions and, in turn, a vigorous preferential energy deposition across the magnetic field, as discussed later on.

According to one possible scenario, the oxygen velocity distribution which is isotropic out to about 1.8 $R_\odot$ becomes anisotropic above this height but with the lowest degree of anisotropy compatible with the UVCS data. The anisotropy ratio, $T_\perp/T_\parallel$, in this case, is minimum throughout the outer corona; it increases from unity to a value of about 10 at 2 $R_\odot$, the maximum being ≈ 20 farther out (open circles in Figure 2). Alternatively, the oxygen temperature along the magnetic field retains the value of the electron, or more likely of the proton, temperature, as in

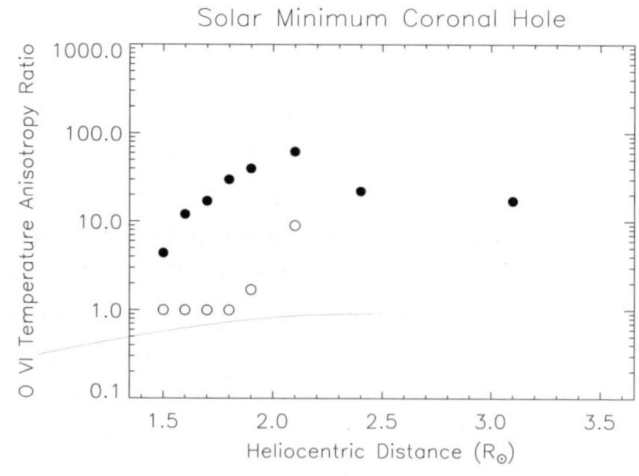

Figure 2. Ratio of the O VI kinetic temperature across and along the coronal magnetic field, $T_\perp/T_\parallel$, in the core of a polar coronal hole; this quantity measures the degree of anisotropy of the oxygen ion velocity distribution (Antonucci, 2002, presented at the Solar Wind 10 conference). Open circles denote isotropy out to 1.8 $R_\odot$ and minimum anisotropy farther out (first hypothesis discussed in the text); full circles denote maximum anisotropy out to 2.1 $R_\odot$ (second hypothesis in the text); the values at 2.4 $R_\odot$ and 3.1 $R_\odot$ correspond to minimum anisotropy (these points are common to the first and second hypothesis formulated in the text).

the innermost corona and the highly anisotropic distribution relaxes, above 1.8 $R_\odot$ toward the lowest anisotropy compatible with the data (full circles in Figure 2). This situation correspond to high anisotropy ratios, of the order of 60 around 2 $R_\odot$. In the extreme case still compatible with the data maximum anisotropy is maintained throughout the observed region.

If the anisotropy ratio is assumed to be the lowest throughout the outer corona, ions are first heated to a few 10^7 K according to a maxwellian distribution, and then above 1.8 $R_\odot$ a strong preferential heating across the field is established, which results in a bimaxwellian distribution, with a temperature about 10^7 K along the magnetic field (Antonucci *et al.*, 2000a). In the second case (full circles in Figure 2), acceleration occurs always preferentially across the field, but the energy is redistributed more efficiently in the direction along the field above 2.1 $R_\odot$. If instead maximum anisotropy is maintained throughout the range of observation, energy is not redistributed and the anisotropy ratio remains close to the value reached around 2 solar radii.

Since very large anisotropy ratios are quite difficult to be explained in terms of the physical processes that can act in the outer corona, the first scenario, which implies lower anisotropy ratios, appears to be the most plausible. In this case, however, one should explain the reason for the transition from a maxwellian to a bi-maxwellian distribution when the wind propagates across 1.8 $R_\odot$. This might reside in a difference in either the deposition or the redistribution of the energy. The above results on anisotropy are obtained without correcting the broadening of OVI lines for the effects of the outflow speed of the solar wind. Telloni *et al.* (2006) clearly show that this effect does not alter the importance of the anisotropy that is established in the outer corona, that has recently been questioned in a paper by Raouafi and Solanki (2004).

The preferential acceleration of the oxygen ions across the magnetic field, revealed by their anisotropic temperature, is an indubitable signature of the energy addition required to create the fast wind. It is well known that the fast polar wind requires a supply of energy, in addition to that provided by thermal conduction, in the region of the supersonic flow in order to attain the high asymptotic speed observed '*in situ*' (Leer and Holzer, 1980). The observations indeed prove that energy is added in an extended region just above the sonic point, that in a coronal hole is reached roughly at 1.6-1.7 $R_\odot$ (Figure 3), as required in order to accelerate the wind without increasing the mass flux of the wind itself.

As a likely candidate for accelerating the ions across the magnetic field Cranmer *et al.*, 1999 have suggested ion cyclotron resonance, a process of collisional wave-particle interaction that can damp energy of the Alfvén waves assumed as propagating in the corona from the base of the solar atmosphere. The energy of ions perpendicular to the field direction is increased at the expense of the energy of the waves. One of the main issues in this scenario is how to create a wave spectrum that can be absorbed by the coronal ions over an extended region of the outer corona. The hypotheses put forward up to now to explain the needed wave spectrum are

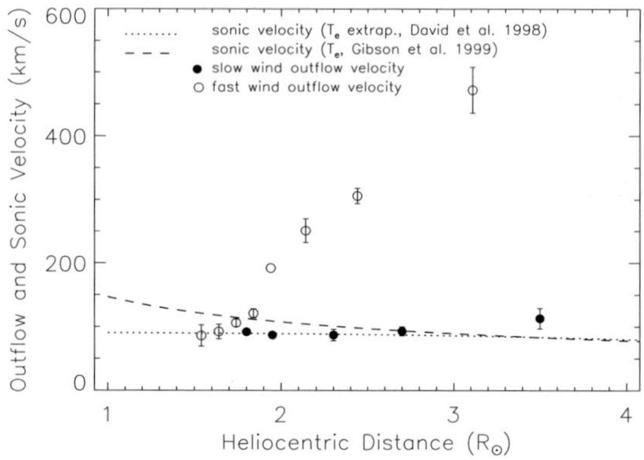

Figure 3. Sound and expansion outflow speed in the core of a polar coronal hole for the fast wind (open circles) and the slow wind (full dots). The outflow velocities are derived from Antonucci *et al.* (2004, 2005). The plasma temperature from David *et al.* (1998) for coronal holes and Gibson *et al.* (1999) close to the streamer belt. (Courtesy of Daniele Telloni.)

summarized and discussed by Kohl *et al.* (2006). They range from the idea that high frequency oscillations are generated in small–scale reconnection events at the base of a coronal hole to the scenario of the local wave generation which implies that low-frequency waves generated at the coronal base are gradually converted into ion cyclotron waves in the outer corona.

Another crucial aspect is to create a model of energy deposition on the basis of a physical process which can generate the observed oxygen temperature ratios which however, as shown in the previous paragraph, cannot be univocally defined. Up to now efforts have been directed to explain the extremely large anisotropy ratios invoked by Cranmer *et al.* (1999). Less attention has been paid to the fact that the data do not provide a unique curve of anisotropy ratios versus heliocentric distance. Equally valid from the observational point of view is a polar region with an oxygen component with a relatively small temperature anisotropy which at maximum reaches approximately a value of 10–20 as shown in Figure 2. Therefore efforts should be devoted to model a wave particle interaction process that causes a temperature across the field $\approx$10 times larger than that along the field. In addition a model should fully explain also the variations of the anisotropy ratio with height due to the interplay between an energy deposition perpendicular to the field, extended over more than 3 solar radii (Figure 2), and the transfer of energy to the direction along the field lines.

Energy deposition has the effect of a rapid increase of the speed of the oxygen component of the solar wind as shown in Figure 4, where the results obtained by Telloni *et al.* (2006), show that the wind has reached 500 km s^{-1} at 5 $R_\odot$.

The behavior of the main wind component formed by protons, can be traced by the neutral hydrogen atoms as far as charge exchange is sufficiently efficient in an ambient becoming less and less dense with height, that is out to about 3–4

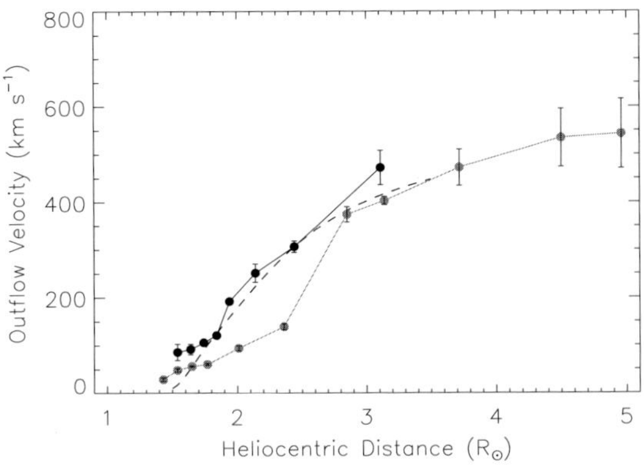

Figure 4. Outflow velocity of the oxygen component of the fast solar wind in the core of a polar hole according to the results by Telloni *et al.* (2006), grey dots; Antonucci *et al.* (2004), blueblack dots; and Cranmer *et al.* (1999), dashed line; computed for maximum anisotropy of the oxygen ions. (Courtesy of Daniele Telloni.)

solar radii. The acceleration of protons leads to lower temperatures, $\leq 3 \times 10^6$ K, than the oxygen ions. The anisotropy of the proton velocity, indicating preferential acceleration across the field is likely to occur (Cranmer *et al.*, 1999) but it cannot be fully proved (Antonucci *et al.*, 2000b). At the same time there are indications that the outflow speed of the proton component might be lower than that of the oxygen component (Cranmer *et al.*, 1999).

Although the oxygen component has a negligible influence on the physics of the solar wind, due to the low oxygen abundance in the solar plasma, the knowledge that can be achieved from the results obtained by its observation is crucial in order to understand the physical processes in a collisionless coronal plasma and to pave the way for a full understanding of the acceleration of the main proton component of the solar wind.

3. Slow Expansion at the Edges of the Large Polar Coronal Holes and Between Sub-Streamers

The identification of the sources and acceleration regions of the slow solar wind is less straightforward than that of the fast wind which emanates from the core of the large polar coronal holes characterizing the Sun during the activity minimum. The solar minimum '*in situ*' measurements obtained with Ulysses clearly indicate the existence of two basic solar wind regimes, the fast and slow wind, divided by a sharp boundary, e.g. von Steiger *et al.* (2000). The slow wind in this phase of the cycle is confined in the equatorial belt. The main questions then are whether in corona there are two acceleration regions with distinct characteristics for the slow and fast wind, separated by a sharp boundary as observed in the heliosphere and whether there might be more than one coronal source of the slow wind.

The anomalous depletion of the oxygen relative to hydrogen abundance, observed with UVCS in the core of large quiescent solar minimum streamers, has been immediately related to a possible signature of the slow wind in the outer corona. However two alternative suggestions were put forward. The first one relates the slow wind to the oxygen depletion in the streamer core (Noci *et al.*, 1997). The second one suggests the bright regions existing inside the streamer at the sides of the core as sources of the slow wind (Raymond, 1997). Noci *et al.* (1997) proposed to interpret the core dimming by relating it to the existence of open field lines inside magnetically complex large streamers. Such open field lines are capable of channelling plasma flowing at low speed into the heliosphere. If a streamer is formed by multiple sub–streamers, these are separated by open flux tubes characterized by an expansion factor which does not increase monotonically with height. Such flux tubes indeed widen and then shrink before reaching the critical point with a consequent decrease of the outflow speed relative to a radial flow. Thus in the flux tube the proton velocity decreases whereas the density remains the same. The reduction of the proton flux implies a reduction of the dragging force exerted by the protons on the oxygen ions that can lead to the depletion of oxygen observed in the streamer core. On the other hand, the finding that the low first ionization potential (FIP) elements, such as Si and Fe, are enhanced in the bright regions of the streamer, observed at the sides of the OVI dim core, has led to the suggestion that the majority of the slow wind originates from these regions (Raymond, 1997). In the bright regions the elemental abundance, at least in the case of oxygen, decreases with height (Marocchi *et al.*, 2001). Due to the fainter emission of Si and Fe it is not known whether their abundance varies with height and whether the observed enhancement of low FIP elements persists where the bright regions of streamers extend outward, thus possibly contributing to the slow wind.

The slow wind in corona has also been investigated by analyzing the dynamics of the plasma in the streamer belt. The streamer boundary can be considered either as a possible dividing line between the two regimes of fast and slow wind or simply the separation between the open field lines of the polar holes bordering the closed magnetic field of the streamer (notwithstanding the fact that the streamer might indeed contain open flux tubes in the inner part, as discussed above). If the boundary is identified with the line marking the $1/e$ decrease of the peak intensity inside the streamer, plasma at low speed is observed to flow in a lane, 15°–20° wide, situated outside and bordering the streamer. In this flux tube the outflow speed varies from about 90 km s^{-1} at 1.8 $R_\odot$ to 110 km s^{-1}at 3.5 $R_\odot$ (Abbo and Antonucci, 2002; Antonucci *et al.*, 2005). On the other hand, the ultraviolet emission averaged inside the streamer boundaries does not show any significant Doppler dimming, hence the streamer plasma is predominantly static. Outflows can exist but at very low speed or confined to tiny regions. Since the bright regions at the sides of the dim core dominate the streamer emission, the logical consequence is that these regions cannot be the main contributors to the slow wind (Antonucci *et al.*, 2005). Frazin *et al.* (2003) also exclude significant outflow velocities inside streamers. Expansion at low speed is also observed along the streamer axis approximately at, or above, the streamer cusp, where the interplanetary current sheet forms. Flows originate between 2.7 $R_{\odot}$ and 3.5 $R_{\odot}$ as shown by Abbo and co-authors (Abbo and Antonucci, 2002; Abbo *et al.*, 2003; Antonucci *et al.*, 2005), and farther out between 3.6 and 4.1 $R_{\odot}$ in a streamer observed by Strachan *et al.* (2002). The plasma flowing along the streamer axis has a tendency to be slower, 90 km s^{-1}, than that observed outside the streamer boundaries. The results obtained for the slow wind by studying the oxygen outflows are also representative of the proton behavior, since in the coronal regions under investigation the expansion time of the coronal plasma does not exceed the proton/oxygen thermalization time out to ≈ 2.7 $R_{\odot}$ (Antonucci *et al.*, 2006).

The dynamics of the coronal plasma in the streamer belt and its adjacent regions suggests the existence of two components of the slow wind: an outer component which flows first along the open filed lines surrounding the streamer boundary and farther out the heliospheric current sheet; an inner slower component which is detected where the current sheet forms and flows along it. The most plausible explanation (Antonucci *et al.*, 2005) is that the inner slower wind component, according to the hypothesis put forward by Noci *et al.* (1987), is formed by the plasma channelled along the flux–tubes separating sub–streamers, and extending along the streamer axis. Recently G. Noci has derived the contour marking the minima of the intensity ratio of the OVI 1037 Å to the OVI 1032 Å line in a streamer. Below this line the coronal plasma cannot expand. Noci finds that this line reaches its minimum height in the dim core of a streamer, whilst its distance from the Sun peaks near to the brightest features at the sides of the streamer core (Figure 5). This

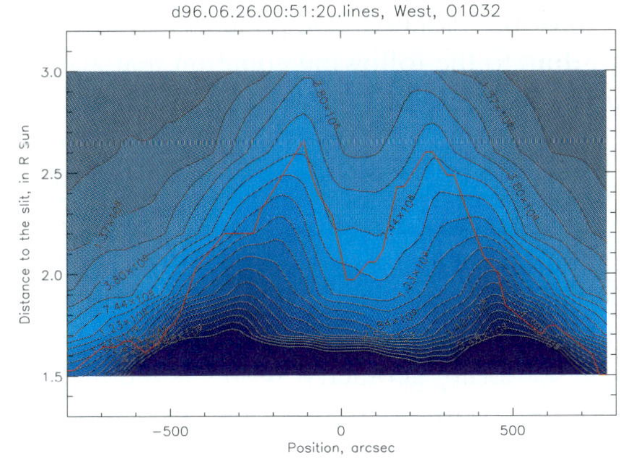

Figure 5. OVI 1032 intensity contours (colored lines) in a large quiescent equatorial streamer compared with the dark line marking the region where the plasma is static. (Courtesy of Giancarlo Noci, presented at the UVCS meeting, Giardini Naxos, 13 May 2006).

proves that the bright regions observed inside a streamer are predominantly regions of plasma confined by closed magnetic fields. Thus they cannot be plausible sources of slow wind. Furthermore it proves that the dim core, where the open magnetic field lines separating sub–streamers converge to extend into the heliosphere, can be the only source of slow wind inside a quiescent streamer. If the bright regions are indeed characterized by a closed field than the decrease of their oxygen abundance with height might be due to gravitational settling (Marocchi *et al.*, 2001).

The scenario derived on the basis of the UVCS results is consistent with the two–component solar wind proposed by Wang *et al.* (2000) by studying the flow of the small inhomogeneities, or coronal blobs, observed in the visible light coronal images, which form at the streamer cusp and move along the current sheet.

The conclusions that can be drawn on the slow wind in corona are confirmed by an analysis of the absolute abundance of oxygen in the regions where the wind outflows are observe. A comparison of this quantity to that detected '*in situ*' in the slow wind streams indeed substantiates the finding that the slow wind is predominantly generated at the edges of the solar minimum polar coronal holes, close to the streamer boundary (Antonucci *et al.*, 2006).

In the upper chromosphere abundances are affected by an atom-ion separation process depending on the element first ionization potential (FIP), leading to an enrichment of low FIP elements ($<$10 eV) more enhanced in the slow than in fast wind (von Steiger *et al.*, 2000). Oxygen however is an element with high first ionization potential (13.6 eV), the same as hydrogen, thus its wind abundance should retain the photospheric value. Yet the SWICS instrument onboard Ulysses reveals a tendency to a systematic variation of the oxygen abundance between 5.3×10^{-4} (8.7) and 6.3×10^{-4} (8.8), as the wind speed varies between 400 km s^{-1} and 800 km s^{-1} typical values of the slow and fast wind streams, respectively (von Steiger *et al.*, 1995). This variation is then probably not related to the FIP effect but to an effect occurring during the wind propagation. The oxygen abundance can be derived from the ratio of the radiative components of the OVI 1032 and HI 1216 line intensities, according to the following equation that accounts for the Doppler dimming where outflows are present:

$$\left(\frac{N_{\rm O}}{N_{\rm H}}\right) \sim \frac{I_{r,\rm OVI}}{I_{r,\rm HI}}\frac{b_{\rm HI}}{b_{\rm OVI}}\frac{B_{12,\rm HI}}{B_{12,\rm OVI}}\frac{\lambda_{0,\rm HI}}{\lambda_{0,\rm OVI}}\frac{n_{\rm HI}/n_{\rm H}}{n_{\rm OVI}/n_{\rm O}}\frac{\langle\int_\Omega \Phi(\delta\lambda_{0,\rm HI})p(\varphi)\,d\omega\rangle}{\langle\int_\Omega \Phi(\delta\lambda_{0,\rm OVI})p(\varphi)\,d\omega\rangle}, \tag{1}$$

where b is the branching ratio, B_{12} the Einstein coefficient, $\frac{n_{\rm HI}/n_{\rm H}}{n_{\rm OVI}/n_{\rm O}}(T_e)$ the ratio of HI atoms and OVI ions concentrations as a function of the electron temperature T_e, $p(\varphi)$ accounts for the scattering geometry, $\Phi(\delta\lambda)$ is the integral of the product of the intensity $I_{ex}(\lambda)$ of the exciting spectrum and $\Psi(\lambda, \mathbf{n})$ the normalized coronal absorption profile along the direction of the incident radiation $\mathbf{n}$, and $\delta\lambda$ is the wavelength shift due to the outflow velocity (Antonucci *et al.*, 2006).

Inside streamers the average oxygen abundance rapidly falls with altitude below the slow wind composition values (Figure 6), according to the tendency noted by

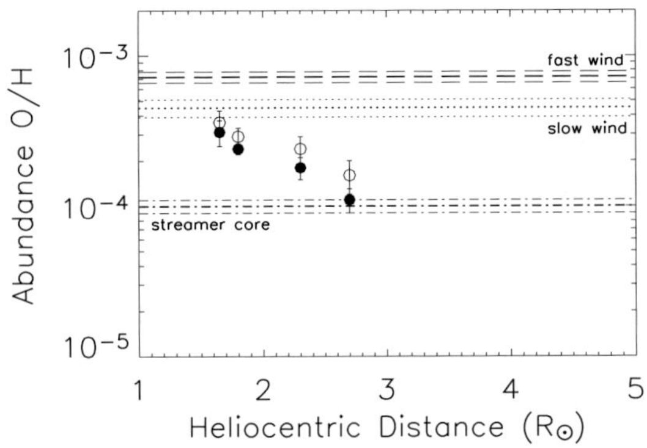

Figure 6. Average oxygen abundance inside streamers. Full dots refer to the results obtained in the assumption of anisotropic ion velocity distribution, open dots to the case of isotropy (Antonucci *et al.*, 2006).

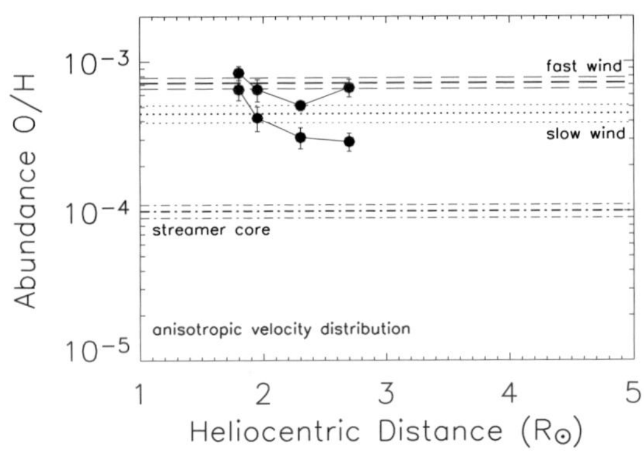

Figure 7. Oxygen abundance of slow wind in corona outside streamers, obtained for anisotropic velocity distribution of the oxygen ions. The upper and lower curves delimit the region of acceptable results. The width of this region is related to the uncertainty in the local electron temperature (Antonucci *et al.*, 2006).

Marocchi *et al.* (2001). This confirms that the brightest regions of the streamers have to be excluded as sources of the slow wind. On the other hand, the oxygen abundance of the coronal plasma flowing at low speed, ≥ 90 km s^{-1}, just outside the boundaries of a solar minimum streamer is consistent with the slow wind composition observed in the heliosphere (Figure 7). It is interesting to note that at about 1.8 $R_{\odot}$, the oxygen abundance of the slow wind is close to the value found for the the fast wind in corona (Antonucci and Giordano, 2001) and in the heliosphere (von Steiger *et al.*, 1995). These findings support the conclusion that the edges of a coronal hole can be the sources of the slow wind during solar minimum. It is to be noted that close to the Sun the abundances of the slow and fast wind are in agreement with the photospheric value, 6.7×10^{-4}, determined by Grevesse and Sauval (1998), and not with the much lower value, 4.6×10^{-4}, found by Asplund *et al.* (2004).

The main result then is that the major component of the slow wind is observed along the open field lines outside the streamer boundary. This implies that both the fast and slow wind are originating in the large polar coronal holes which are characteristic of solar minimum. The fast wind emanates from the core of coronal holes and the slow wind toward the edges of the polar holes, in the regions bordering the streamer. At the base of the corona the plasma emanating from the edges of coronal holes indeed has similar properties, e.g. density and oxygen abundance, to that emanating from the core. Its physical properties however differentiate during propagation (Antonucci *et al.*, 2005, 2006).

The question concerns then what causes different conditions of propagation when spanning the coronal hole from the core to the edge. The most obvious difference resides in the topology of the magnetic field lines. The coronal plasma flows in flux tubes with expansion factors which show a strong variation from the core to the edge, as pointed out by Wang and Sheeley (1990), who proposed the magnetic topology as an element of strong influence on the wind speed. The variation of the expansion factor with latitude was therefore also invoked by Antonucci *et al.* (2005) to explain the results obtained with UVCS indicating that the slow wind is emanating from the edges of coronal holes.

Large expansion factors characterize the coronal segment of the flow tubes far from the axis of a coronal hole. Furthermore close to and at the interface with the streamer the expansion factor peaks and then decreases before the wind reaches the critical point (Wang and Sheeley, 1990). The last conditions are the same ones postulated by Noci *et al.* (1997) for the field lines separating sub–structures in streamers.

Therefore we can extend the Noci *et al.* idea, put forward to interpret the oxygen dimming in the streamer core, to the case of the plasma expansion in the flux tubes bordering the streamer boundary, in order to explain both the slow flow and the abundance decrease with height observed in Figure 7. The narrowing of the flux tube indeed can cause the low speed of expansion and at the same time result in a depletion of the oxygen abundance. The argument is the following. The characteristics of the critical solution in flux tubes narrowing before the critical point can be derived in comparison with the solution for a wind which expands radially. The equation for the critical point is

$$\frac{2kT}{m}\left(\frac{2}{r}+\frac{1}{F}\frac{dF}{dr}-\frac{1}{T}\frac{dT}{dr}\right)-\frac{GM}{r^2}=0, \tag{2}$$

T the temperature, G the gravitational constant, M the solar mass, $F(r) = \frac{A(r)}{r^2}$ describes the deviation of the expansion factor from the radial case and $A(r)$ is the cross section of the flux tube. In the radial expansion case, $\frac{dF}{dr} = 0$, the left hand side of Equation 2 is zero at the critical point $r = r_c$. If the flux tube narrows below the critical point then $\frac{dF}{dr} < 0$ at r_c as defined in the radial case. Since the left hand side of the equation is negative below the critical point, the conclusion is that a decrease of $A(r)$ with height pushes the critical point outward thus lowering the outflow speed below it. Noci *et al.* have proposed this effect to explain why the outflow speed of the plasma which moves between sub–streamers (in flux tubes converging in a multiple current sheet) is smaller than that flowing where no narrowing of the flux tube occurs. The same reasoning can be applied to the coronal hole flux tubes close to and at the interface with a streamer, since these are affected by a narrowing of the cross section before the critical point (Wang and Sheeley, 1990), that for the slow wind is found at about 2.8–3 $R_\odot$ (Figure 3). This argument is valid only when the flux tubes have a negative $\frac{dF}{dr}$ in corona. Departing from the streamer boundary $\frac{dF}{dr}$ becomes positive but still significantly larger than in the core of coronal holes in the range of coronal heights of interest (Wang and Sheeley, 1990).

The Noci *et al.* discussion indicates that the propagation in corona can modify the abundance of the elements. Therefore it would be necessary a quantitative analysis of the effect of a reduced dragging force on the different ion species (Geiss *et al.*, 1970) to assess if this effect could also explain the relative enhancement in the low FIP elements which in the slow wind is found to be larger than in the fast wind (von Steiger *et al.*, 2000).

The consistency of coronal and heliospheric oxygen abundance shown in the analysis by Antonucci *et al.* (2006) is ensured by imposing the constraint that in the region where the slow wind is accelerated the oxygen ion velocity distribution is anisotropic. In analogy with the fast wind case, we thus obtain indirect information on energy deposition which also in the slow wind occurs preferentially across the field. Thus it can be further suggested that the same heating process acting in the fast wind, also occurs in the slow wind. The process however is less effective, since in the slow wind at 2.0–2.7 $R_\odot$ the oxygen kinetic temperature, of the order of 1×10^7 K, remains approximately one order of magnitude lower than in the core of coronal holes (Figure 8). The lower kinetic temperatures in the slow wind suggest that the rate of energy deposition itself might be related to the local magnetic topology, as the outflow speed. Energy deposition by wave–particle interaction can indeed vary according to the divergence of the open field lines that determine the ion cyclotron frequency. It is however necessary to investigate in detail whether and how the rate of energy deposition is quantitatively related to the local magnetic field topology in different regions of a coronal hole. That is, whether the geometry of the flux tube is indeed influencing energy deposition in such a way to vary its rate from the core to the edge of coronal holes.

In polar regions the outflow velocity reaches the sonic value at about 1.7 $R_\odot$. Beyond the sonic point acceleration is indeed increasing significantly as does the observed oxygen kinetic temperature, indicating the deposition of energy. In the slow wind the sonic point occurs around 3 $R_\odot$ (Figure 3), thus energy is deposited over an extended region in the subsonic flow with the effect that the mass flux of the slow wind should increase, in agreement with the results obtained '*in situ*'. It is however true that both winds reach approximately the same kinetic temperature in the subsonic region. Hence only an accurate quantitative assessment of the energy

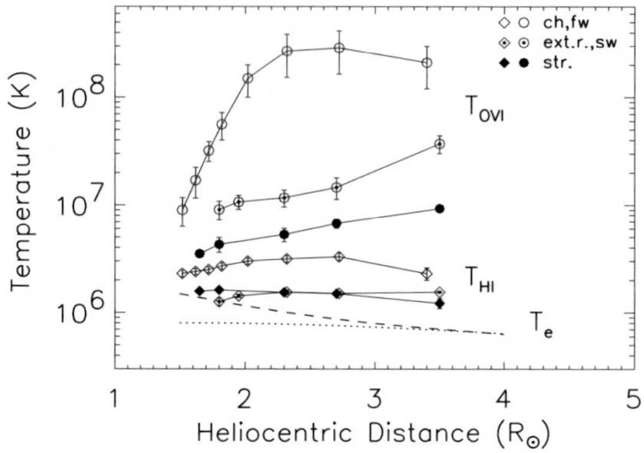

Figure 8. Kinetic temperature variation of the OVI ions (open circles, coronal holes; dot–circle, slow wind at the edge of a polar hole; full circles, streamers), HI neutral hydrogen (open diamonds, coronal holes; dot–diamond, slow wind at the edge of a polar hole; full diamond, streamers) and electron temperature (dotted line, coronal holes; dashed line, streamers) in the outer corona from the core of polar holes to streamers (Antonucci *et al.*, 2005).

deposited in the acceleration region of the fast and slow wind can relate on a quantitative basis the energy addition to the asymptotic speed and mass flux reached by the fast and slow wind.

The discussion on the oxygen abundance of the plasma expanding at low speed in proximity of streamers leads then to the conclusion that energy is deposited not only in the fast but also in the slow coronal wind and that the efficiency of the process is probably related to the local magnetic topology, being more efficient in less divergent magnetic field lines.

4. Conclusions

In conclusion during solar minimum polar coronal holes are identified as the sources not only of the fast wind which emanates from their core and reaches a speed of 500 km s^{-1} around 5 $R_\odot$. Polar holes are also the most likely sources of the majority of the slow wind which originates toward their edges close to the interface with streamers and reaches 110 km s^{-1}around 3.5 $R_\odot$. It is suggested that the varying magnetic topology of the holes from their core to the edges is the clue to understand the formation of the fast and slow wind. Moving away from the core of a coronal hole the expansion factor of the flux tubes increases and at the interface with streamers the flux tubes widen and then shrink still in the subsonic region, with the effect of slowing down the flow. This process has been first proposed by Noci *et al.* (1997) to suggest that also open field lines separating the closed loops of sub–streamers, and converging outward to form a multiple current sheet, might channel slow wind. This hypothesis has indeed been recently confirmed, thus weakening the suggestion that the bright regions at the sides of the oxygen dim core of streamers might be the sources of the slow wind. The wind properties, such as density, abundance and outflow speed, have a tendency to vary without abrupt changes from the edges to the core of coronal holes indicating that in corona a gradual variation from the slow to the fast wind is more likely. This is in contrast with the heliospheric sharp transition existing between the two wind regimes. Therefore we suggest that the change from gradual to sharp transition from the fast to the slow wind is indeed a result of the wind propagation.

Energy is supplied at a high rate in the oxygen component of the fast wind above the sonic point, as required by the theory to reach the heliospheric asymptotic value of the outflow speed. The mechanism of energy deposition has to be found in processes of wave particle interaction which act across the magnetic field, such as ion cyclotron resonance, in order to induce the observed anisotropy in the oxygen ion velocity distribution. Less stringent constraints on the models needed to explain energy deposition are posed by assuming an anisotropy ratio of the ion velocity distribution that at maximum reaches values of the order of 10–20. This however implies that below 1.7 $R_{\odot}$ the oxygen kinetic temperature, of the order of 10^7 K, should be isotropic. According to the observations energy is supplied also to the slow wind and there is evidence for preferential deposition across the field also in this case. For this reason the same deposition process it is probably at work in both types of wind. The addition of energy is observed in the subsonic slow wind, which has the effect to increase its mass flux. As the outflow speed is influenced by the topology of the magnetic field which guides the expansion in this region of the solar corona, also energy deposition is suggested to be influenced by the geometry of the field lines.

These conclusions are reached on the basis of the results on the outflow velocity, density and abundance of the wind obtained by analyzing the observations of the outer corona performed with UVCS during solar minimum.

Acknowledgements

The author wishes to thank Prof. Giancarlo Noci who has given the permission to refer to his recent results prior to their publication; Dr. Lucia Abbo and the Ph. D. student Daniele Telloni for their enthusiasm in the research activities on the solar wind based on the UVCS observations; Daniele Telloni for his assistance in preparing some of the figures. The support of the Agenzia Spaziale Italiana, ASI, through the contract ASI/I/035/05/0 is acknowledged.

References

Abbo, L. and Antonucci, E.: 2002, *ESA–SP* **508**, 477.

Abbo, L., Antonucci, E., Dodero, M. A., and Benna, C.: 2003, in *Proceedings SolarWind 10, AIP* **679**, 238.

Antonucci, E.: 1999, *ESA SP* **466**, 53.
Antonucci, E., Giordano, S., and Dodero, M. A.: 2000a, *Adv. Space Res.* **25**, 1923.
Antonucci, E., Dodero, M. A., and Giordano, S.: 2000b, *Solar Phys.* **197**, 115.
Antonucci, E., and Giordano, S.: 2001, in *Solar Galactic Composition*, *AIP* **598**, 77.
Antonucci, E., Dodero, M. A., Giordano, S., Krishnakumar, V., and Noci, G.: 2004, *Astron. Astrophys.* **416**, 749.
Antonucci, E., Abbo, L., and Dodero, M. A.: 2005, *Astron. Astrophys.* **435**, 699.
Antonucci, E., Abbo, L., and Telloni, D.: 2006, *Astrophys. J.* **643**, 1239.
Asplund, M., Grevesse, N., Sauval, A.J., Allende Prieto, C., and Kiselman, D.: 2004, *Astron. Astrophys.* **417**, 751.
Cranmer, S.R., Kohl, J.L., Noci, G., *et al.*: 1999, *Astrophys. J.* **511**, 481.
David, C., Gabriel, A.H., Bely-Dubau, F., *et al.*: 1998, *Astron. Astrophys.* **336**, L90.
Dodero, M.A., Antonucci, E., Giordano, S., and Martin, R.: 1998, *Solar Phys.* **183**, 77.
Esser, R., Fineschi, S., Dobrzycka, D., Habbal, S.R., Edgar, R.J., Raymond, J.C., Kohl, J.L., and Guhathakurta, M.: 1999, *Astrophys. J.* **510**, L63.
Frazin, R.A., Cranmer, S.R., and Kohl, J.L.: 2003, *Astrophys. J.* **597**, 1145.
Geiss, J., Hirt, P., and Leutwyler, H.: 1970, *Solar Phys.* **12**, 458.
Gibson, S.E., Fludra, A., Bagenal, F., *et al.*: 1999, *J. Geophys. Res.* **104**, 9691.
Grevesse, N., and Sauval, A.J.: 1998, *Space Sci. Rev.* **85**, 161.
Kohl, J.L., Noci, G., Antonucci, E., *et al.*: 1997, *Solar Phys.* **175**, 613.
Kohl, J.L., Noci, G., Cranmer, S.R., and Raymond, J.C.: 2006, *Astron. Astrophys. Rev.* **13**, 31.
Leer, E., and Holzer, T.E.: 1980, *J. Geophys. Res.* **85**, 4681.
Li, X., Habbal, S.R., Kohl, J.L., and Noci, G.: 1998, *Astrophys. J.* **501**, L133.
Marocchi, D., Antonucci, E., and Giordano, S.: 2001, *Ann. Geophys.* **19**, 135.
Noci, G., Kohl, J.L., and Withbroe, G.L.: 1987, *Astrophys. J.* **315**, 706.
Noci, G., Kohl, J.L., Antonucci, E., *et al.*: 1997, in Wilson, A. (ed.), *Fifth SOHO Workshop, The Corona and Solar Wind near Minimum Activity*, *ESA SP*-**404**, Noordwijk, ESA, 75.
Raouafi, N.E., and Solanki, S.K.: 2004, *Astron. Astrophys.* **427**, 725.
Raymond, J.C., Kohl, J.L., Noci, G., *et al.*: 1997, *Solar Phys.* **175**, 645.
Strachan, L., Suleiman, R., Panasyuk, A.V., Biesecker, D.A., and Kohl, J.L.: 2002, *Astrophys. J.* **571**, 1008.
Telloni, D., Antonucci, E., and Dodero, M.A.: 2006, *Astron. Astrophys.*, submitted.
von Steiger, R., Wimmer-Schweingruber, R.F., Geiss, J., and Gloeckler, G.: 1995, *Adv. Space Res.* **15**, 3.
von Steiger, R., Schwadron, N.A., Geiss, J., *et al.*: 2000, *J. Geophys. Res.* **105**, 27217.
Wang, Y.H., and Sheeley, N.R.: 1990, *Astrophys. J.* **355**, 726.
Wang, Y.H., Sheeley, N.R., Socker, D.G., Howard, R.A., and Rich, N.B.: 2000, *J. Geophys. Res.* **105**, 25133.
Withbroe, G.L., Kohl, J.L., Weiser, H., and Munro, R.H.: 1982, *Space Sci. Rev.* **33**, 17.

SOLAR WIND SOURCES AND THEIR VARIATIONS OVER THE SOLAR CYCLE

R. SCHWENN
Max-Planck-Institut für Sonnensystemforschung, Max-Planck-Str. 2, 37191 Katlenburg-Lindau, Germany
(E-mail: schwenn@mps.mpg.de)

(Received 25 August 2005; Accepted in final form 4 September 2006)

Abstract. In this paper I will briefly summarize the present status of our knowledge on the four different sorts of solar wind, their sources and their short- and long-term variations. First: the fast solar wind in high-speed streams that emerges from coronal hole regions. Second: the slow solar wind emerging from the non-active Sun near the global heliospheric current sheet above helmet streamers and underlying active regions. Third: the slow solar wind filling most of the heliosphere during high solar activity, emerging above active regions in a highly turbulent state, and fourth: the plasma expelled from the Sun during coronal mass ejections. The coronal sources of these different flows vary dramatically with the solar activity cycle.

Keywords: corona, solar wind, high-speed streams, streamers, differential rotation, coronal mass ejections (CMEs), solar activity cycle

1. Introduction

From eclipse observations it has been well known that the solar corona is highly structured and dynamic. It changes its shape enormously during the solar activity cycle. Hence, it was no great surprise when both these properties (spatial structure and temporal variability) were found to be reproduced in the corona's offspring, i.e., the solar wind.

It was not until the Skylab era in 1973/1974 when coronal holes were discovered to be the sources of long-lived solar wind high speed streams (Krieger *et al.*, 1973). Coronal holes are usually located above inactive parts of the Sun, where "open" magnetic field lines prevail, e.g., at the polar caps around activity minima. In contrast, the more active near-equatorial regions on the Sun are most often associated with "closed" magnetic structures, such as bipolar loop systems and helmet streamers on top. It is important to note that both: the coronal holes as well as their offspring, the high-speed solar wind streams are representatives of the inactive or "quiet" Sun. Thus, the only state of the solar wind that may deserve the label "quiet" is the high speed wind, rather than the more variable slow wind from above active regions. This perception, first described by Feldman *et al.* (1976) and Bame *et al.* (1977), caused a major paradigm change. No longer could the slow wind be

DOI: 10.1007/s11214-006-9099-5

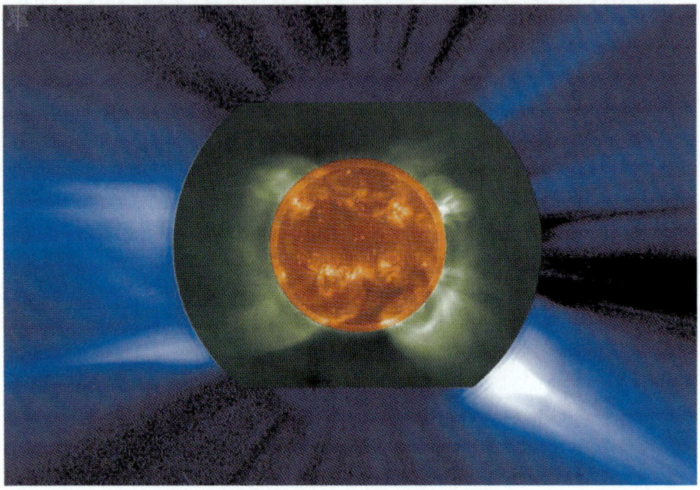

Figure 1. Montage of images from 3 different telescopes on SOHO: from EIT the EUV disk image (at 19.5 nm), from LASCO-C1 the inner corona (at 530.3 nm), from LASCO C2 the outer corona (in white light). Note, in particular, the dark area on the disk near the south pole which is a good example of a large coronal hole.

considered the "quiet" or "ground state" type although it would fit much better to the famous model of a thermally driven solar wind as derived by Parker (1958).

Figure 1 exhibits the two states of the near-minimum corona in 1998 and the associated solar wind rather nicely. The parts of this combined image were taken almost simultaneously from the EIT and LASCO telescopes on SOHO. This figure shows the two states by their different brightness in the corona. Note also how well separated from each other they are, both in the low corona as well as in the extended corona, i.e., in the solar wind.

In fact, the existence of sharp boundaries between solar wind streams (in longitude as well as in latitude) had already been noticed by Rosenbauer *et al.* (1977) and Schwenn *et al.* (1978) on the basis of *in-situ* measurements from the Helios solar probes that went as close as 0.3 AU to the Sun. These two basic types of quasi-steady solar wind differ markedly by their main properties and by the location and magnetic topology of their sources in the corona, thus probably in their acceleration mechanism.

In addition to these two basic states the slow solar wind filling most of the heliosphere during high solar activity has to be considered a third category. It emerges above active regions distributed over large parts of the Sun, far from the heliospheric current sheet, and in a highly turbulent state. It differs in some aspects from the minimum type of slow solar wind. Finally, we regard the plasma expelled from the Sun during huge coronal mass ejections as a category on its own, because of some fundamental differences to be described later.

The status of knowledge on the solar wind before 1972 had been very well summarized in the textbook by Hundhausen (1972). Then, from the mid 1970s on, a new class of space missions (Skylab, Helios, Voyager, and Ulysses) equipped with a new generation of instruments had initiated a new epoch in solar and heliospheric research. Numerous important discoveries were made and are documented in a multitude of scientific papers. Major reviews can be found, e.g., in Zirker (1977), Schwenn and Marsch (1990), Schwenn and Marsch (1991), (Marsden, 1986, 1995, 2001), Kohl and Cranmer (1999), Balogh *et al.* (2001). A comparable step forward occurred in the mid 1990s when the Yohkoh, SOHO, Wind, ACE, and TRACE spacecraft went into operation. Reviews of early results can be found, e.g., in Fleck and Svestka (1997), Brekke *et al.* (2001), and a series of SOHO Workshop Proceedings published in the ESA SP series.

Because of the wealth of the already existing literature I will only briefly summarize in this review the main results concerning solar wind types, sources and evolution. Rather, I try to guide the interested reader to pertinent references, i.e., usually to the first relevant paper plus sometimes some recent papers and reviews.

2. The Two Basic States of Solar Wind

In this chapter I will present the main characteristics of the two basic states of solar wind (high-speed streams, called HSS, and low speed wind of minimum type, called LSM), discuss in detail their differences and similarities and some other peculiarities.

In Table I the more specific values taken from Schwenn (1990) (called S90 in this paper) are given for both types separately:

TABLE I

Average solar wind parameters at 1 AU, for the time around solar activity minimum, compiled by S90.

	Low speed wind (LSM)	Fast wind (HSS)
Flow speed v_p	250–400 km s^{-1}	400–800 km s^{-1}
Proton density n_p	10.7 cm^{-3}	3.0 cm^{-3}
Proton flux density $n_p v_p$	3.7×10^8 cm^{-2}s^{-1}	2.0×10^8 cm^{-2}s^{-1}
Proton temperature T_p	3.4×10^4 K	2.3×10^5 K
Electron temperature T_e	1.3×10^5 K	1×10^5 K
Momentum flux density	2.12×10^8 dyn cm^{-2}	2.26×10^8 dyn cm^{-2}
Total energy flux density	1.55 erg cm^{-2}s^{-1}	1.43 erg cm^{-2}s^{-1}
Helium content n_α/n_p	2.5%, variable	3.6%, stationary

The most pronounced differences between LSM and HSS:

- The flow speed was found to be a superior criterion for classifying solar wind types.
- The proton density differs by a factor of 3.5 between fast and slow wind.
- The proton flux density (equivalent to the total solar wind mass flux density) differs by a factor of 2 between fast and slow wind.
- The helium content in slow wind (LSM) is markedly lower than in the fast wind (HSS). It is particularly low near the heliospheric current sheet (Borrini *et al.*, 1981). This is an important result in that it indicates a larger release height of the slow wind, consistent with the idea of a gravitationally stratified corona, where the heavier helium would be relatively enriched at lower levels. Another plausible explanation for the low helim abundances in slow wind in terms of insufficient Coulomb drag with protons (as defined by Geiss *et al.*, 1970) was given by von Steiger *et al.* (1995). In any case, note that even the 3.6% abundance in HSS is less than the half the photospheric abundance (Meyer, 1993; Grevesse and Sauval, 1998).
- The helium content variability in HSS is remarkably low and always alike. In LSM, it is highly variable, from almost zero near the HCS to a steady 4% in HSS and up to 20% or even more in ICMEs (S90).
- The measured coronal source temperatures (derived from the O^{7+}/O^{6+} abundance ratio) for HSS are 1.2×10^6 K compared to a value of 1.7×10^6 K for LSM (Geiss *et al.*, 1995).
- The FIP effect (Meyer, 1993) is significantly less pronounced in LSM, as indicated by the difference in the Mg/O element abundance ratio (Geiss *et al.*, 1995).
- The temperatures of protons are generally much higher in HSS than in LSM.
- The proton temperature in slow wind drops with distance from the Sun as $R^{-1.21}$ (Schwenn *et al.*, 1981). That is about what would be expected for an adiabatic expansion. Extrapolation back to the corona leads right to the observed coronal temperatures of about 2×10^6 K. That means that this type of solar wind is a real adiabatic flow, i.e. without any energy input or loss beyond the corona.
- In contrast, the fast wind cools off as $R^{-0.69}$, thus indicating substantial additional heat input in interplanetary space. It is due to heat conduction and local dissipation of convected wave energy (Marsch *et al.*, 1982b; Cranmer, 2002).
- In HSS, there is usually a strong anisotropy in the proton velocity distribution functions leading to $T_{perp} > T_{par}$ relative to the local magnetic field direction (Marsch *et al.*, 1982b). This anisotropy is rapidly decreasing with increasing distance from the Sun.
- The temperatures of electrons are lower in HSS than in LSM. Further, only in HSS the electron velocity distributions show a striking anisotropy for suprathermal electrons ($E \gtrsim 50$ eV). This "Strahl" of suprathermal electrons is a

beam of electrons directed away from the Sun along the local the magnetic field (Rosenbauer *et al.*, 1977; Pilipp *et al.*, 1987).

- The electron temperature in HSS is much lower than the proton temperature, as first noted by Feldman *et al.* (1976).
- The transport of the Sun's angular momentum into space by the escaping solar wind is almost exclusively restricted to LSM (Pizzo *et al.*, 1983). That indicates that the slow plasma may partly corotate up to much larger distances before it is finally released. This observation allows inferences on the release height of the different flow types (Marsch and Richter, 1984).
- In HSS, the helium ions and all other ions travel faster than the protons by about the local Alfvén velocity (Marsch *et al.*, 1981, 1982a). The differential speed is always aligned with the local magnetic field. Therefore, the absolute speeds become nearly equal in cases where the field happens to be tangentially oriented, e.g. in the outer heliosphere.
- In HSS, there are always strong perpendicular Alfvénic fluctuations present (Belcher and Davis, 1971) that are well discernible by the correlated or anti-correlated deflections of the magnetic field and the proton flow.

These latter two facts lead to a striking effect: The helium ions behave like surfers near an ocean beach who travel ahead of and in phase with a big wave, not taking part in the waves' up and down motion. Similarly, the helium ions do not participate in the transverse excursions of the solar wind flow and move at rather constant, purely radial speed (Marsch *et al.*, 1981).

Some striking similarities between LSM and HSS:

- The momentum flux which is equivalent to the ram pressure the solar wind is exerting on an obstacle appears to be an invariant of the solar wind flow state within 7% (Steinitz, 1983, S90). Using the data from the Helios solar probe mission in the inner heliosphere that same result was found to also apply for latitudinal stream structures up to at least 30° in heliomagnetic latitude (Bruno *et al.*, 1986).
- The total energy flux density in the solar wind is the sum of two main components: the kinetic energy flux and the potential energy flux (basically the work done in moving the solar wind out of the solar gravitational potential). Further contributions by the enthalpy and wave energy fluxes are negligible. Although all these contributions vary substantially between the two basic solar wind states, the sum of them is always found to be invariant within 8%.

There is no explanation yet for these strange invariances. It remains valid what was stated in S90: "...this uniformity is fortuitous, but it is certainly fair to say that we are far from understanding it." We have reasons to suspect that a crucial clue to the understanding of the solar wind phenomenon may be hidden here.

We should mention here that the results concerning the latitudinal invariances are not fully consistent with those obtained by the Ulysses mission traversing the outer heliosphere up to polar latitudes (Goldstein *et al.*, 1996; McComas *et al.*, 2000). Potential reasons: the Helios results were obtained well inside 1 AU where not much interaction between the streams has occurred; Ulysses went to much higher heliographic/heliomagnetic latitudes; the data were collected in different solar cycles; there are differences in the instruments and the data evaluation techniques (for further details see the Ulysses books by Marsden, 1995, 2001; Balogh *et al.*, 2001).

3. The Fast Wind in HSS and its Sources

The fast solar wind in HSS from coronal holes is an off-spring of the inactive Sun. Accordingly, its final speed, density, helium content and other characteristics are remarkably stable and do not vary much from stream to stream.

The advent of EUV spectrometers on solar telescopes orbiting in space has lead to a major breakthrough in understanding solar wind generation. More than 1000 papers have already been published in the last 8 years based alone on data of the SUMER spectrograph on SOHO! Hassler *et al.* (1999) identified the source regions of the fast solar wind by means of radiance and Doppler shift maps in the polar coronal hole, and Xia *et al.* (2003) in equatorial coronal holes. They found a relation between the Doppler shift of a NeVIII line and the chromospheric network. Apparently, there is outflow very low down, and it is strongest near the network lanes and their intersections of the chromospheric network. Wiegelmann *et al.* (2005) recently found the Doppler shift pattern to be related with the magnetic field topology derived from a simultaneous magnetogram by a potential field extrapolation. Tu *et al.* (2005) could even determine that the fast solar wind starts flowing out of the corona at heights between 5000 and 20000 km above the photosphere with an initial speed of 10 km/s. The detailed magnetic structure of the source was revealed as coronal "funnels" right above the network lanes and their intersections. The concept of such funnels which are magnetic structures expanding in the corona with a narrow neck in the photosphere was first described by Gabriel (1976) and formed the basis of the magnetic "furnace" model by Axford and McKenzie (1997). The solar wind plasma is now considered to be supplied by plasma stemming from the many small magnetic loops with only a few thousand kilometers in height, crowding the funnels. Through magnetic reconnection plasma is fed from all sides to the funnel, where it may be accelerated and finally form the fast solar wind.

There is still controversy about such basic issues as, e.g., the role of polar plumes in the wind. These bright, ray-like extended structures sticking out from coronal holes and outlining the magnetic field lines have been well-known from eclipse pictures. Spectroscopic analyses revealed that the plume plasma is cooler than the space between plumes, and it does not participate in the flow, like blades of grass sprouting from a river floor (Wilhelm *et al.*, 2000). Supporting evidence was given by Feldman *et al.* (2005), who found strong elemental abundance variations in plumes which are not present in high-speed solar wind. However, other authors reached opposite conclusions (Gabriel *et al.*, 2003); the final answer is still pending.

Some remnants of an originally filamentary structure of HSS were found in the solar wind, both close to the Sun (around 0.3 AU with Helios) and beyond 1 AU at high latitudes. Thieme *et al.* (1990) analyzed Helios plasma data taken near 0.3 AU and found evidence for many small structures with a diameter of about 5° in longitude that eventually merge to form a HSS. In these structures, differences in the plasma and magnetic field pressure balance each other such that the total pressure is rather constant. McComas *et al.* (1996) found similar "pressure-balanced structures" (PBS) in Ulysses data taken at high latitudes. The "microstreams" identified by McComas *et al.* (1995) and Neugebauer *et al.* (1995) by some characteristic compressional increases of plasma and magnetic pressures are probably of a different nature.

The PBS as well as the microstreams have probably a solar origin. Thieme *et al.* (1990) argued that the PBS are signatures of the underlying supergranulation structure, because the scale sizes are consistent. Also, relations with polar plumes are being considered. Unique associations with *in situ* solar wind measurements are difficult since interactions between different structures wipe out the differences and cause flux tubes to merge and form the well-known rather uniform pattern of HSS. Only at locations with little stream-stream interaction some remnants of the fine structure patterns may be preserved, e.g., close to the Sun or at high latitudes.

The spectroscopic measurements by UVCS on SOHO revealed surprising results about the fast wind between 1.5 and 4 R_s. The outflow speed increases dramatically with distance (see Figure 2), the speed of O^{5+} ions being twice that of the protons. While the proton velocity distributions are only mildly anisotropic, the O^{5+} ions are strongly anisotropic, with T_{perp} approaching 2×10^8 K (Kohl *et al.*, 1998). That effect is being interpreted in terms of ion cyclotron heating, as had been suggested by Marsch *et al.* (1982b) on the basis of *in situ* data. Note though that the velocity distributions deduced from UVCS may not be real since they depend strongly on the adopted density profile (Raouafi and Solanki, 2004).

Figure 2 taken from Esser *et al.* (1997) shows a compilation of several studies that report observed, or inferred, or modeled profiles of high speed flow in the whole range up to 1 AU. Clearly, the flow has reached substantial speeds as low down as 3 R_s, and by 10 R_s or below it has reached its final state. These observations have strong implications for solar wind heating and acceleration because the classical Alfvén wave models fail to explain it. These models require that the waves travel quite some distance prior to their energy deposition. Instead, if the models include proton heating to very high temperatures low down, as suggested in the funnel model by Tu *et al.* (2005), then they would yield high speed flow close to the Sun, as is observed.

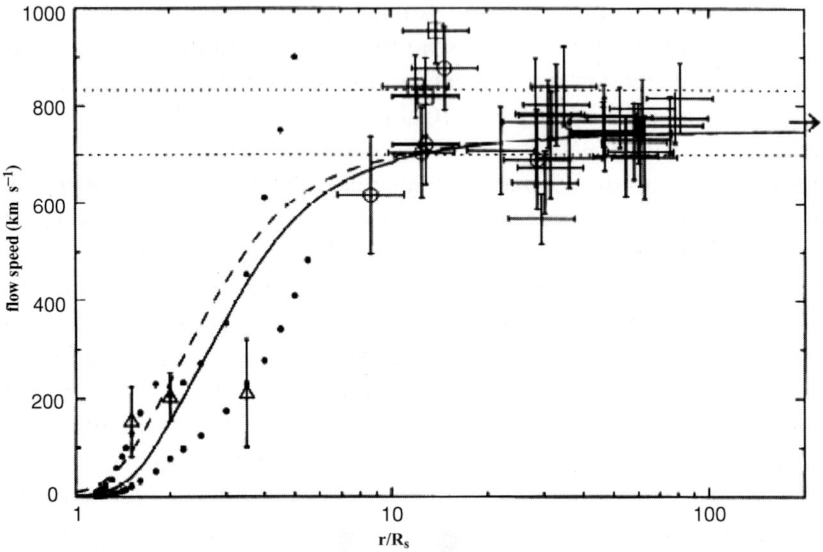

Figure 2. Speed profiles of the fast solar wind from various observations and models. Open triangles: from Lyman alpha Doppler dimming technique (Strachan *et al.*, 1993). Solid dots: flow speeds derived from the assumption of constant mass flux. Open circles and squares inferred from IPS measurements (Grall *et al.*, 1996). Dotted horizontal lines: range of flow speeds measured by Ulysses in high speed wind at high latitudes. Also plotted are the flow speeds calculated for the two models by Esser *et al.* (1997).

Recent EUV measurements showed that the electrons are rather "cold" in coronal holes. According to David *et al.* (1998) the electron temperatures remain well below 10^6 K and fall rapidly off with distance. This is consistent with *in situ* measurements from Helios which showed that electrons in HSS are generally cold and their radial temperature gradients are flat (Marsch, 1991).

At the end of this section let me recommend for further reading the articles in Kohl and Cranmer (1999), and by Neugebauer (2001) and Cranmer (2002).

4. The Low Speed Solar Wind of Minimum Type (LSM) and its Sources

Streamers and interfaces between streamers have been regarded as the most plausible sources of slow solar wind, but the details are still unclear. Around solar activity minimum, there is usually a streamer "belt" encircling the Sun. It encloses the remaining sunspots and activity centers. Also, it contains the warped magnetic neutral line that separates the two polarities of the global magnetic dipole and can therefore be regarded as the heliomagnetic equator. Here emerges the heliospheric current sheet which rotates with the Sun and extends throughout the whole heliosphere like a giant ballerina skirt (Alfvén, 1977). The streamer belt is the source region of the

LSM. It fills a latitude range of $\pm 15^\circ$ around the heliospheric current sheet. The LSM region is confined by HSS from the large polar coronal holes, separated from them by very sharp boundaries (Schwenn *et al.*, 1978; Bruno *et al.*, 1986, see also S90).

Figure 3 presents a view of the extended minimum corona seen by the LASCO C1 and C2 coronagraphs (Schwenn *et al.*, 1997). In this phase of very low solar activity, bright, apparently closed loop systems are almost permanently present, centered at latitudes of 30° to 45° in both hemispheres. Their centers are usually found at the position of a magnetic neutral line in the photosphere. Their helmet-like extensions are bent towards the equator plane. Further out, they merge into one large equatorial streamer sheet that is clearly discernible as far out as 32 R_s (the edge of the LASCO C3 field of view). They remain stable on time scales of days. Above the equator, there is usually a more diffuse pattern visible, well separated from the high-latitude loops and with a very pronounced variability.

It is interesting to compare the minimum corona with the magnetic field model presented by Banaszkiewicz *et al.* (1998) as shown in Figure 3. For their model they took into account an equatorial current sheet plus a magnetic dipole and quadrupole. There is an overall good agreement between the observed electron density distribution made visible by the white light images and the inferred morphology of the plasma-confining magnetic field. Note further, how similar both patterns look to the one shown in the lower panel of Figure 3 that was obtained from an eclipse observation near activity minimum (Loucif and Koutchmy, 1989).

In all three panels of Figure 3, we note substantial non-radial field components in the innermost corona that are strongest near the coronal base and disappear by about 5 R_S. The "expansion factor" of the magnetic field between the coronal base and the outer regions varies from 7 to 10 above coronal holes. But note that the field expansion is significantly stronger within the equatorial streamer belt (expansion factor $\sim$30), because of the multipole components in the low active corona that do not reach out very far. Wang and Sheeley (1990) had first noticed this significant inverse correlation between the rate of magnetic flux-tube expansion and the solar wind speed at 1 AU. This empirical relation allows to predict pretty well solar wind speeds at Earth from a current-free extrapolation of observed photospheric fields into the corona, as has been proven many times (see, e.g., Whang *et al.*, 2005).

The patterns in Figure 3 lead to the impression as if the over-expanding field from the polar coronal holes is squeezing the equatorial plasma into this narrow sheath (with a total width of about 30° in latitude) above the streamer belt (with a width of some 90° near the Sun). Note in passing, that flow from coronal holes does indeed deflect adjacent flow, as was clearly seen in case of CMEs near solar minimum (Cremades *et al.*, 2004; see also Section 6)

The plasma flow in this sheath is rather inhomogeneous and full of density fluctuations. That's what renders this type of solar wind its visibility and even traceability in coronagraph images. Animations of LASCO images have demonstrated this very impressively (see, e.g., http://lasco-www.nrl.navy.mil/images/aaas/xmas_c3.mpg

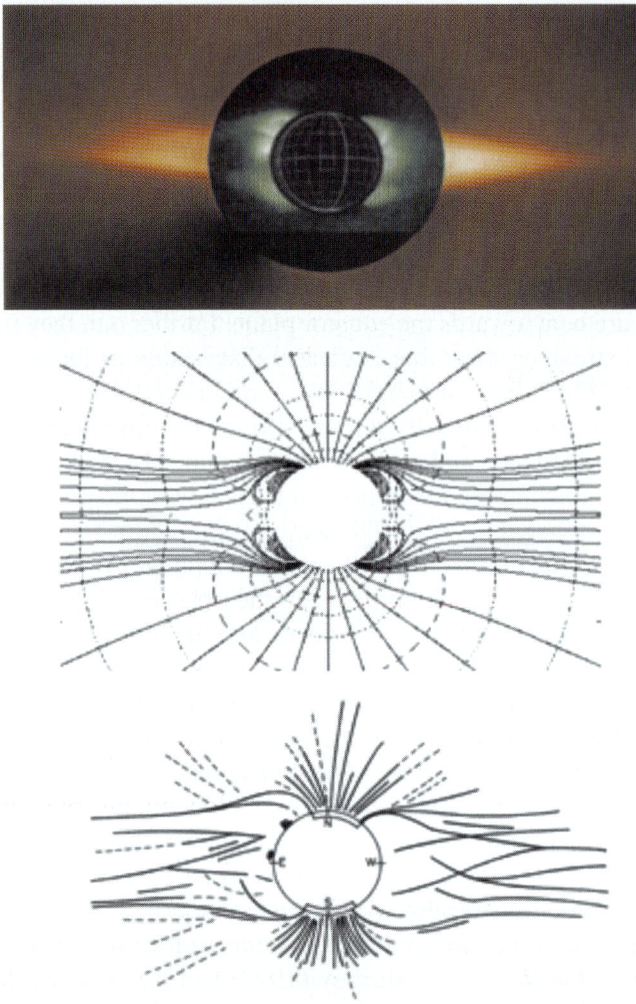

Figure 3. Upper panel: A coronagraph view of the extended minimum corona (on February 1st, 1996), composed by a green-line emission image and a white-light image taken by the LASCO coronagraphs C1 and C2 onboard SOHO. From Schwenn *et al.* (1997). Middle panel: A simple analytic model for the magnetic field in the solar corona and interplanetary space which is appropriate to solar minimum conditions. The model combines an azimuthal current sheet in the equatorial plane with an axisymmetric multipole field representing the internal magnetic field of the Sun. The radial component of the field filling interplanetary space is approximately monopolar at large heliocentric distances as observed. These open field lines connect to the polar regions of the Sun and define the polar coronal holes which are prevalent at solar minimum and which are the source of the fast solar wind. From Banaszkiewicz *et al.* (1998). Lower panel: A typical solar minimum corona, observed during the eclipse on October 23, 1976 and processed by Loucif and Koutchmy (1989).

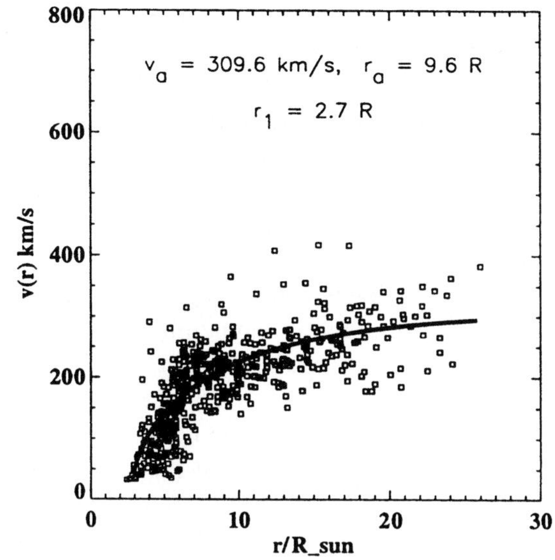

Figure 4. Speed/height measurements for 40 moving density enhancements as tracers of the slow solar wind flow. The solid line is the best fit to the unweighted data points. From Sheeley *et al.* (1997).

<http://lasco-www.nrl.navy.mil/images/aaas/xmas_c3.mpg> or http://star.mpae.gwdg.de/release/movie/mpg/9612c3xm.mpg <http://star.mpae.gwdg.de/release/movie/mpg/9612c3xm.mpg>).

Sheeley *et al.* (1997) tried to quantify the outflow by observing distinct density blobs on their passage through the coronagraphs' fields of view. They claim that these blobs trace out the slow solar wind flow like "leaves in the wind". Statistical analysis of many such leaves allows deriving a speed profile as shown in Figure 4. Apparently, the blobs begin to accelerate not closer than at around 3 R_s, and by 10 R_s they are barely faster than 200 km/s, and 300 km/s at 30 R_s. This profile is fairly consistent with an isothermal expansion at a temperature of 1.1×10^6 K and an Alfvén point at 5 R_s, in good agreement with the simple model by Parker (1958). It also fits well to the *in situ* measurements from Helios at 60 R_s of about 350 km/s. Note how markedly this profile differs from the one for the HSS shown in Figure 2!

These blobs represent only a small component of the total LSM flux, and the questions now are: how are these blobs generated and how representative are they? How does all the other LSM emerge? Where does the consistent outflow of the slow wind really begin? Unfortunately, quantitative diagnostics using a spectrographic coronagraph in the critical distance range from 2 to 5 R_s are very difficult. The intensities of relevant spectral lines fall off with distance very rapidly such that they are soon swamped by both: the continuum corona (from Thompson scattering on coronal electrons); and the instrumental straylight that is plaguing present days coronagraphs. The LASCO-C1 coronagraph allowed spectral analysis of the green coronal emission line at 530.3 nm wavelength out to $\sim 2R_s$ (Mierla *et al.*, 2004). For an isolated streamer they measured the Doppler shift of the green line above the limb and determined a line-of-sight velocity of 5 km/s for the range from 1.2 to 1.5 R_s. Since they were able to locate the streamer's source region on the disk, the projection effect could be corrected and a true radial speed of 9 km/s be calculated. This is roughly consistent with the model by Chen *et al.* (2004) and with measurements of Strachan *et al.* (2002) and Frazin *et al.* (2003) who used data from UVCS on SOHO.

Wang *et al.* (1998) suggested that the blobs might be the product of reconnection between distended helmet streamer loops and neighboring open field lines. During such exchange of foot points, no new open magnetic flux is created and the helmet streamers continue to remain in quasi-steady state. However, such processes can only account only for a small component of the LSM.

A different approach towards the origin of LSM was given by Noci *et al.* (1997). In their model, the slow wind originates from within streamers, rather than from their flanks. The plasma moves along open field lines that are located in between the multiple loops that form the large-scale streamer (see Figure 3). They conclude this from the depletion of FIP elements that occurs similarly in streamers (observed with UVCS on SOHO) and in the slow solar wind (Geiss *et al.*, 1995).

A study by Wiegelmann *et al.* (1998) shows that small-scale eruptions at the cusp of the helmet streamer may accelerate small amounts of plasma, thereby contributing to the non-stationary LSM. In their numerical simulations they assumed a triple streamer configuration as observed by LASCO-C1 (Schwenn *et al.*, 1997, see also Figure 3) and found that reconnection will release small plasmoids from the streamer stalk above the cusp.

We see increasing evidence for transient processes to be involved in generating the slow solar wind. However, we have to realize that the distant heliospheric current sheet (that is embedded in the LSM) as well as the streamer sheet are permanent features and remain rather stable on large scales, and so do the large polar coronal holes. The general flatness (at minimum) and stability of the streamer sheet are probably due to the persistence of the mid-latitude streamers. In other words: the sheet is determined by mid-latitude phenomena rather than by the near-equatorial activity belt, as was often thought in the past. The mid-latitude streamers are bent towards the solar equator by the over-expanding polar coronal holes and the HSS emerging from there. After all, that would mean that the heliospheric streamer sheet near the ecliptic plane is also being shaped by forces originating in the Sun's polar regions!

In this context, it is interesting to study some strange effects concerning solar rotation. From sunspot observations the differential rotation of the photosphere has long been known (see, e.g., Howard, 1984). However, the overlying corona rotates almost rigidly, and so do all stream patterns (Sime *et al.*, 1989; Stenborg *et al.*, 1999). The coronal rotation rate was found to be 27.2 days (synodic) which is very close to both: the equatorial photospheric rate and the rate of the solar core (Kosovichev *et al.*, 1997). In other words: the heliospheric stream pattern (including very high latitude high-speed streams) appears to be "attached" to the equatorial photosphere or to the solar core, respectively!

There is a considerable mismatch between the rigid rotation of, e.g., coronal holes with their open magnetic filed lines and the differentially rotating photosphere. Fisk (1996) offered a plausible explanation (see also Posner *et al.*, 2001; Fisk, 2005) that is consistent with the concept of interchange reconnection at coronal hole boundaries as suggested, e.g., by Wang *et al.* (1998), Crooker *et al.* (2002), Lionello *et al.* (2005). However, experimental confirmation is still lacking, and further surprises are to be expected.

5. The Low Speed Solar Wind at High Solar Activity (LSA) and its Sources

The slow solar wind filling most of the heliosphere during high solar activity has to be considered a separate category. It emerges above active regions distributed over large parts of the Sun, probably far from any heliospheric current sheet. In fact, even at high latitudes there is now always slow wind prevailing, in sharp contrast to

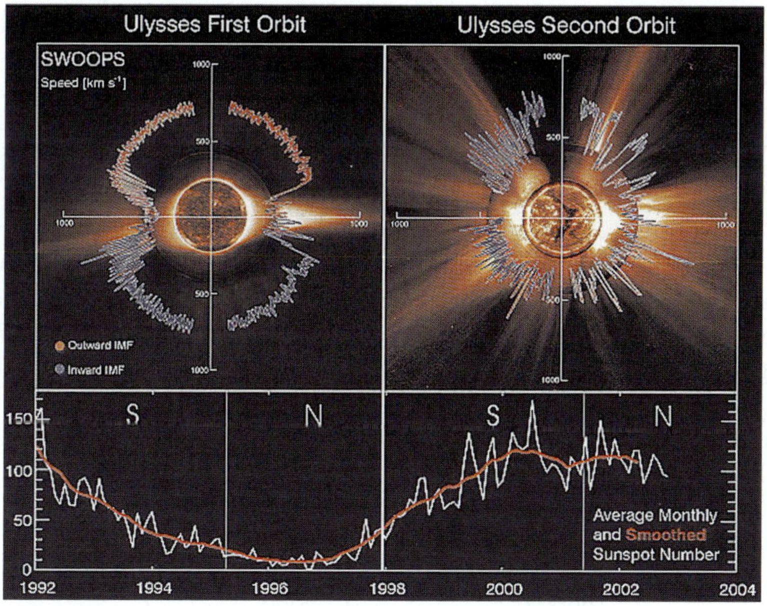

Figure 5. Polar plots of solar wind speed as a function of latitude of Ulysses' first two orbits. Sunspot number (bottom panel) shows that the first orbit occurrred through the solar cycle declining phase and minimum while the second orbit spanned solar maximum. Both are plotted over solar images characteristic of solar minimum (8/17/96) and maximum (12/07/00); from the center out, these images are from the Solar and Heliospheric Observatory (SOHO) Extreme ultraviolet Imaging Telescope (Fe XII at 195 Å), the Mauna Loa K-coronameter (700–950 nm), and the SOHO C2 Large Angle Spectrometric Coronagraph (white light). From McComas *et al.* (2003).

activity minimum times. This is impressively demonstrated in Figure 5 taken from McComas *et al.* (2003).

The most remarkable difference between the two types of slow solar wind is the helium abundance A. In the course of the solar activity cycle, A was found to vary in slow wind from 2.5% at minimum to 4% at maximum, while in HSS the value of A remained almost constant (S90). This was taken as indication that the slow wind, (at least as it is encountered in the ecliptic plane) changes its character dramatically during the solar cycle. At minimum, gravitational stratification or insufficient Coulomb drag between helium ions and protons allows A to become very low at the release altitude of the LSM which is rather high. It appears as if that depletion is progressively disturbed with increasing activity and the disappearance of stable large-scale helmet streamers and current sheets.

Indeed, the LSA is usually encountered in a highly turbulent state which is in part due to the many shocks from solar transient events plowing through it.

After all, it is by no means clear yet how the slow solar wind (both types) is released at all, since apparently it emerges from coronal regions with mainly closed magnetic field loops underneath. The transition from there to an open field topology (that is required to allow plasma outflow altogether) requires some kind of magnetic reconnection, be it in a steady-state process or involving transient phenomena. Here we are confronted with another crucial problem in coronal and solar wind physics that is still waiting for its solution.

6. Solar Wind in Context with Transients at the Sun

At times, gigantic explosions (flares, coronal mass ejections (CMEs), prominence eruptions) make the whole Sun's atmosphere shudder. These spectacular processes form an exciting topic of current research, and a rich literature has been piled up. In the context of this review on solar wind sources, I restrict myself to CMEs since they cause gigantic plasma clouds to leave the Sun. In Figure 6 a snapshot of a typical CME as observed by the coronagraph LASCO-C3 onboard the SOHO spacecraft is shown. More than half of all CMEs have a similar 3-part structure: a bright outer loop, followed by a dark void and finally by a bright kernel, often with substantial fine structure. For further information, the reader is referred to St. Cyr *et al.* (2000) and Gopalswamy (2004).

Some authors claim that there are two (or more) kinds of coronal mass ejections (e.g., Sheeley *et al.*, 1999; Srivastava *et al.*, 1999a,b): (1) Gradual CMEs, with balloon-like shapes, accelerating slowly and over large distances to speeds in the range 300 to 600 km/s, and (2) Impulsive CMEs, often associated with flares, accelerated already low down to extreme speeds (sometimes more than 2000 km/s). It is not clear yet whether these are really fundamentally different processes or whether they represent just the extrema of an otherwise continuous spectrum of CME properties.

Figure 6. The coronal mass ejection on Feb. 27, 2000, observed by the LASCO C3 coronagraph. The Sun is drawn to scale. The image is a difference between two exposures taken about 10 hours apart from each other. From Schwenn (2005).

Zhang *et al.* (2001, 2004) described the initiation of CMEs in a three-phase scenario: the initiation phase, the impulsive acceleration phase and the propagation phase. The initiation phase (taking some tens of minutes) always occurs before the onset of an associated flare, and the impulsive phase coincides well with the flare's rise phase. The acceleration ceases with the peak of soft X-ray flares. It is interesting to notice that some of the theoretical CME models begin to postulate different phases of acceleration (see, e.g. Chen and Krall, 2003).

Right at the launch time of several CMEs, Kaufmann *et al.* (2003) discovered rapid solar spikes at submillimeter wavelengths that might be representative of an early signature of CME onset. The role of some other observed processes is also still unclear: coronal "dimmings" (Hudson *et al.*, 2003), Moreton waves (Thompson and Murdin, 2000), EIT waves (Thompson *et al.*, 1998), the various types of radio bursts (Reiner *et al.*, 2001), coronal inflow (Wang *et al.*, 1999; Sheeley and Wang, 2002; Tripathi *et al.*, 2005).

In order to disentangle the various processes around CME initiation new observations with significantly better resolution, spatially and in time and even supported by spectroscopic diagnostics are needed, as was demonstrated by Innes *et al.* (2001) and Balmaceda *et al.* (2003).

The fast CMEs often drive large-scale density waves out into space which eventually steepen to form collisionless shock waves. The shock front is the outer boundary of a plasma sheath (see, e.g., Tsurutani *et al.*, 1988) that results from compression, deflection, and heating of the ambient solar wind by the ensuing ejecta. The sheath may contain substantial distortions of the interplanetary magnetic field due to field line draping (McComas *et al.*, 1989) around the ejecta cloud pressing from behind. Figure 7 gives a good example of a typical fast ICME event (ICME stands for Interplanetary counterparts of CMEs), as observed by *in-situ* instrumentation on an interplanetary spaceprobe.

The ejecta themselves (called "piston gas" or "driver gas" in earlier papers) are often separated from the sheath plasma by a tangential discontinuity. Their very different origin is discernible from their different elemental composition (Hirshberg *et al.*, 1970), ionization state (Bame *et al.*, 1979; Schwenn *et al.*, 1980; Henke *et al.*, 1998; Rodriguez *et al.*, 2004), temperature depressions (Gosling *et al.*, 1973; Montgomery *et al.*, 1974; Richardson and Cane, 1995), cosmic ray intensity decreases ("Forbush decreases", see, e.g., Cane *et al.*, 1994), the appearance of bi-directional distributions of energetic protons and cosmic rays (Palmer *et al.*, 1978) and supra-thermal electrons (Gosling *et al.*, 1987). In many ejecta, major over-abundances of Helium are observed, up to 30%, as first noted by Hirshberg *et al.* (1970). This indicates that this ejecta material originates from low layers in the solar atmosphere, where dynamical accumulation and gravitational stratification allow substantial enrichment of heavy ions.

For about one third of all shocks driven by ICMEs, the succeeding plasma shows to an *in-situ* observer the topology of magnetic clouds (Burlaga *et al.*, 1981; Burlaga, 1991). Smooth rotation of the field vector in a plane vertical to the propagation direction, mostly combined with very low plasma beta, i.e. low plasma densities and strong magnetic field with low variance give evidence of a flux rope topology (Marubashi, 1986) of these magnetic clouds. That is consistent with the concept of magnetic reconnection processes in coronal loop systems in the course of prominence eruptions at the Sun (Priest, 1988).

Most of these ICME signatures can be seen in the event shown in Figure 7. Usually, only a fraction of the criteria for identifying ejecta is encountered in individual events, and to this day a trained expert's eye is needed to tell what is ejecta and what not. The situation is additionally complicated by the class of very slow CMEs found to take off more like balloons rather than as fast projectiles (Srivastava *et al.*, 1999a, b). After many hours of slow rise, they finally float along in the ambient slow solar wind. Naturally, they do not drive a shock wave. Only in rare cases, a few of their ejecta signatures (e.g., composition anomalies, magnetic cloud topology) remain and disclose their origin.

The compressed sheath plasma behind shocks and the ejecta clouds may both exhibit substantial deviations of the magnetic field direction from the usual Parker spiral, including strong out-of-the-ecliptic components. These, in turn, are the main ingredients for space weather effects impacting the Earth system (Tsurutani *et al.*, 1988). That is why we have to keep in mind that the sources of magnetic field deflections in the sheath plasma and the ejecta are of fundamentally different origin:

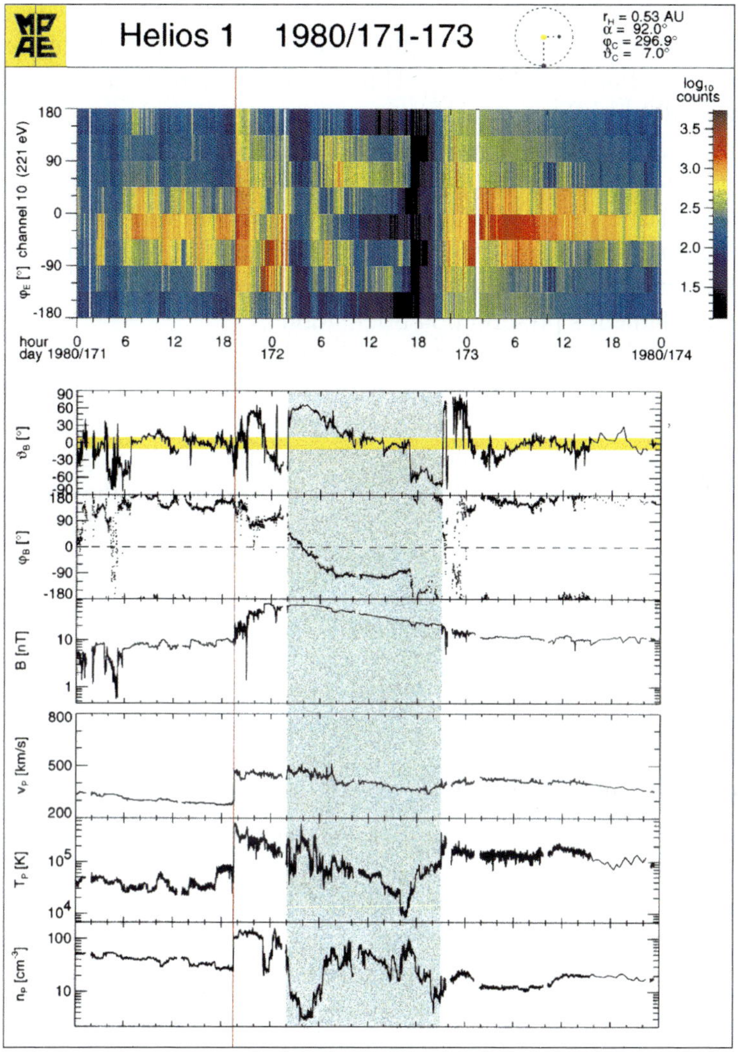

Figure 7. This shock event was observed by the Helios 1 solar probe during 3 days in 1981, at a distance from the Sun of 0.53 AU. The panels show the solar wind parameters (from bottom to top): proton density, temperature, flow speed, magnetic field magnitude and its azimuthal and elevation angles. The upper panel shows the azimuthal flow direction of suprathermal electrons (at 221 eV), the direction away from the Sun being at 180°. The jumps in all parameters and the sudden widening of the electron angular distribution at 19:20 UT on day 171 denotes the arrival of a fast shock wave. The time between 02:00 and 19:00 UT (shaded area) on day 172 denotes the passage of a magnetic cloud, with its characteristic change of the field direction in the sense of a magnetic flux rope, and the simultaneous appearance of oppositely flowing (bi-directional) suprathermal electron streams. Note also the mono-directional electron flow before and after the event series. From Schwenn (2005).

– The field line deflection in the sheath due to draping depends on the orientation of the ejecta relative to the heliospheric current sheet and to the observer sitting, say, near the Earth's bow shock. Thus, the field orientations in the sheath vary dramatically from case to case.
– The magnetic field inside an ejecta cloud, is generally thought to follow a flux rope topology. In fact, an observer who is passed by such a cloud observes a characteristic rotation of the field vector that allows him to determine the orientation of the flux rope. Bothmer and Schwenn (1998) found that most often the configuration of the ejected magnetic cloud within the ICME reflects well the filament's pre-eruption orientation.

Most CMEs are originating near the heliographic equator (Howard *et al.*, 1985; St. Cyr *et al.*, 2000). Sometimes CMEs are seen at very high latitudes (for example the one shown in Figure 6). Usually these are CMEs originating at mid latitudes but directed near the Sun-Earth line, such that in projection they look as if they were poleward pointed (Burkepile *et al.*, 2004). Cremades and Bothmer (2004) corrected this projection effect for some 200 CMEs observed between 1996 and 2002 and determined their "true" center latitudes. At solar minimum, they clearly peak at the solar equator. But their source regions are centered in two belts at around 25° northern or southern latitude. That means that these CMEs must have been deflected from their mid-latitude sources towards the solar equator. Cremades *et al.* (2004) found that the deflection is proportional to both: the proximity and the size of nearby coronal holes. At times of high solar activity, in absence of the big polar coronal holes, there was no net deflection found. The latitudinal distribution of source regions and CME center latitudes were generally broader, but high latitude CMEs were observed rarely even then.

After all, we have to admit that some fundamental questions about CMEs are still unsolved. Most importantly: What causes a CME to erupt in the first place? Many researchers around the world are intensely tackling this problem. However, the essential ingredients for CME onset are not yet identified. Some candidates: the proximity of a CME site to coronal holes (Bravo *et al.*, 1999), magnetic shear (Mikic and Linker, 1997), filament helicity (Martin, 2003; Rust, 2003), sigmoids (Rust and Kumar, 1996; Moore *et al.*, 2001).

The uncertainty about CME initiation has a serious consequence: We do not know the signatures of an imminent eruption, i.e., our predictive capabilities are very limited. There are more questions waiting to be answered. For example: Where in the well-known three-part structure of a CME as shown in Figure 6 would the shock front be located? Where would the magnetic flux rope be located? How and where is this pattern transformed into the well-known two part structure of an ICME (just sheath and ejecta cloud) as shown in Figure 7? What is the role of magnetic reconnection for CME onset and evolution: trigger, driver, or sequel? Important hints are probably hidden in the wealth of observational data from the SOHO mission (Plunkett *et al.*, 2002). Further on, we are looking forward to the upcoming STEREO mission with its instrumentation that is particularly dedicated to the study of CMEs.

7. Variations over the Solar Activity Cycle

One major manifestation of the solar activity cycle is certainly the dramatic shrinking, disappearing, and rebirth of the polar coronal holes from one minimum to the next. The changes in shape of coronal holes are always caused by the birth and emergence of new bipolar magnetic regions which then interact with the pre-existing fields. In particular, the emergence of new magnetic flux with rising activity at high latitudes shrinks the polar holes and simultaneously leads to the formation of isolated near-equatorial holes. The total area of coronal holes is lowest around activity maximum. The total amount of magnetic flux on the Sun is increased by a factor of 3 (Howard and Labonte, 1981), but the interplanetary magnetic filed is increased only by 40% or less (Slavin *et al.*, 1984). This means that the percentage of solar magnetic flux "leaking out" into the heliosphere is much less. Further: a significant fraction of the Sun's fluxes (both: magnetic field and solar wind) must originate outside coronal holes.

Around solar activity minimum, the structures of the corona and the solar wind are rather simple and remain stable for several months. The current sheet separating inward and outward going magnetic field lines is very flat (see Figure 3) and the warps in the current sheet do not extend beyond 15° (Hoeksema *et al.*, 1982; Sanderson *et al.*, 2003). With increasing activity the warps in the shirt begin to grow. They finally break up and turn over around the activity maximum (Bilenko, 2002). The current sheet begins reorganizing itself according to the overall magnetic field of the Sun, which has just undergone a polarity reversal. A full magnetic cycle of the Sun takes 22 years, i.e. two activity cycles, until the original polarity is restored.

According to the structural changes of the corona, the solar wind also changes. The remarkably stable shapes of both: coronal holes and solar wind HSS, in the years before minimum activity are giving way to increasingly irregular patterns. The few remaining fast streams are smaller in size and amplitude, and the whole structure is additionally confused by abrupt disturbances due to interplanetary shock waves driven by CMEs (Sheeley and Harvey, 1981).

One would expect to find some considerable variations of the average solar wind properties going in parallel with solar activity changes. It turns out, though, that this is actually not the case, at least not for the plane of the ecliptic to which observations have been constrained so far. There is not much change: the average speed dropped from some 500 km/s in early 1975 to 400 km/s in 1980 (S90). This is due to the disappearance of the big high speed streams from minimum to maximum. However, other quantities such as particle densities, mass, momentum, and energy fluxes are modulated by not more than about 20%. Note that they reach their minima at that time of peak solar activity, despite the many CME and shock events (S90)!

Outside the ecliptic plane and especially at very high solar latitudes the situation is completely different. At solar minimum we encounter nothing but the fast type of solar wind which is characterized by its high speed and very low density (see Figure 5). With increasing activity and the associated disappearance of the polar coronal holes the share of slow type solar wind grows, with its high values of particle and mass flux densities, and strong modulation of these quantities at high latitudes during the solar cycle can be expected. There is in fact direct evidence from optical observations. The instrument dedicated to determining the zodiacal light brightness on the Helios probes was sensitive enough to detect and separate the Thompson scattered light from solar wind electrons integrated along the line of sight at high latitudes. “The conclusion is hard to escape that for this region inside 0.4 AU the plasma density nearly doubles from minimum to maximum”, as Leinert *et al.* (1982) noted. Further evidence was derived from measurements of resonant scattering of solar Lyman-alpha emission, which is a result of charge exchange collisions between interstellar neutral hydrogen atoms and the solar wind proton flux (Lallement *et al.*, 1985; Nakagawa *et al.*, 2003).

Coronal evolution during the solar cycle and the resulting modulation of the solar wind does only weakly affect another important quantity: the momentum flux. The observed invariance of the momentum flux with the stream structure also applies with respect to heliomagnetic latitudes (Bruno *et al.*, 1986, S90) up to at least 30°. We have no reason to expect major deviations from that invariance at higher latitudes. The momentum flux modulation above the poles does probably not exceed the one observed in the ecliptic plane (+/−25%). This conclusion implies important consequences for the three-dimensional shape, size, and the cyclic modulation of the heliospheric termination shock.

The occurrence rate of CMEs is ∼3 per day at activity maximum and decreases to about the tenth of this value at minimum (Webb and Howard, 1994). The absolute number of the CME rate depends somewhat on the sensitivity and dynamic range of the instrument used for CME detection and on the exact definition of what a CME really is. That’s probably the reason why (Gopalswamy, 2004) found higher rates form SOHO-LASCO data than (Webb and Howard, 1994) did in the pre-SOHO era. However, the relative change during the solar cycle is about the same. It is important to note that the quality of the CMEs (angular size, speed, mass content etc.) does not change significantly during the solar cycle. For further details see Gopalswamy (2004).

8. Summary

Parker’s calculations (Parker, 1958) of a thermally driven continuous outflow of solar material yielding the solar wind led to surprisingly good agreements with observations at the time. However, the more detailed data were collected and the more the models were refined, the more discrepancies showed up. At present, we can differentiate between four different types of solar wind, and they all may result from different acceleration mechanisms.

The fast solar wind in high speed streams (HSS) emerges from coronal holes which are representative of the inactive Sun, i.e. the "quiet" Sun. This type of solar wind remains fairly steady throughout a high speed stream (apart from the commonly present Alfvénic fluctuations). Also, the properties of all HSS are much alike. The plasma in HSS is brought up to speed very low down in the corona, and even inside 10 R_s it has reached its final state. Present theories have still difficulties in explaining the high values of the flow speed in HSS.

The slow solar wind apparently originates from above the more active regions on the Sun. In fact, there are actually two different types of slow solar wind: the low speed wind of minimum type (LSM) is typical for times around activity minimum. It is constrained to the warped streamer "belt" of about 30° width in latitude which encircles the Sun close to the heliographic equator. This is the regime of bright coronal streamers where the heliomagnetic current sheet is "attached". The strikingly low helium content ($<2\%$) in this type of solar wind indicates a larger release height in the gravitationally stratified solar atmosphere. Consistent outward motion of the LSM begins only at $\sim 3R_s$.

In contrast, at times of high solar activity, the low speed wind at maximum activity (LSA) is found to emerge all over the Sun from large areas that are often located far from any current sheet. It is highly variable and usually contains a significant fraction (some 4%) of helium.

For both types of slow wind, it is not clear yet how it can be released after all, since apparently it emerges from coronal regions with mainly "closed" magnetic field loops underneath.

The slow and fast streams are separated from each other by sharp boundaries that can be traced back well into the corona (see Figure 1) Apparently, there are not only fundamentally different states of solar wind, but also of their coronal sources. We cannot see how the models explaining acceleration of fast wind from coronal funnels (Tu *et al.*, 2005) that are presently en vogue could be extended to cover slow wind generation as well. That supports our suspicion that the acceleration mechanisms for the different solar wind types might also be of fundamentally different nature.

As a fourth type of solar wind we consider the coronal mass ejecta that are produced in the course of major explosions in the solar atmosphere. They can easily be discerned by the characteristic disturbances they impose on the ambient solar wind. In many cases, they attract notice by an unusually high helium percentages (up to some 30%) that is observed in the driver gas behind many CME driven shock waves. Unfortunately, it remains unclear after all what causes a CME to erupt and what are the signatures of an imminent eruption.

This summary illustrates that our understanding of the solar wind and its sources is not yet on firm ground. We have to admit that even some fundamental problems in coronal and solar wind physics are still waiting for their solution.

Kohl, J. L., and Cranmer, S. R.: 1999, in *Proceedings of the SOHO-7 Workshop Northeast Harbor*, Maine, Kluwer Academic Publishers.

Kosovichev, A. G., Schou, J., Scherrer, P. H., Bogart, R. S., Bush, R. I., Hoeksema, J. T., *et al.*: 1997, *Solar Phys.* **170**, 43.

Krieger, A. S., Timothy, A. F., and Roelof, E. C.: 1973, *Solar Phys.* **29**, 505.

Lallement, R., Bertaux, J. L., and Kurt, V. G., 1985, *J. Geophys. Res.* **90**, 1413.

Leinert, C., Richter, I., and Planck, B., 1982, *Astronom. Astrophys.* **110**, 111.

Lionello, R., Riley, P., Linker, J. A., and Mikić, Z.: 2005, *Astrophys. J.* **625**, 463, doi: 10.1086/429268.

Loucif, M. L., and Koutchmy, S.: 1989, *Astron. Astrophys. Suppl. Ser.* **77**, 45.

Marsch, E.: 1991, in Schwenn R., and Marsch, E. (eds.), *Physics of the Inner Heliosphere II*, Springer-Verlag Berlin Heidelberg New York, p. 45.

Marsch, E., Rosenbauer, H., Schwenn, R., Mühlhäuser, K.-H., and Denskat, K. U.: 1981, *J. Geophys. Res.* **86**, 9199.

Marsch, E., Rosenbauer, H., Schwenn, R., Mühlhäuser, K.-H., and Neubauer, F. M.: 1982, *J. Geophys. Res.* **87**, 35.

Marsch, E., Rosenbauer, H., Schwenn, R., Mühlhäuser, K.-H., Pilipp, W., and Neubauer, F. M.: 1982, *J. Geophys. Res.* **87**, 52.

Marsch, E., and Richter, A. K.: 1984, *J. Geophys. Res.* **89**, 5386.

Marsden, R. G. (ed.): 1986, *Astrophys. Space Sci. Library* **123**, D. Reidel Publishing Company Dordrecht, The Netherlands.

Marsden, R. G. (ed.): 1995, The High Latitude Heliosphere, 28th ESLAB Symposium, Kluwer, Dordrecht, The Netherlands.

Marsden, R. G. (ed.): 2001, The 3-D Heliosphere at Solar Maximum, 34th ESLAB Symposium, Kluwer, Dordrecht, The Netherlands.

Marubashi, K.: 1986, *Adv. Space Res.* **6**, 335.

Martin, S. F.: 2003, *Adv. Space Res.* **32**, 1883, doi: 10.1016/S0273-1177(03)90622-3.

McComas, D. J., Gosling, J. T., and Bame, S. J.: 1989,*J. Geophys. Res.* **94**, 1465.

McComas, D. J., Barraclough, B. L., Gosling, J. T., Hammond, C. M., Phillips, J. L., Neugebauer, M., *et al.*: 1995, *J. Geophys. Res.* **100**, 19893, doi: 10.1029/95JA01634.

McComas, D. J., Hoogeveen, G. W., Gosling, J. T., Phillips, J. L., Neugebauer, M., Balogh, A., *et al.*: 1996, *Astron. Astrophys.* **316**, 368.

McComas, D. J., Barraclough, B. L., Funsten, H. O., Gosling, J. T., Santiago-Muñoz, E., Skoug, R. M., *et al.*: 2000, *J. Geophys. Res.* **105**, 10419, doi: 10.1029/1999JA000383.

McComas, D. J., Elliott, H. A., Schwadron, N. A., Gosling, J. T., Skoug, R. M., and Goldstein, B. E.: 2003, *Geophys. Res. Lett.* **30**, 24, doi: 10.1029/2003GL017136.

Meyer, J. P.: 1993, *Adv. Space Res.* **13**, 377, doi: 10.1016/0273-1177(93)90509-A.

Mierla, M., Schwenn, R., Stenborg, G., Teriaca, L., and Podlipnik, B.: 2004, *35th COSPAR Scientific Assembly*, p. 2275.

Mikic, Z., and Linker, A. J., 1997, *AGU Geophysical Monograph Series* **99**, 57.

Montgomery, M. D., Asbridge, J. R., Bame, S. J., and Feldman, W. C.: 1974, *J. Geophys. Res.* **79**, 3103.

Moore, R. L., Sterling, A. C., Hudson, H. S., and Lemen, J. R.: 2001, *Astrophys. J.* **552**, 833, doi: 10.1086/320559.

Nakagawa, H., Fukunishi, H., Takahashi, Y., Watanabe, S., Taguchi, M., Bertaux, J.-L., *et al.*: 2003, *J. Geophys. Res.* **108**, 10, doi: 10.1029/2003JA009882.

Neugebauer, M., Goldstein, B. E., McComas, D. J., Suess, S. T., and Balogh, A.: 1995, *J. Geophys. Res.* **100**, 23389, doi: 10.1029/95JA02723.

Neugebauer, M.: 2001, in Balogh, A., Marsden, R. G., and Smith, E. J. (eds.), *The Heliosphere near Solar Minimum. The Ulysses Perspective*, Praxis Books in Astrophysics and Astronomy, Springer London, pp. 43–106.

Noci, G., Kohl, J. L., Antonucci, E., Tondello, G., Huber, M. C. E., *et al.*: 1997, *The Corona and Solar Wind Near Minimum Activity*, Fifth SOHO Workshop, ESA SP-404, p. 75.

Palmer, I. D., Allum, F. R., and Singer, S.: 1978, *J. Geophys. Res.* **83**, 75.

Parker, E. N.: 1958, *Astrophys. J.* **128**, 664.

Pilipp, W. G., Mühlhäuser, K.-H., Miggenrieder, H., Rosenbauer, H., and Schwenn, R.: 1987, *J. Geophys. Res.* **92**, 1103.

Pizzo, V., Schwenn, R., Marsch, E., Rosenbauer, H., Mühlhäuser, K.-H., and Neubauer, F. M.: 1983, *Astrophys. J.* **271**, 335, doi: 10.1086/161200.

Plunkett, S. P., Michels, D. J., Howard, R. A., Brueckner, G. E., St. Cyr, O. C., Thompson, B.J., *et al.*: 2002, *Adv. Space Res.* **29**, 1473.

Posner, A., Zurbuchen, T. H., Schwadron, N. A., Fisk, L. A., Gloeckler, G., Linker, J. A., *et al.*: 2001, *J. Geophys. Res.* **106**, 15869, doi: 10.1029/2000JA000112.

Priest, E. R.: 1988, *Astrophys. J.* **328**, 848.

Raouafi, N.-E., and Solanki, S. K.: 2004, *Astron. Astrophys.* **427**, 725, doi: 10.1051/0004-6361:20041203.

Reiner, M. J., Kaiser, M. L., and Bougeret, J.-L.: 2001, *J. Geophys. Res.* **106**, 29989, doi: 10.1029/2000JA002228.

Richardson, I. G., and Cane, H. V.: 1995,*J. Geophys. Res.* **100**, 23397.

Rodriguez, L., Woch, J., Krupp, N., Fränz, M., von Steiger, R., Forsyth, R. J., *et al.*: 2004, *J. Geophys. Res.* **109**, 1108, doi: 10.1029/2003JA010156.

Rosenbauer, H., Schwenn, R., Marsch, E., Meyer, B., Miggenrieder, H., Montgomery,, M.D., *et al.*: 1977, *J. Geophys.* **42**, 561.

Rust, D. M.: 2003,*Adv. Space Res.* **32**, 1895, doi: 10.1016/S0273-1177(03)90623-5.

Rust, D. M., and Kumar, A.: 1996, *Astrophys. J. Lett.* **464**, L199, doi: 10.1086/310118.

Sanderson, T. R., Appourchaux, T., Hoeksema, J. T., and Harvey, K. L.: 2003, *J. Geophys. Res.* **108**, 7-1, doi: 10.1029/2002JA009388.

Schwenn, R.: 1990, in: Schwenn, R., and Marsch, E. (eds.), *Physics of the Inner Heliosphere I*, Springer-Verlag Berlin Heidelberg New York, p. 99.

Schwenn, R.: 2005, in *Geophysics and Geochemistry, from Encyclopedia of Life Support Systems (EOLSS)*, Eolss Publishers, Oxford, UK [Online].

Schwenn, R., and Marsch, E. (eds.): 1990, *Physics of the Inner Heliosphere I, Large-Scale Phenomena*, Springer-Verlag Berlin Heidelberg New York.

Schwenn, R., and Marsch, E. (eds.): 1991, *Physics of the Inner Heliosphere II, Particles, Waves and Turbulence*, Springer-Verlag Berlin Heidelberg New York.

Schwenn, R., Rosenbauer, H., Miggenrieder, H., and Meyer, B.: 1976, *Space Res.*, 671.

Schwenn, R., Montgomery, M. D., Rosenbauer, H., Miggenrieder, H., Mühlhäuser, K. H., Bame, S. J., *et al.*: 1978, *J. Geophys. Res.* **83**, 1011.

Schwenn, R., Rosenbauer, H., and Mühlhäuser, K. H.: 1980, *Geophys. Res. Lett.* **7**, 201.

Schwenn, R., Mhlhäuser, K. H., Marsch, E., and Rosenbauer, H.: 1981, *Solar Wind Conference*, p. 126.

Schwenn, R., Inhester, B., Plunkett, S. P., Epple, A., Podlipnik, B., Bedford, D. K., *et al.*: 1997, *Solar Phys.* **175**, 667.

Sheeley, N. R., and Harvey, J. W.: 1981, *Solar Phys.* **70**, 237.

Sheeley, N. R., and Wang, Y.-M.: 2002,*Astrophys. J.* **579**, 874, doi: 10.1086/342923.

Sheeley, N. R., Wang, Y.-M., Hawley, S. H., Brueckner, G. E., Dere, K. P., Howard, R. A., *et al.*: 1997, *Astrophys. J.* **484**, 472, doi: 10.1086/304338.

Sheeley, Jr., N. R., Walters, J. H., Wang, Y.-M., and Howard, R. A.: 1999, *J. Geophys. Res.* **104**, 24739.

Sime, D. G., Fisher, R. R., and Altrock, R. C.: 1989, *Astrophys. J.* **336**, 454, doi: 10.1086/167024.

Slavin, J. A., Smith, E. J., and Thomas, B. T.: 1984, *Geophys. Res. Lett.* **11**, 279.

Srivastava, N., Schwenn, R., Inhester, B., Stenborg, G., and Podlipnik, B.: 1999a, *Space Sci. Rev.* **87**, 303.

Srivastava, N., Schwenn, R., Inhester, B., Stenborg, G., and Podlipnik, B.: 1999b, in: Habbal, S. R., Esser, R., Hollweg, J. V., and Isenberg, P. A. (eds.), *Solar Wind Nine*, AIP Conference proceedings 471, Woodbury, New York, pp. 115–118.

Srivastava, N., Schwenn, R., Inhester, B., Martin, S. F., and Hanaoka, Y.: 2000, *Astrophys. J.* **534**, 468.

St. Cyr, O. C., Plunkett, S. P., Michels, D. J., Paswaters, S. E., Koomen, M. J., Simnett, G. M., *et al.*: 2000, *J. Geophys. Res.* **105**, 18169, doi: 10.1029/1999JA000381.

Steinitz, R.: 1983, *Solar Phys.* **83**, 379.

Stenborg, G., Schwenn, R., Inhester, B., and Srivastava, N.: 1999, On the Rotation Rate of the Emission Solar Corona, ESA SP-448: Magnetic Fields and Solar Processes, 1107.

Strachan, L., Suleiman, R., Panasyuk, A. V., Biesecker, D. A., and Kohl, J. L.: 2002, *Astrophys. J.* **571**, 1008, doi: 10.1086/339984.

Strachan, L., Kohl, J. L., Weiser, H., Withbroe, G. L., and Munro, R. H.: 1993, *Astrophys. J.* **412**, 410, doi: 10.1086/172930.

Thieme, K. M., Marsch, E., and Schwenn, R.: 1990, *Ann. Geophysicae* **8**, 713.

Thompson, B., and Murdin, P.: 2000, *Encyclopedia of Astron. Astrophys.*, doi: 10.1888/0333750888/2575.

Thompson, B. J., Plunkett, S. P., Gurman, J. B., Newmark, J. S., St. Cyr, O. C., and Michels, D. J.: 1998, *Geophys. Res. Lett.* **25**, 2465, doi: 10.1029/98GL50429.

Tripathi, D., Bothmer, V., Solanki, S. K., Schwenn, R., Mierla, M., and Stenborg, G.: 2005, *SOHO/EIT Observation of a Coronal Inflow*, IAU Symposium, pp. 133–134.

Tsurutani, B. T., Gonzalez, W. D., Tang, F., Akasofu, S. I., and Smith, E. J.: 1998, *J. Geophys. Res.* **93**, 8519.

Tu, C.-Y., Zhou, C., Marsch, E., Xia, L.-D., Zhao, L., Wang, J.-X., *et al.*: 2005, *Science* **308**, 519, doi: 0.1126/science.1109447.

von Steiger, R., Wimmer-Schweingruber, R. F., Geiss, J., and Gloeckler, G.: 1995, *Adv. Space Res.* **15**, 3.

Vourlidas, A., Buzasi, D., Howard, R. A., and Esfandiari, E.: 2002, in *Solar Variability: From Core to Outer Frontiers* ESA SP-506, pp. 91–94.

Wang, Y.-M., Sheeley, N. R., Walters, J. H., Brueckner, G. E., Howard, R. A., Michels, D. J., *et al.*: 1998, *Astrophys. J. Lett.* **498**, L165, doi: 10.1086/311321.

Wang, Y.-M., Sheeley, N. R., Howard, R. A., St. Cyr, O. C., and Simnett, G. M.: 1999, *Geophys. Res. Lett.* **26**, 1203, doi: 10.1029/1999GL900209.

Wang, Y.-M., and Sheeley, N. R.: 1990, *Astrophys. J.* **355**, 726, doi: 10.1086/168805.

Whang, Y. C., Wang, Y.-M., Sheeley, N. R., and Burlaga, L. F.: 2005, *J. Geophys. Res.* **110**, 3103, doi: 10.1029/2004JA01087B.

Webb, D. F., and Howard, R. A.: 1994, *J. Geophys. Res.* **99**, 4201.

Wiegelmann, T., Schindler, K., and Neukirch, T.: 1998, *Solar Phys.* **180**, 439.

Wiegelmann, T., Xia, L. D., and Marsch, E.: 2005, *Astron. Astrophys.* **432**, L1, doi: 10.1051/0004-6361:200500029.

Wilhelm, K., Lemaire, P., Curdt, W., Schuhle, U., Marsch, E., Poland, A. I., *et al.*: 1997, *Solar Phys.* **170**, 75.

Wilhelm, K., Dammasch, I. E., Marsch, E., and Hassler, D. M.: 2000, *Astron. Astrophys.* **353**, 749.

Zhang, J., Dere, K. P., Howard, R. A., Kundu, M. R., and White, S. M.: 2001, *Astrophys. J.* **559**, 452.

Zhang, J., Dere, K. P., Howard, R. A., and Vourlidas, A.: 2004, *Astrophys. J.* **604**, 420.

Zirker, J.: 1977, *Coronal Holes and High Speed Solar Wind Streams*, Boulder, Colorado University Press.

Xia, L. D., Marsch, E., and Curdt, W.: 2003, *Astron. Astrophys.* **399**, L5.

HELIOSPHERIC PHYSICS: LINKING THE SUN TO THE MAGNETOSPHERE

THOMAS H. ZURBUCHEN
Department of Atmospheric, Oceanic and Space Sciences, University of Michigan, USA
(E-mail: thomasz@umich.edu)

(Received 4 May 2006; Accepted in final form 8 November 2006)

Abstract. Research into the heliospheric structure and its relation to the solar boundary is at an impasse. After successful predictions by Parker about the zeroth-order behavior of the heliospheric magnetic field and the solar wind, the heliospheric community struggles to make substantive progress toward a predictive model describing the connections between the Sun and its space environment, between the closed corona and the open corona extending to the planets. This is caused by our lack of understanding of the basic processes heating the corona and transporting open magnetic field. We detail the models used to describe this connectivity, from potential field source surface models to full MHD techniques. We discuss the current limitations of both approaches. Finally, we address a recent attempt to advance our understanding beyond these limitations. At this point in time the proposed theory remains controversial in the community, but it addresses important shortcomings of current approaches outlined above.

Keywords: solar wind, solar corona, coronal heating

1. Introduction

The heliosphere is an exciting place connecting the Sun and its atmosphere to the planets of our solar system and to the galactic environment beyond. It provides the solar plasma, which determines the space environment of all bodies in the solar system including the Earth. The inner heliosphere and the Earth's space environment are dominated by the solar corona, which in many ways has been very elusive to our investigations: We are not currently able to observe the driving agent of the corona, its magnetic field. Furthermore, we have sampled the heliosphere mostly at 1 AU. The *Helios* spacecraft are still the only inner heliospheric probes to date to reach an approximate heliospheric distance of 0.29 AU, or approximately 62 solar radii (R_s). *Helios* revealed a strong radial evolution in the solar wind (Schwenn, 1990). Structures on small spatial scales dominate the solar wind up close, but get washed out as the dynamic evolution of the solar wind carries it out into the solar system. But the most important dynamic processes occur closer to the Sun than the *Helios*' orbit. For comparison, solar wind heating and acceleration occurs within 5 R_s, and the Alfvén radius, at which the supersonic solar wind finally decouples from the Sun, is at 10–20 R_s. Most energetic particles are also accelerated within the Alfvén point (Parker, 2001, and references therein). The key processes dominating

Space Science Reviews (2006) 124: 77–90
DOI: 10.1007/s11214-006-9130-x

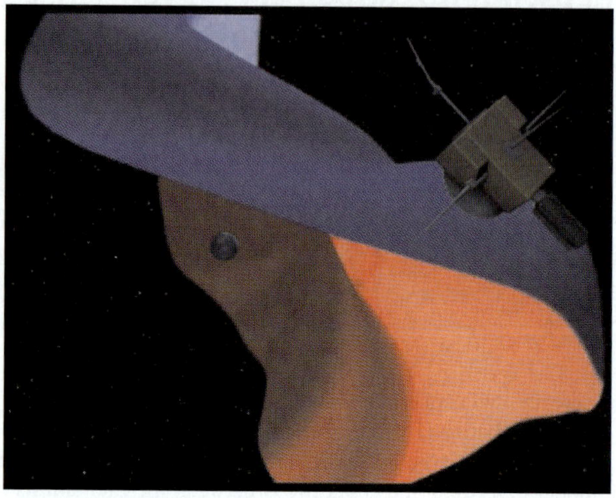

Figure 1. Artist's representation of the *Ulysses* spacecraft probing the heliospheric current sheet at high latitudes. The current sheet was computed from photospheric field distributions by Riley *et al.* (2002).

the heliosphere therefore remain unknown. This was exposed by out-of-ecliptic measurements made by the *Ulysses* spacecraft (Figure 1) (Wenzel *et al.*, 1992), where fundamental discrepancies with predictions were found: For example, the solar wind speed was higher than predicted, coronal-hole solar wind was found to have a different composition than slow solar wind, there were unexpected particles that propagated from low to high latitudes, and the galactic cosmic ray modulation was larger than previously anticipated.

There are, of course, important white-light and radio observations of solar wind plasmas in that near-solar range, but they often result only in column densities and are therefore not very sensitive to the physical processes that are of key interest here.

Even with these limitations, crucial constraints on the corona have been derived (Guhathakurta and Sittler, 1999; Wang *et al.*, 1999). Furthermore, current heliospheric radio observations either have frequency resolution or position resolution on the Sun (Bastian *et al.*, 1998). It is therefore very difficult to interpret them in the context of physical processes that have important spatial, frequency, and temporal signatures. Furthermore, modern UV spectroscopic observations of the Sun have unraveled significant departures of the corona from a locally thermal distribution (Kohl *et al.*, 1998), but the interpretation of these data in the context of coronal physics is not straightforward (Hollweg, 2000; Antonucci *et al.*, 2004).

Due to this lack of direct observations of key phenomena, proper modeling tools are essential to test our understanding of the Sun-heliosphere connections (Schatten *et al.*, 1969; Wang *et al.*, 1989; Bravo *et al.*, 1998; Fisk *et al.*, 1999). These models have to predict the heliospheric magnetic field, its topology, and its large-scale configurations. They also have to be consistent with other constraints from heliospheric observations and observations of their coronal sources. But, by necessity, the models often use approximations that are sometimes difficult to defend, and care has to be taken to only use the models within their realm of applicability.

We first define this problem and then briefly discuss its classic solutions, the potential field source surface (PFSS) models. We then discuss MHD models, their successes and important current limitations. Finally, we look at novel approaches for addressing the important limitations of models to date.

2. Problem Definition

The Sun-heliospheric magnetic field problem can be defined as a combination of two important connections which are intimately related, but have often been dealt with separately. The first connection relates to the solar-wind plasma energization, or the "coronal heating problem," and the second relates to the magnetic connectivity of this solar wind and its transport properties or, the "heliospheric field transport problem."

This is best formulated using the MHD force equation.

$$\rho\left(\frac{\partial \mathbf{v}}{\partial t} + (\mathbf{v}\cdot\nabla)\mathbf{v}\right) = -\nabla p + \rho\mathbf{g} + \mathbf{J}\times\mathbf{B}. \tag{1}$$

Here, ρ denotes the mass density, $\mathbf{v}$ the plasma velocity, and p the plasma pressure. Also, $\mathbf{g}$ is gravity, and the last term is the Lorentz force of a magnetic field $\mathbf{B}$, with $\mathbf{J} = \nabla \times \mathbf{B}/\mu_0$. The evolution of the magnetic field is described by the induction equation, and p has to be calculated from some energy-conservation law.

In principle, Equation (1) describes the plasma in both the closed and open corona. In the closed corona, the plasma is captured in loops. Therefore, at the tops of these loops, the radial velocity disappears, and magnetic and gas pressure terms are balanced by magnetic tension and gravity. The corona is heated by processes currently not fully understood. Magnetic fields most likely have a very important role in generating this heat and dissipating it in the corona (Klimchuk and Porter, 1995). Due to a combination of increased heat dissipation, increasing the thermal pressure p, and successive changes in magnetic fields, these loop plasmas can open and the centrifugal term, $(\mathbf{v}\cdot\nabla)\,\mathbf{v}$, exceeds the magnetic field tensional forces and gravity, stretching the magnetic loop into the heliosphere. This process has been described as a major source of open flux in the heliosphere, directly relating coronal heating to heliospheric magnetic flux. But, such opinions are often based on models and have difficulty explaining the heliospheric behavior of that flux.

However, there is another way to frame this connection problem, addressing magnetic field transport in the corona at multiple spatial scales. The corona is, on average, magnetically dominated, but is rooted in the photosphere which is constantly in motion, driven by convection below. In steady state, the overall global-scale magnetic field will tend to evolve to its lowest energy state, the potential field configuration, with $\nabla \times \mathbf{B} = 0$, eliminating large-scale currents and hence all magnetic forces. The inner boundary is given by photospheric observations, but the choices for the outer boundary are less clear. The simplest version of an outer boundary is that of a conductor at a given heliocentric distance, r_s. This analysis bypasses the coronal heating problem immediately arising from the first formulation, but introduces uncertainties that are not easily overlooked. The most obvious one is the choice of r_s, which is ill constrained. One would have guessed that r_s should relate to the Alfvén radius, where the solar wind magnetically decouples, but the most successful choices of r_s are only at approximately one-tenth of the Alfvén radius. In fact, the best choices are even within the critical radius, where the solar wind passes through the sonic point. Secondly, this approximation is not useful for describing temporal changes of the coronal field, for instance arising from motions of magnetic footpoints on the surface. These can be addressed, for example, using the full induction equation.

Such footpoint motions have a range of consequences, depending on whether the motions are slow or fast compared to the Alfvén speed in the corona. For fast motions, these perturbations propagate along the magnetic field and can be interpreted as a series of Alfvén waves. For slow motions, large-scale magnetic field perturbations are introduced in space and can be calculated from the induction equation (see Hollweg and Lee, 1989). For example, for motions perpendicular to the radial direction, with velocity $\mathbf{u}_\perp$, and if the solar-wind velocity v is radial, the heliospheric magnetic field is

$$B_r = B_s \left(\frac{r_s}{r}\right)^2,$$
$$\mathbf{B}_\perp = -\mathbf{u}_\perp \frac{B_r}{v} \frac{r}{r_s}. \tag{2}$$

Here, B_r is the radial component of the magnetic field, directly integrated from $\nabla \cdot \mathbf{B} = 0$, r_s is the radius of the magnetic source, and B_s is the magnetic field strength at this source radius. Also, $\mathbf{B}_\perp$ is the magnetic field perpendicular to the radial direction. For example, a flow field given by a simple rotation in the φ direction at r_s, with an angular velocity of Ω, results in a flow velocity given by $\mathbf{u}_\perp = \Omega r_s \sin\theta \mathbf{e}_\phi$, resulting in non-zero currents in contradiction to the assumption of $\nabla \times \mathbf{B} = 0$ (see Fisk and Schwadron, 2001). This transport can also be random in nature, as pointed out by Leighton (1964), and has been used for the explanation of heliospheric phenomena using arguments very similar to the ones outlined above (Schrijver and DeRosa, 2003; Giacalone and Jokipii, 2004). There are also important magnetic interactions that occur in the closed corona. Due to their coupling to the convective zone, magnetic fields in the corona are constantly twisted and significant dissipation occurs (Parker, 1993). This small-scale magnetic interaction is therefore, for the thermal properties of the plasma, directly linking magnetic transport and coronal heating.

In summary, the corona-heliosphere plasma connection problem therefore falls into two specific themes which are intimately linked. First, the energy available for heating the corona determines the topology of the open corona, how the magnetic field locally evolves, and how the solar wind can be accelerated. The nature of this heating and dissipation process is therefore crucial, and directly related to the second process, the transport properties of the magnetic field in the solar atmosphere. It reflects the evolution of the solar corona through an interaction of many spatial scales. The description of the large-scale magnetic field does a remarkable job of describing the global topology of the heliospheric field. But there are no obvious ways to address time dependences, which are important on small scales.

In this paper, we organize this discussion along the lines of arguments focused on the second question ("magnetic field transport problem"), and link back to the first question ("coronal heating") at various points. This approach has some important advantages, which will become clear throughout this discussion. This is different from common lines of inquiry, which tend to focus on coronal heating.

3. Potential Field Source Surface (PFSS) Models

PFSS models are perhaps the most useful Sun-heliosphere connection models to date and have many applications (Arge *et al.*, 2002; Luhmann *et al.*, 2002; DeRosa and Schrijver, 2002). They use the potential field assumption mentioned above and hence decouple the coronal heating problem and the open-flux connection problem by making the inner corona exclusively magnetically dominated and time-independent. For a given choice of source surface radius, r_s, the magnetic field potential then can be written as a sum of spherical harmonics.

The PFSS models reveal important characteristics of the Sun-heliosphere connection, which have been successfully confirmed. The first conclusion from this model is also demonstrated in Figure 2, namely that open flux organizes itself into volumes of open field with a given polarity. The interface between these regions of given polarity is a current layer which extends from the outer corona into the heliosphere to form a global heliospheric current sheet. Figure 2 shows the solar magnetic field and the heliospheric extension of this current sheet for two specific conditions. The top-left panel shows photospheric magnetic fields at solar minimum, for Carrington rotation 1913. The top-right panel shows the global structure of the heliospheric current sheet arising from this field configuration. Clearly, the dipole moment is the most dominant part of multipole expansion. The dipole is slightly tilted and also shows some additional warping, which can lead to multiple crossings of a spacecraft at a given heliospheric latitude. The bottom panels of Figure 2 show the equivalent figures near solar maximum conditions, for Carrington rotation 1969. Again, there is only one single-current sheet with interesting characteristics. First, the current sheet extends over all latitudes and shows a large

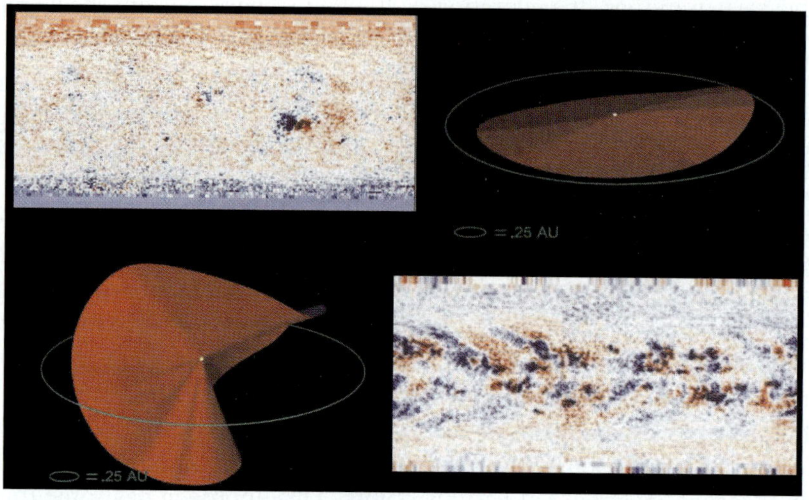

Figure 2. Photospheric distribution of magnetic field, from Wilcox solar observatory, and corresponding current-sheet orientation for Carrington rotations 1913 and 1969. Both solar minimum and solar maximum conditions lead to a single current sheet which extends to high latitudes near solar maximum. Computations from Riley *et al.* (2002).

degree of warping induced from higher-order multipoles that become dominant during this period of the solar cycle.

Model predictions for the location of the current sheet have been successfully tested throughout the previous solar cycles, using in-ecliptic measurements near Earth, but also, for the first time, using high-latitude measurements from *Ulysses*, as indicated in Figure 1. These combined measurements were shown to be in remarkable agreement with predictions. The models have also been used successfully to describe the magnetic topology of the closed corona (DeRosa and Schrijver, 2002), even though deviations from potential fields are discernable from detailed observations (Schrijver and van Ballegooijen, 2005). PFSS models, without any additional correction terms, also predict a latitude-dependent magnitude of the total magnetic flux emerging into the heliosphere, which is inconsistent with observations (Smith *et al.*, 2001). Such correction terms were proposed by several authors (Wang and Sheeley, 1995 and references therein), but are generally not included for practical use, such as the computation of magnetic field expansion factors.

As discussed above, PFSS models are generally unsuitable to describe the time evolution of the global magnetic field. They have nonetheless been used in the fashion of cartoons, "many images make a movie," far beyond the realm of physical applicability of PFSS models. This approach, of course, neglects crucial physics, but its physical meaning can be shown to be equivalent to a situation in which the timescale for magnetic dissipation, or reconnection effects, τ_{Rec}, outpaces the typical timescale of the evolution of this structure, τ_{Evo}, or,

$$\tau_{\mathrm{Rec}} \ll \tau_{\mathrm{Evo}}. \tag{3}$$

Under these assumptions, the evolution of the heliospheric current sheet from one solar minimum to the next is well approximated by a slow, but highly irregular rotation of this current sheet. During its 11 years (on average), the current sheet simply rotates through 180° to complete its field reversal.

It should be pointed out that there may still be benefits to the application of PFSS models in such a time-dependent mode, as shown by Arge and Pizzo (2000), with respect to the predictions of the polarity of the heliospheric magnetic field and the solar wind speed. These physical quantities appear to be determined by large-scale topology successfully modeled by PFSS models. But, this approach is limited: Large-scale field-deviations from the average Parker model, for example, result from evolutionary aspects of the field, and cannot be derived from a PFSS approach. The knowledge of this large-scale field is required for the prediction of energetic particles, for example.

Some MHD models, such as Riley *et al.* (2002), predict multiple disconnected current sheets during sole Carrington rotations. Thus far this has not been directly observed (Smith *et al.*, 2001), but we have limited data. We should also remember the important model assumptions, and the limitations, of observing magnetic fields only from near the Earth, which clearly limits the knowledge of the boundary conditions for any heliospheric model calculations. However, their fundamental shortcomings are their neglect of plasma time dependence and plasma interactions, which have to be addressed in a more rigorous way.

4. Magnetohydrodynamic (MHD) Models

The PFSS models have now been surpassed by MHD models that have more capability to address the important connection between coronal heating and the field and plasma ranging into the heliosphere (Linker *et al.*, 1999; Riley *et al.*, 2002; Roussev *et al.*, 2003, 2004). These computations solve all MHD equations, typically using a simple energy equation with additional *a priori* assumptions about heating profiles or boundary conditions.

MHD models have important advantages over PFSS models because they *can* address the interactions between fields and plasmas in a self-consistent way. Adding the full MHD force equation now relates the magnetic connection problem to the coronal heating problem, which is unfortunately not understood. This is demonstrated in Figure 3, which compares two heliospheric models with two slightly different assumptions for the thermodynamic properties of the coronal plasma (Roussev *et al.*, 2003, 2004). In this solution, *a priori* information about the field topology from PFSS models is being used. The large image in Figure 3 shows a very reasonable coronal topology, in qualitative agreement with heliospheric data. Magnetic fields open into the heliosphere near coronal holes, and remain closed near active regions. But there are several *a priori* assumptions for this. For example,

are rather abrupt, with the tendency to discredit the diffusive picture (Roelof, 2006 personnel communication).

The proposed large-scale transport of the magnetic field leads to important consequences for the low-latitude corona. As the open flux gets transported to low latitudes, it has to interact with topologically closed regions that give rise to the slow solar wind (Fisk *et al.*, 1999). The transport properties there should no longer be expected to be dominated by differential rotation and magnetic expansion, but should be associated with interactions between open and closed flux, through magnetic reconnection. Such interactions were predicted prior to the 1990s, often in conjunction with energetic particles observed in the corona (see a more recent summary by Reames, 2002). Most theories for impulsive particle acceleration involve a closed magnetic topology, but the particles need to escape onto open field lines for us to see them. That requires the same kind of reconnection predicted in this theory.

This proposed transport would lead to a very different character for the slow wind when compared to coronal-hole-associated fast wind. Slow wind would be intrinsically time dependent and composed of streams from localized release events in the corona, initiated by reconnection. Indeed, in-situ analysis of the slow solar wind near the Sun reveals important structures that tend to wash out before they arrive at 1 AU. But they are still visible in the compositional pattern of the solar wind (Zurbuchen *et al.*, 2001), as predicted. There is no current explanation for such behavior based on other PFSS-based views of the Sun-heliosphere connection. This explanation also clearly delineates the fundamental differences between fast and slow solar wind, also clearly indicated by *in-situ* data (von Steiger *et al.*, 2000).

The challenge then turns to the theoretical description of this magnetic field transport and its relation to coronal heating. This is a formidable task and requires new approaches addressing both large- and small-scale transport properties of flux on the Sun. An approach was chosen which is comparable to the one used for decades to describe cosmic-ray particles in the large-scale heliosphere (Parker, 1958). Transport properties have to be calculated based on the properties of coronal fields from the smallest scales to the large-scale (Fisk, 2005; Fisk and Zurbuchen, 2005, 2006). We will not address the details of these calculations, which predict the transport of open field and its distribution on the Sun.

The expansion of this field can be predicted and compared with PFSS models. A new mathematical method has been devised by Gilbert *et al.* (2006), using a diffusive process to redistribute the flux to be consistent with the observation of approximately constant magnetic flux in the heliosphere, as discussed above. The open flux in the photosphere is redistributed in the entire heliosphere, while maintaining the shape and location of the current sheet. Figure 4 summarizes the results of such a calculation (red dashed lines), and compares them with PFSS calculations (black lines). The method can be shown to be consistent with magnetically dominated MHD solutions. There are important differences between this mapping technique

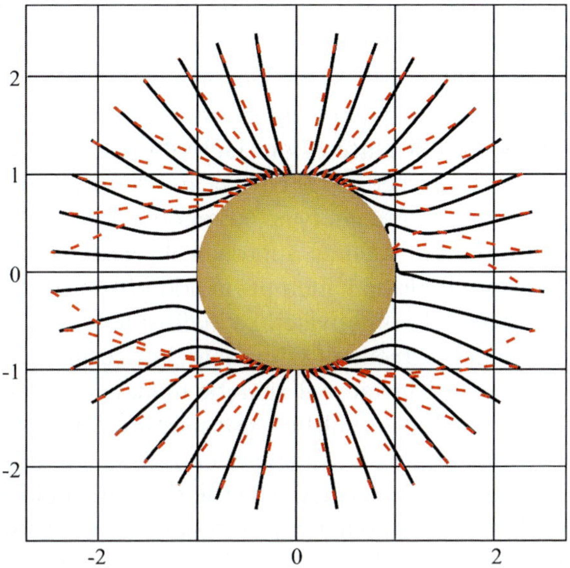

Figure 4. Comparison between PFSS expansions based on a model that includes expansion to constant radial field (Gilbert *et al.*, 2006). Red dashed lines indicate field lines in a PFSS expansion, and black lines are from an expansion model consistent with heliospheric observations by Smith *et al.* (2001). A small component of open flux was included emerging from topologically closed regions, as predicted by Fisk *et al.* (1999).

and traditional PFSS maps. Much has been said about the relation between PFSS expansion factors and the solar wind speed. In a time-stationary solar wind model, expansion factors directly affect the dynamics of the solar wind through the continuity equation (Holzer, 2005, and references therein). There are two questions associated with these expansion factors. The vast majority of the factors used in the literature are computed from an incorrect outer boundary condition. Such expansion factors therefore have the same inconsistencies as the field expansions from which they are calculated, caused by the outer boundary conditions that are inconsistent with MHD models and heliospheric data. The Gilbert *et al.* (2006) method allows for a simple calculation of expansion factors with outer-boundary conditions consistent with the heliospheric observations. Furthermore, it allows for the inclusion of open flux in the photosphere away from coronal holes calculated using PFSS methods. There are expectations of such sources, based on in-situ measurements of energetic particles from the Sun, and based on the properties of the slow solar wind (Zurbuchen *et al.*, 2002; Fisk and Zurbuchen, 2005, 2006).

This mapping method is interesting by itself, but when coupled with the open field transport, has the potential to address the coronal-hole heating problem. There have been initial attempts at this integrated model based on compositional signatures for such connections (Gloeckler *et al.*, 2003). But this is still being worked on.

6. Summary

This is an exciting time in heliospheric physics. We have had a number of successes describing the zeroth-order behavior of the Sun-heliosphere connection. But we are unable to move beyond that because of our lack of understanding of the physical processes that govern the interface of the corona and the heliosphere. This situation is rather embarrassing and unfortunately no commonly accepted strategic approach is in place. We have proposed such an approach. It uses magnetic field transport processes as a guiding principle and attempts to address coronal heating as a result of that transport. This model is motivated by heliospheric observations and is in qualitative and quantitative agreement with many key results. But it remains controversial on many grounds. The current analytic models will have to be combined into MHD simulations and tested with self-consistent calculations. Only then will we be able to decide whether this novel approach is in fact living up to its promise to become a predictive model for the heliospheric magnetic field and its solar wind. It is quite possible that we will find that this approach is not an exact match to what we need as a community. Other approaches may then turn out to be even more promising. But a successful and predictive model of the heliosphere will address, in conjunction, the coronal heating problem and the transport of the magnetic field in the corona, with a highly interdisciplinary view of an interface between the Sun and our heliosphere. (Zurbuchen and the TR&T team, 2006).

Finally, we should point out that there is a need for novel observations of the corona. Emerging techniques in position and frequency-resolved radio-observations (Bastian *et al.*, 1998) and also of novel spectroscopic observations of the corona (Lin *et al.*, 2004) have the potential to revolutionize our understanding of the Sun-heliosphere connection. These approaches are currently in reach and hopefully will materialize during the next solar cycle.

Acknowledgements

This work was supported, in part, by NASA grants NAGR-10975 and NNG05 GB43G, and JPL contract 1237843. We thank ISSI for its hospitality and flawless organization of this meeting. THZ thanks Len Fisk, Jon Linker and Spiro Antiochos for useful conversations.

References

Antonucci, E., Dodero, M. A., Giordano, S., Krishnakumar, V., and Noci, G.: 2004, *Astron. Astrophys.* **416**, 749.

Arge, C. N., and Pizzo, V. J.: 2000, *J. Geophys. Res.* **105**(A5), 10,465.

Arge, C. N., Hildner, E., Pizzo, V. J., and Harvey, J. W.: 2002, *J. Geophys. Res.* **107**(A10), 1319, doi: 10.1029/2001JA000503.

Bastian, T., Benz, A. O., and Gary, D. E.: 1998, *Ann. Rev. Astron. Astrophys.* **36**, 131.
Bravo, S., Stewart, G. A., and Blanco-Cano, X.: 1998, *Solar Phys.* **179**, 223.
DeRosa, M. L., and Schrijver, C. J.: 2002, *Solar Phys.* **212**, 165.
Fisk, L. A.: 1996, *J. Geophys. Res.* **101**(A7), 15,547.
Fisk, L. A.: 2005, *Astrophys. J.* **626**(1), 563.
Fisk, L. A., and Schwadron, N. A.: 2001, *Astrophys. J.* **590**, 425.
Fisk, L. A., and Zurbuchen, T. H.: 2005, in B. Fleck and T. H. Zurbuchen (eds.), *Proceedings of Solar Wind 11 – SOHO 16 "Connecting Sun and Heliosphere"*, ESA SP-592, p. 227.
Fisk, L. A., and Zurbuchen, T. H.: 2006, *J. Geophys. Res.* **111**(A9), 28.
Fisk, L. A., Zurbuchen, T. H., and Schwadron, N. A.: 1999, *J. Geophys. Res.* **104**, 19765.
Giacalone, J., and Jokipii, J. R.: 2004, *Astrophys. J.* **616**, 573.
Gilbert, J., Zurbuchen, T. H., and Fisk, L. A.: 2006, *Astrophys. J.*, submitted.
Gloeckler, G., Zurbuchen, T. H., and Geiss, J.: 2003, *J. Geophys. Res.* **108**, doi: 10.1029/2002JA009286.
Guhathakurta, M., and Sittler, E.: 1999, *Astrophys. J.* **523**, 812.
Hollweg, J. V.: 2000, *J. Geophys. Res.* **105**, 15699.
Hollweg, J. V. and Lee, M. A.: 1989, *Geophys. Res. Lett.* **16**, 919.
Holzer, T.: 2005, in B. Fleck and T. H. Zurbuchen (eds.), *Proceedings of Solar Wind 11 – SOHO 16 "Connecting Sun and Heliosphere"*, ESA SP-592, p. 115.
Klimchuk, J. A., and Porter, L. J.: 1995, *Nature* **377**, 131.
Kohl, J. L., *et al.*: 1998, *Astrophys. J.* **501**, L127.
Kóta J., and Jokipii, J. R.: 1998, *Space Sci. Rev.* **83**, 137.
Leighton, R. B.: 1964, *Astrophys. J.* **140**, 1547.
Lin, H., Kuhn, J. R., and Coulter, R.: 2004, *Astrophys. J.* **613**, L177.
Linker, J. A., *et al.*: 1999, *J. Geophys. Res.* **104**, 8909.
Lionello, R., Riley, P., Linker, J. A., and Mikic, Z.: 2005, *Astrophys. J.* **625**, 463.
Luhmann, J. G., Li, Y., Arge, C. N., Gazis, P. R., and Ulrich, R.: 2002, *J. Geophys. Res.* **107**, doi: 10.1029/2001JA007550.
Neugebauer, M., *et al.*: 1998, *J. Geophys. Res.* **103**, 14587.
Pagel, C., Crooker, N. U., and Larson, D. E.: 2005, *Geophys. Res. Lett.* **32**, L14105.
Parker, E. N.: 1958, *Phys. Rev.* **110**, 1445.
Parker, E. N.: 1993, *Astrophys. J.* **407**, 342.
Parker, E. N.: 2001, in J. A. M. Bleeker, J. Geiss, and M. C. E. Huber (eds.), *Century of Space Science*, vol. 1. Kluwer Academic Publishers, Dordrecht.
Pizzo, V., and Gosling, J. T.: 1994, *Geophys. Res. Lett.* **21**, 2063.
Reames, D. V.: 2002, *Astrophys. J.* **571**, L63.
Riley, P., Linker, J. A., and Mikic, Z.: 2002, *J. Geophys. Res.* **107**, doi: 10.1029/2001JA000299.
Roussev, I. I., *et al.*: 2003, *Astrophys. J.* **595**, L57.
Roussev, I. I., Sokolov, I. V., Forbes, T. G., Gombosi, T. I., Lee, M. A., and Sakai, J. I.: 2004, *Astrophys. J.* **605**, L73.
Schatten, K. H., Wilcox, J. M., and Ness, N. F.: 1969, *Solar Phys.* **6**, 443.
Schrijver, C. J., and DeRosa, M. L.: 2003, *Solar Phys.* **212**, 165.
Schrijver, C. J., and van Ballegooijen, A. A.: 2005, *Astrophys. J.* **630**, 552.
Schwenn, R.: 1990, in R. Schwenn and E. Marsch (eds.), *Physics of the Inner Heliosphere I*, Springer-Verlag, Berlin, p. 99.
Shay, M. A., Drake, J. F., Rogers, B. N., and Denton, R. E.: 2001, *J. Geophys. Res.* **106**, 3759.
Smith, E. J., Balogh A., Forsyth, R. J., and McComas, D. J.: 2001, *Geophys. Res. Lett.* **28**, 4159.
von Steiger, R., *et al.*: 2000, *J. Geophys. Res.* **105**, 27217.
Wang, Y-M., Nash, A. G., and Sheeley, N.: 1989, *Science* **245**, 712.
Wang, Y.-M., and Sheeley, N.: 1995, *Astrophys. J.* **447**, L143.

Wang, Y.-M., Howard, R. A., Rich, N. B., and Lamy, P. L.: 1999, *Geophys. Res. Lett.* **26**, 1349.
Wenzel, K.-P., Marsden, R. G., Page, D. E., and Smith, E. J.: 1992, *Astron. and Astrophys. Suppl.* **92**, 207.
Zurbuchen, T. H., and the Heliospheric Magnetic Field TR&T team: 2006, *EOS*, in press.
Zurbuchen, T. H., Schwadron, N. A., and Fisk, L. A.: 1997, *J. Geophys. Res.* **102**, 24175.
Zurbuchen, T. H., Hefti, S., Fisk, L. A., Gloeckler, G., Schwadron, N. A., Smith, C. W., *et al.*: 2001, *J. Geophys. Res.* **106**, 16,001.
Zurbuchen, T. H., Fisk, L. A., Gloeckler, G., and von Steiger, R.: 2002, *Geophys. Res. Lett.* **29**, doi: 10.1029/2001GL013946.

SUBSTORMS AND THEIR SOLAR WIND CAUSES

RUMI NAKAMURA
Space Research Institute, Austrian Academy of Sciences, A-8042 Graz, Austria
(E-mail: rumi.nakamura@oeaw.ac.at)

(Received 31 August 2005; Accepted in final form 29 November 2006)

Abstract. Consequences of the solar wind input observed as large scale magnetotail dynamics during substorms are reviewed, highlighting results from statistical studies as well as global magnetosphere/ionosphere observations. Among the different solar wind input parameters, the most essential one to initiate reconnection relatively close to the Earth is a southward IMF or a solar wind dawn-to-dusk electric field. Larger substorms are associated with such reconnection events closer to the Earth and the magnetotail can accumulate larger amounts of energy before its onset. Yet, how and to what extent the magnetotail configuration before substorm onset differs for different solar wind driver is still to be understood. A strong solar wind dawn-to-dusk electric field is, however, only a necessary condition for a strong substorm, but not a sufficient one. That is, there are intervals when the solar wind input is processed in the magnetotail without the usual substorm cycle, suggesting different modes of flux transport. Furthermore, recent global observations suggest that the magnetotail response during the substorm expansion phase can be also controlled by plasma sheet density, which is coupled to the solar wind on larger time-scales than the substorm cycle. To explain the substorm dynamics it is therefore important to understand the different modes of energy, momentum, and mass transport within the magnetosphere as a consequence of different types of solar wind-magnetosphere interaction with different time-scales that control the overall magnetotail configuration, in addition to the internal current sheet instabilities leading to large scale tail current sheet dissipation.

Keywords: solar wind, substorm, magnetotail

1. Introduction

The Earth's magnetosphere is formed as a consequence of the interaction between the solar wind and the terrestrial magnetic field. Whereas the dayside magnetosphere is compressed by the solar wind, the nightside magnetosphere is stretched out into a long magnetotail. The interaction between the solar wind and magnetospheric field lines drives the global convection in the magnetosphere known as Dungey convection cycle and is illustrated in Figure 1a (from Baumjohann and Treumann, 1996). When the interplanetary magnetic field (IMF) is southward, the merged field lines on the dayside are transported toward the nightside, reconnected in the distant tail and then transported back toward the dayside, creating inward and Earthward flow in the central plasma sheet and causing the two cell flow-current pattern in the ionosphere. As a consequence, enhanced eastward (westward) auroral electrojet is observed in the duskside (dawnside) high-latitude ionosphere. The flux transport rates on the dayside and the nightside are, however, balanced only in an average sense and unbalanced transport is the ultimate cause of a substorm.

Space Science Reviews (2006) 124: 91–101
DOI: 10.1007/s11214-006-9131-9

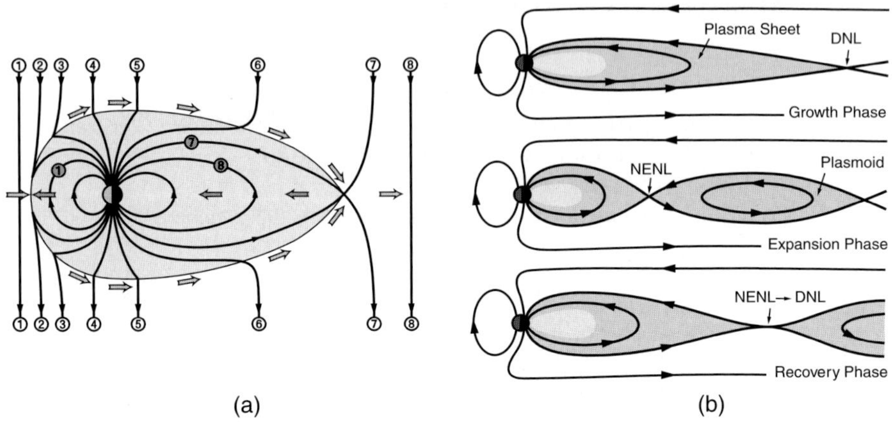

Figure 1. (a) Dungey convection cycle during southward IMF period (b) Reconfiguration of the plasma sheet during a substorm. *Uppermost panel:* Growth phase *Middle panel:* Expansion phase *Lowermost panel:* Recovery phase (From Baumjohann and Treumann, 1996).

The reconfiguration of the plasma sheet during a substorm is illustrated in Figure 1b. When the dayside merging rate is enhanced and exceeds the nightside rate, the magnetic field energy becomes stored in the magnetotail and the magnetotail becomes more stretched during the growth phase, as illustrated in the upper panel of Figure 1b. The tail current sheet becomes thin, often thinner than the compression of the current sheet caused by the accumulated flux in the magnetotail. How and how far the current sheet thins before substorm onset is one of the key problems for understanding the substorm onset mechanism. During the expansion phase, illustrated in the middle panel in Figure 1b, this stored energy is released in an explosive way and can be observed as a variety of signatures, involving both the ionosphere and magnetosphere: enhanced field-aligned currents, strong auroral precipitation, enhanced westward auroral electrojet flowing in the ionosphere in the midnight sector, and particle acceleration and subsequent fast plasma flows in the magnetosphere.

Although how and where the onset instability sets in is yet a debated issue, general consensus is that energy in the thin tail current sheet is dissipated explosively during the expansion phase. There are mainly two regions considered to be important in these processes: One is the inner tail region where field-aligned current pairs (a substorm current wedge) are formed, initiating auroral breakup in the conjugate ionosphere. The other is the midtail region where near-Earth reconnection takes place, creating fast plasma jets directed Earthward and tailward. As illustrated in the middle panel of Figure 1b, in a global sense, the magnetic field changes toward a dipolar configuration on the Earthward side of the reconnection region, whereas a closed flux region, called plasmoid, forms at the tailward side of the reconnection region. The latter eventually moves tailward, removing a part of the plasma sheet. During the recovery phase, shown in the bottom panel in Figure 1b, the field configuration recovers to that of the quiet level.

This simple illustration of Figure 1b, however, describes the changes in the overall global magnetotail configuration and only serves as an average picture. In reality, the global processes involve disturbances with different spatial scales. For example, in spite of the global substorm disturbance, the initial onset region is considered to be concentrated in a very small region. Plasma sheet flows in the magnetosphere actually consist of transient, localized high speed flows, called bursty bulk flow (BBF) (Baumjohann *et al.*, 1990; Angelopoulos *et al.*, 1992), and slow background flows. Reconnection involves processes from MHD scales down to the electron scale. Essential points in substorm physics are therefore to clarify the dissipation process of the tail current sheet at different scales in different key regions and to explain the causal relationships among these different disturbances in the larger context of the magnetosphere/ionosphere system and beyond, namely the solar wind magnetosphere interaction.

While a number of recent studies based on four-point observations by Cluster in the midtail region obtained important results on local processes leading to the understanding of physics of the tail current sheet associated with near-Earth reconnection (e.g., Runov *et al.*, 2003), fast plasma flows (e.g., Nakamura *et al.*, 2002) and flux ropes/plasmoids (e.g., Slavin *et al.*, 2003), studies investigating relationships between solar wind parameters and magnetospheric parameters are also important to understand the substorms in a global context, considering the fact that the ultimate driver of the process is the solar wind. During the ISTP era many events have been accumulated when simultaneous monitor of solar wind were available to study responses in the magnetosphere. Particulary, these studies allowed to investigate magnetotail responses at different regions during a variety of solar wind conditions. Furthermore, fortuitous configuration of multi-spacecraft enabled to study the large-scale response of the magnetototail simultaneously at a wide region. In this paper, three topics on large scale magnetotail dynamics during substorms are selected to discuss the consequences of the solar wind chain on Geospace: (1) Solar wind input control of near-Earth reconnection, (2) Different modes of the magnetotail flux transport as a response to solar wind input, and (3) Solar wind density influences on substorms.

2. Solar Wind Input Control of Near-Earth Reconnection

Determining where and when reconnection takes place is of fundamental importance to understand the substorm onset mechanism as well as to examine where and how the thin current sheet is formed. Although there is a general consensus that near-Earth reconnection takes place outside the dipolar region, it was a long lasting issue of debate at what exact location reconnection occurs, since there were disagreements among the results from different missions. Using an extensive data energy associated with substorms can be observed as increase and recover of the energy density (total pressure) in the midtail (Caan *et al.*, 1973). The mode of dissipation of the accumulated energy observed in the tail, however, does not seem always the same. That is, it is not a simple enhancement of the energy and then a decrease toward the quiet level. For example, at times the accumulated energy is rapidly dissipated during a substorm, whereas at other times the dissipation can occur more gradually during ongoing magnetic activity (Fairfield *et al.*, 1981). During the latter cases, the energy supplied by the solar wind may even exceed that being dissipated, thus causing the tail energy to increase. On the other hand, there are times when convection dominates and no loading/unloading process are observed. Such periods are called steady magnetospheric convection (SMC) (Sergeev *et al.*, 1977), or more recently called continuous magnetospheric dissipation (CMD) emphasizing the nearly constant magnetotail pressure profile (Tanskanen *et al.*, 2005). Based on the global multi-point observations from solar wind, magnetotail and ionosphere, these different modes of magnetospheric transport in association with enhanced solar wind energy input were quantitatively compared by Nakamura *et al.* (1999).

The left panels in Figure 3 show the relationship between the solar wind electric field and the associated changes of the midtail parameters and the auroral electrojet. The total pressure in the midtail started to increase in association with the solar wind electric field enhancement and reached the maximum near the onset of a substorm at 1502 UT (line (iii) in Figure 3). This onset also coincided with the dawn-to-dusk magnetospheric electric field enhancement ($-(V \times B)_Y$ electric field, or flux transport rate). The close time scales of the period of enhanced solar wind electric field and magnetospheric electric field (if shifted in time of $\tau \sim$40 min) indicate a direct control of the solar wind input to the magnetotail response. The relationship can be examined quantitatively by taking into account the day-night reconnection processes (see Figure 1). That is, the change in the midtail magnetic flux, $\partial F/\partial t$, can be expressed as the balance between the magnetic flux stripped from the dayside (and added to the tail), $\Delta\Phi_d$, due to dayside reconnection, and the flux transported away from the midtail region, $\Delta\Phi_n$, by nightside reconnection processes in the distant tail as well as in the near-Earth tail, such as $\partial F/\partial t = \Delta\Phi_d - \Delta\Phi_n$. The obtained relationship among these three quantities are shown in the right panels of Figure 3 (for details see Nakamura *et al.*, 1999). Here $\Delta\Phi_d$ was obtained from the solar wind and ionospheric data, whereas F from the midtail data. In case of steady state convection as illustrated in Figure 1a, $\partial/\partial t = 0$ is expected. As a reference value for this mode, a simple shift of $\Delta\Phi_d$ by the time difference, τ, is shown as a dotted line in the right third panel of Figure 3. $\Delta\Phi_n$ estimated from the observed values, F and $\Delta\Phi_d$, and thereby taking into account the magnetotail substorm process (Figure 1b) is shown as solid line in the same panel.

Because of the near-Earth reconnection during the substorm, $\Delta\Phi_n$ increases significantly removing the magnetic flux from the midtail, either tailward or earthward, at a value higher than expected from steady state convection. This unloading phase the plasma sheet. During the recovery phase, shown in the bottom panel in Figure 1b, the field configuration recovers to that of the quiet level.

This simple illustration of Figure 1b, however, describes the changes in the overall global magnetotail configuration and only serves as an average picture. In reality, the global processes involve disturbances with different spatial scales. For example, in spite of the global substorm disturbance, the initial onset region is considered to be concentrated in a very small region. Plasma sheet flows in the magnetosphere actually consist of transient, localized high speed flows, called bursty bulk flow (BBF) (Baumjohann *et al.*, 1990; Angelopoulos *et al.*, 1992), and slow background flows. Reconnection involves processes from MHD scales down to the electron scale. Essential points in substorm physics are therefore to clarify the dissipation process of the tail current sheet at different scales in different key regions and to explain the causal relationships among these different disturbances in the larger context of the magnetosphere/ionosphere system and beyond, namely the solar wind magnetosphere interaction.

While a number of recent studies based on four-point observations by Cluster in the midtail region obtained important results on local processes leading to the understanding of physics of the tail current sheet associated with near-Earth reconnection (e.g., Runov *et al.*, 2003), fast plasma flows (e.g., Nakamura *et al.*, 2002) and flux ropes/plasmoids (e.g., Slavin *et al.*, 2003), studies investigating relationships between solar wind parameters and magnetospheric parameters are also important to understand the substorms in a global context, considering the fact that the ultimate driver of the process is the solar wind. During the ISTP era many events have been accumulated when simultaneous monitor of solar wind were available to study responses in the magnetosphere. Particulary, these studies allowed to investigate magnetotail responses at different regions during a variety of solar wind conditions. Furthermore, fortuitous configuration of multi-spacecraft enabled to study the large-scale response of the magnetototail simultaneously at a wide region. In this paper, three topics on large scale magnetotail dynamics during substorms are selected to discuss the consequences of the solar wind chain on Geospace: (1) Solar wind input control of near-Earth reconnection, (2) Different modes of the magnetotail flux transport as a response to solar wind input, and (3) Solar wind density influences on substorms.

2. Solar Wind Input Control of Near-Earth Reconnection

Determining where and when reconnection takes place is of fundamental importance to understand the substorm onset mechanism as well as to examine where and how the thin current sheet is formed. Although there is a general consensus that near-Earth reconnection takes place outside the dipolar region, it was a long lasting issue of debate at what exact location reconnection occurs, since there were disagreements among the results from different missions. Using an extensive data

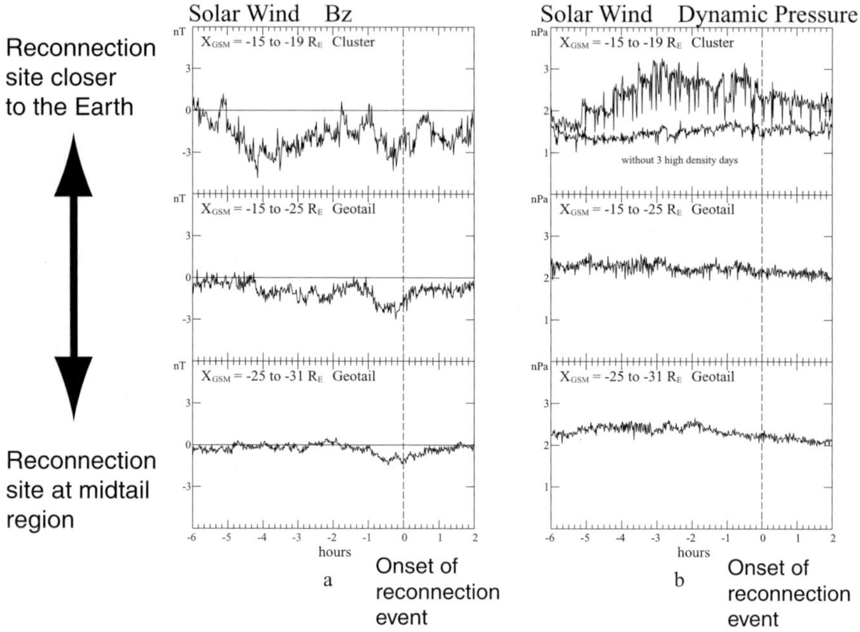

Figure 2. Relationships between the solar wind input and reconnection site. Average variations of IMF Bz (*left three panels*) and solar wind pressure (*right three panels*) for the period between 6 before and 2 hours after the reconnection when the observed reconnection site was $X_{\mathrm{GSM}} = -15$ to $-19R_E$ (*top panels*), $X_{\mathrm{GSM}} = -15$ to $-25R_E$ (*middle panels*), and $X_{\mathrm{GSM}} = -25$ to $-31R_E$ (*bottom panels*) (After Nagai *et al.*, 2005).

base from Geotail and Cluster, which surveyed the near-Earth plasma sheet at radial distances of 10–31 R_E, Nagai *et al.* (2005) investigated the solar wind condition for the reconnection events associated with substorm onset. Reconnection events were identified using the criterion of strong electron acceleration and also using tailward flow events, which obtained similar statistical results. Various solar wind parameters prior to each reconnection event were examined in order to find the factor controlling the magnetic reconnection site. Figure 2 shows the average temporal variations of the IMF B_Z (left panels) and the dynamic pressure (right panels) around the reconnection events identified from the tailward flows sorted by the reconnection events observed at different radial distances from the Earth. The upper panels show the solar wind variations when the reconnection events were identified from Cluster, meaning that these reconnection events took place between $X_{\mathrm{GSM}} = -15$ and $-19R_E$. The middle and the bottom panels show the solar wind profiles when the reconnection events were observed by Geotail at $X_{\mathrm{GSM}} = -15$ to $-25R_E$ and $X_{\mathrm{GSM}} = -25$ to $-31R_E$, respectively. Hence, the uppermost panels represent the solar wind profile around the reconnection event which took place close to the Earth, whereas the bottom panels represent those of the midtail reconnection events. It can be seen that the southward IMF is significantly larger for reconnection events closer to the Earth, while less clear differences can be seen among the three left plots of the dynamic pressure (if the high-density period in Cluster is removed in the upper panel). Among the different solar wind parameters, it was found that the most important factor in the solar wind energy input can be expressed by $-V_X \times B_S$, where V_X is the X component of the solar wind velocity and B_S is the southward component of the interplanetary magnetic field. The effect of solar wind dynamic pressure was found to be minor by Nagai *et al.* (2005).

This results clearly showed that the solar wind energy input strongly controls magnetic reconnection in the magnetotail. The reconnection events closer to the Earth occur for large energy input (average level as well as total amount), while the midtail events for low energy input levels. On the other hand, although larger dynamic pressure is thought to make the current thinner, expecting to produce favorable conditions to initiate reconnection, no clear dynamic pressure effect is found in the present analysis. The results imply that there is no threshold level in the accumulated energy for which the reconnection can be initiated, but the magnetotail seems to behave differently between strong and weak solar wind driver cases by changing its configuration and thereby changing the favorable site of the reconnection. Nagai *et al.* (2005) also showed that the solar wind input control of reconnection results in solar cycle dependence of tail magnetic reconnection site and thereby explains the previous different results in terms of the location of the near-Earth neutral line obtained from different missions during different phase of the solar cycle.

IMF dependence is also observed in the global auroral observation by Polar in which stronger auroral precipitation covering lower latitude regions were detected during southward IMF period (Shu *et al.*, 2001). It has been known that the expansion onset of a more intense substorm takes place at lower latitudes in the ionosphere (Kamide and Akasofu, 1974). Furthermore, the radial profile of average B_Z obtained from Geotail statistical study showed that the location of the reconnection region is closer to the Earth during intense substorms (defined by auroral electrojet activity; Miyashita *et al.*, 2004). Combining these results it can be concluded that a strong southward IMF B_Z is an essential factor for a larger substorms, where disturbances take place covering lower latitudes in the ionosphere, and magnetotail disturbances take place closer to the Earth. Yet, the onset of the tail current sheet instabilities developing to a substorm is not determined by the amount of the solar wind input but is possibly controlled by some internal processes or by IMF trigger mechanism (e.g., Lyons *et al.*, 1997).

3. Different Modes of the Magnetotail Flux Transport

As illustrated in Figure 1 southward IMF B_Z drives the convection in the magnetotail, thereby transferring solar wind energy into the magnetosphere. In the magnetotail, evidence supporting the storage (loading) and release (unloading) of energy associated with substorms can be observed as increase and recover of the energy density (total pressure) in the midtail (Caan *et al.*, 1973). The mode of dissipation of the accumulated energy observed in the tail, however, does not seem always the same. That is, it is not a simple enhancement of the energy and then a decrease toward the quiet level. For example, at times the accumulated energy is rapidly dissipated during a substorm, whereas at other times the dissipation can occur more gradually during ongoing magnetic activity (Fairfield *et al.*, 1981). During the latter cases, the energy supplied by the solar wind may even exceed that being dissipated, thus causing the tail energy to increase. On the other hand, there are times when convection dominates and no loading/unloading process are observed. Such periods are called steady magnetospheric convection (SMC) (Sergeev *et al.*, 1977), or more recently called continuous magnetospheric dissipation (CMD) emphasizing the nearly constant magnetotail pressure profile (Tanskanen *et al.*, 2005). Based on the global multi-point observations from solar wind, magnetotail and ionosphere, these different modes of magnetospheric transport in association with enhanced solar wind energy input were quantitatively compared by Nakamura *et al.* (1999).

The left panels in Figure 3 show the relationship between the solar wind electric field and the associated changes of the midtail parameters and the auroral electrojet. The total pressure in the midtail started to increase in association with the solar wind electric field enhancement and reached the maximum near the onset of a substorm at 1502 UT (line (iii) in Figure 3). This onset also coincided with the dawn-to-dusk magnetospheric electric field enhancement ($-(V \times B)_Y$ electric field, or flux transport rate). The close time scales of the period of enhanced solar wind electric field and magnetospheric electric field (if shifted in time of τ ~40 min) indicate a direct control of the solar wind input to the magnetotail response. The relationship can be examined quantitatively by taking into account the day-night reconnection processes (see Figure 1). That is, the change in the midtail magnetic flux, $\partial F/\partial t$, can be expressed as the balance between the magnetic flux stripped from the dayside (and added to the tail), $\Delta\Phi_d$, due to dayside reconnection, and the flux transported away from the midtail region, $\Delta\Phi_n$, by nightside reconnection processes in the distant tail as well as in the near-Earth tail, such as $\partial F/\partial t = \Delta\Phi_d - \Delta\Phi_n$. The obtained relationship among these three quantities are shown in the right panels of Figure 3 (for details see Nakamura *et al.*, 1999). Here $\Delta\Phi_d$ was obtained from the solar wind and ionospheric data, whereas F from the midtail data. In case of steady state convection as illustrated in Figure 1a, $\partial/\partial t = 0$ is expected. As a reference value for this mode, a simple shift of $\Delta\Phi_d$ by the time difference, τ, is shown as a dotted line in the right third panel of Figure 3. $\Delta\Phi_n$ estimated from the observed values, F and $\Delta\Phi_d$, and thereby taking into account the magnetotail substorm process (Figure 1b) is shown as solid line in the same panel.

Because of the near-Earth reconnection during the substorm, $\Delta\Phi_n$ increases significantly removing the magnetic flux from the midtail, either tailward or earthward, at a value higher than expected from steady state convection. This unloading phase

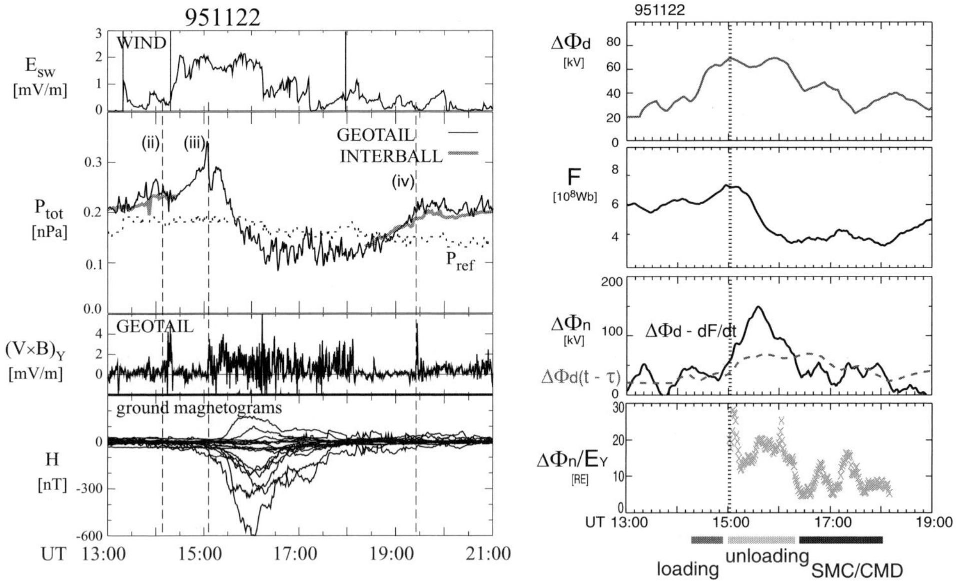

Figure 3. Left panel: Solar wind electric field, E_{SW}, and total pressure, P_{tot}, in the magnetotail from Geotail and Interball (compared to a reference value, P_{ref}), dawn-dusk component of the $-V \times B$ electric field in the magnetotail, and ground magnetogram data H showing the auroral electrojet (from top to bottom). The vertical lines show the start times of the electric field enhancements. *Right panel:* Estimated dayside potential drop, $\Delta\Phi_d$, magnetic flux in the midtail, F, and nightside effective potential drop, $\Delta\Phi_n$, and spatial scale of the flux transport, $\Delta\Phi_n/E_Y$ (from top to bottom). The vertical line shows the onset time of the substorm expansion phase. $\Delta\Phi_n$ is estimated in the following two ways: (1) difference between the dayside potential drop and the rate of change of the magnetic flux in the tail (solid curve), (2) dayside potential drop shifted in time, which is the time difference between the onset of the solar wind electric field enhancement and that of the magnetospheric electric field (dashed curve) (Both after Nakamura *et al.*, 1999).

is characterized by large scale flux transport ranging between 15 and $20R_E$ as seen in the $\Delta\Phi_n/E_Y$ values. Yet, after the expansion phase the flux stayed at a lower level. $\Delta\Phi_n$ obtained from the flux balance (solid curve in the third panel) become in fact comparable to that of the convection level (dotted curve in the third panel) as expected for SMC or CMD period. The magnetospheric electric field (third panel at the left side) was still enhanced during this period. As a result, the scale size of the flux transport becomes quite small (down to $4R_E$). Hence, the nightside flux transport during the low-pressure period (equivalent to the low-flux period) takes place at smaller spatial scales than that during the expansion phase. Interesting to note that although SMC stands for steady magnetospheric convection, the actual magnetospheric flow is known to be very bursty (Sergeev *et al.,* 1996) as can be also seen in the bursty profile of $(-V_X \times B_Y)$ in Figure 3. This indicates that the global balance in flux transport during SMC is not obtained by steady homogeneous flow but could rather consists of localized bursty flows.

The controlling parameters for these various responses of the tail to solar wind input are not fully understood. A recent statistical study investigating the solar wind parameters for unloading/loading periods and CMD periods concluded that the former periods likely have a stronger mean southward IMF $B_Z < -5$ nT than the latter (Tanskanen *et al.*, 2005). Yet, as discussed in previous sections, the magnitude of B_Z can also control the reconnection region related to substorm. Pulkinnen *et al.* (1994) found that during SMC periods, the plasma sheet was thicker in the midtail region, resembling the growth phase of a substorm, while the near tail current sheet was strongly thinned. This peculiar configuration allowed to gives rise to the magnetic field minimum in the near-Earth region which is a favorable configuration of the magnetotail to perform a steady adiabatic convection, without necessity of substrom cycle. Hence, both the magnetotail condition and the solar wind driver would affect the different modes of the flux transfer in the magnetotail.

4. Solar Wind Density Influences on Substorms

So far we have discussed mainly the effect of the IMF Bz on the magnetotail response which can be ascribed to the process illustrated in Figure 1. The input solar wind energy causes the tail to be stressed (loading) or directly drives night side reconnection (SMC) but then the tail relaxes (unloading) to the initial state at some stage so that solar wind input and output will eventually be balanced. Yet, within a single substorm cycle, the input and output is not always balanced, as clearly shown in recent observations by Sergeev *et al.* (2005). In this event, the tail was significantly loaded, but the unloading took place in a rather modest fashion, indicating that the amount of dissipation in the tail during a single substorm can not be determined from the amount of the input energy only. Here another solar wind factor, namely the density which determines the precondition of the plasma sheet was suggested to play a significant role in the magnetospheric response.

The overall profile of the interplanetary magnetic field, magnetotail current signatures observed by multi-spacecraft in the tail and ground magnetograms for the substorm studied by Sergeev *et al.* (2005) are shown in Figure 4. The event was initiated by a IMF southward turning (arriving at the magnetopause around 20 UT, Figure 4a). During the growth phase (starting after 2015 UT) a steady growth of the tail current and stretching of the tail configuration were observed (Figure 4c,d). During that time period the Polar Cap (PC) index, which is an indicator of polar cap convection, also increased (Figure 4b). The loading of magnetic flux and magnetic energy into the magnetotail, starting after 2015 UT, is very distinctly observed in this event. The amount of lobe magnetic field increase is unusually large, as can be seen from the large discrepancy of the Cluster lobe field (B_X in Figure 4c) compared to the empirical model of the lobe field at substorm (unloading) onset by Shukhtina *et al.* (2004). This discrepancy suggests that the energy storage is a factor 3 to 4 higher than in the average substorm. The tail current density obtained

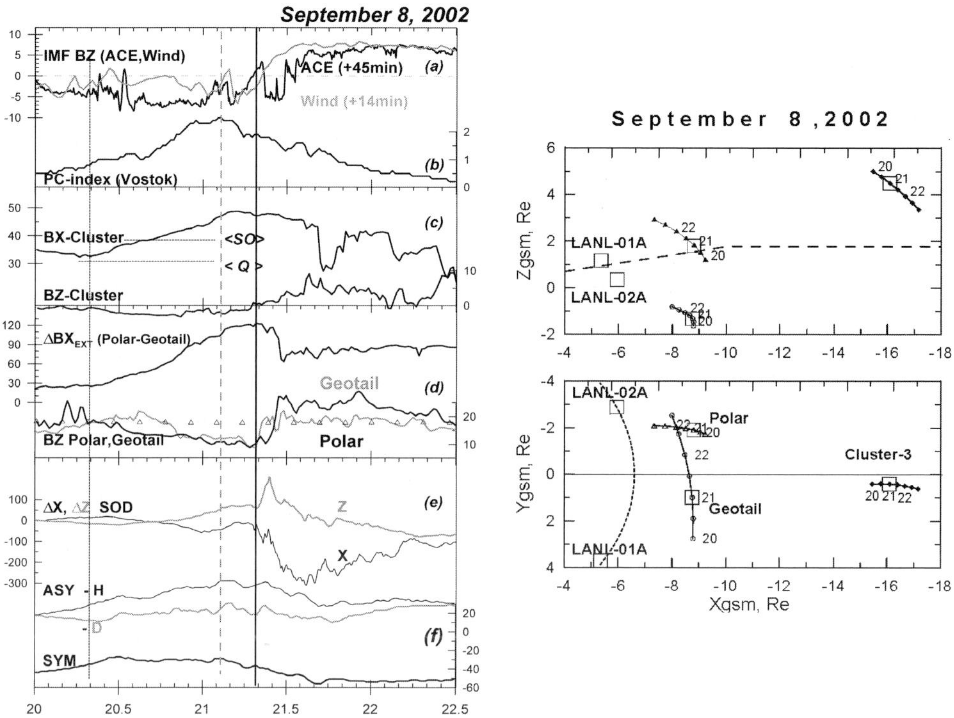

Figure 4. Left panel: Overview of activity at the end of September 8, 2002: From top to the bottom: (a) IMF B_Z-variations at ACE and WIND spacecraft time-shifted to the subsolar magnetopause position, shown are their GSM coordinates and time shift applied; (b) polar cap PC-magnetic index from the Vostok station; (c) B_X and B_Z component variations at Cluster; (d) difference of B_X components of external field at Polar and Geotail spacecraft and corresponding B_Z field components (with T96 model values shown by triangles); (e) Sodankyla (63.8 CGLat); (f) midlatitude ASY/SYM indexes. The growth phase onset and substorm onset (2118UT) are marked by the vertical lines; the average values of the lobe field expected at the Cluster location for quiet conditions (Q) and at the unloading (substorm expansion) onset (SO) (after Shukhtina *et al.*, 2004) are shown by horizontal lines on the panel (c). *Right panel:* Configuration of basic spacecraft on September 8, 2002 in GSM-coordinates. The neutral sheet position is indicated on the XZ cross-section, spacecraft positions at 21 UT are marked by the rectangles (Both from Sergeev *et al.*, 2005).

from Geotail and Polar closer to the Earth also shows a factor of 5 increase during the growth phase. The extraordinary large increase in the tail current results in an unusually large equatorward expansion of the auroral oval and an unusually low-latitude location of the substorm onset at 2118 UT. Yet, the electrojet intensity during the substorm was moderate, reaching only about 300 nT (as can be seen, for example, in the SOD magnetogram in Figure 4e). The total energy precipitated in the auroral oval on the nightside evaluated from FUV Image observations showed also a modest value (not shown) and the standard substorm signature, the injections of energetic electrons into the inner magnetosphere, was absent.

Sergeev *et al.* (2005) interpreted these seemingly inconsistent features, i.e., extraordinary large energy storage and modest dissipation rate, as being due to the effect of a cold and dense plasma sheet (CDPS) during this interval. The effects of CDPS can be, for example, softer precipitation, since less field-aligned acceleration is needed taking into account the Knight conductance. Or, stronger loading and larger tail current is needed to reach typical thresholds of current sheet instability, since the gyroradius or inertial length is smaller for CDPS. Note that CDPS typically appear after many hours long periods of northward IMF and low magnetic activity and suggesting slow entry process of the solar wind cold/dense plasma (e.g. Terasawa *et al.*, 1997). Several recent studies discussed the cause of CDPS to be the double lobe reconnection during northward IMF intervals, enabling the dense magnetosheath plasma to be captured on closed field lines and transported toward tail (e.g., Li *et al.*, 2005) or solar-wind plasma entry via the K-H instability at the LLBL (Hasegawa *et al.*, 2004). Note that the timescale of the formation of the cold and dense plasma sheet is therefore quite different from the processes discussed in the previous sections.

Therefore,together with southward IMF variations as discussed before, the density/temperature of the plasma sheet may considerably influence the substorm manifestations and introduce a large variability in the behavior and appearance of the substorms. Since the plasma sheet density variations is also controlled by the solar wind but at a different time scale, solar wind-magnetosphere interaction controlling the magnetospheric substorms should be considered as a multi-time scale process.

5. Conclusions

Substorms are most fundamental processes in the magnetosphere. The essential solar wind parameter controlling the substorm is the southward IMF for the large scale magnetotail dynamics. Large IMF input produces current sheet disturbances close to the Earth and involves lower latitude region in the ionosphere. The dependence on dynamic pressure, on the other hand, is less apparent, at least for the near-Earth reconnection process. Yet, southward IMF is only a necessary condition for a large substorm, since southward IMF does not necessarily causes a substorm during periods of SMC or CMD. Furthermore, large scale magnetospheric response is likely controlled by the intrinsic condition of the magnetosphere, which is modified by different modes of solar wind magnetosphere interaction other than southward IMF reconnection. To explain substorm dynamics and to predict the magnetospheric response from the solar wind input it is important to understand both the internal current sheet instabilities leading to the large scale tail current sheet dissipation and the different modes of large scale solar wind-magnetosphere interaction.

Acknowledgements

I wish to thank T. Nagai, V. A. Sergeev, and W. Baumjohann for helpful discussions. The organizers of the workshop on "Solar dynamics and its effects on the heliosphere and Earth" and the International Space Science Institute (ISSI) are gratefully acknowledged.

References

Angelopoulos, V., Baumjohann, W., Kennel, C. F., Coroniti, F. V., Kivelson, M. G., Pellat, R., *et al.*: 1992, *J. Geophys. Res.* **97**, 4027.

Baumjohann, W., Paschmann, G., and Luehr, H.: 1990, *J. Geophys. Res.* **95**, 3801.

Baumjohann, W., and Treumann, R. A.: 1996, *Basic Space Plasma Physics*. Imperial College Press, London.

Caan, M. N., McPherron, R. L., and Russell, C. T.: 1973, *J. Geophys. Res.* **78**, 8087.

Fairfield, D. H., Lepping, R. P., Hones Jr., E. W., Bame, S. J., and Asbridge, J. R.: 1981, *J. Geophys. Res.* **86**, 1396.

Hasegawa, H., Fujimoto, M., Phan, T.-D., Reme, H., Balogh, A., Dunlop, M. W., *et al.*: 2004, *Nature* **415**, 755.

Kamide, Y., and Akasofu, S.-I.: 1974, *J. Geophys. Res.* **79**, 3755.

Li, W., Raeder, J., Dorelli, J., Oieroset, M., and Phan, T. D.: 2005, *Geophys. Res. Lett.* **32**, L12S08, doi: 10.1029/2004GL021524.

Lyons, L. R., Blanchard, G. T., Samson, J. C., Lepping, R. P., Yamamoto, T., and Moretto, T.: 1997, *J. Geophys. Res.* **102**, 27039.

Miyashita, Y., Kamide, Y., Machida, S., Liou, K., Mukai, T., Saito, Y., *et al.*: 2004, *J. Geophys. Res.* **109**, A11205, doi: 10.1029/2004JA010588.

Nagai, T., Fujimoto, M., Nakamura, R., Baumjohann, W., Ieda, A., Shinohara, I., *et al.*: 2005, *J. Geophys. Res.* **110**, A09208, doi: 10.1029/2005JA011207.

Nakamura, R., *et al.*: 1999, *J. Geophys. Res.* **104**, 17299.

Nakamura, R., *et al.*: 2002, *Geophys. Res. Lett.* **29**, 1942, doi: 10.1029/2002GL015763.

Pulkkinen, T. I., Sergeev, V. A., Toivanen, P. K., and Pellinen, R. J.: 1994, in Kan, J. R., Craven, J. D. and Akasofu, S.-I. (eds.), *What Can We Learn About Substorms by Studying Steady Convection Events?*, Proc. Int. Conf. Substorms, Univ. Alaska Fairbanks, p. 449.

Runov, A., *et al.*: 2003, *Geophys. Res. Lett.* **30**(11), 1579, doi: 10.1029/2002GL016730.

Shue, J.-H., Newell, P. T., Liou, K., and Meng, C.-I.: 2001, *J. Geophys. Res.* **106**, 5913.

Shukhtina, M. A., Dmitrieva, N. P., and Sergeev, V. A.: 2004, *Ann. Geophys.* **22**, 1019.

Sergeev, V. A.: 1977, *Phys. Solarterr. Potsdam* **5**, 39.

Sergeev, V. A., Pellinen, R. J., and Pulkkinen, T. I.: 1996, *Space Sci. Rev.* **75**, 551.

Sergeev, V. A., *et al.*: 2005, *Ann. Geophys.* **23**, 2183.

Slavin, J., *et al.*: 2003, *Geophys. Res. Lett.* **30**, 1362, doi: 10.1029/2003GL016411.

Tanskanen, E. I., Slavin, J. A., Fairfield, D. H., Sibeck, D. G., Gjerloev, J., Mukai, T., *et al.*: 2005, *J. Geophys. Res.* **110**, A03216, doi: 10.1029/2004JA010561.

Terasawa, T., Fujimoto, M., Mukai, T., *et al.*: 1997, *Geophys. Res. Lett.* **24**, 935.

MODELING OF THE MAGNETOSPHERIC RESPONSE TO THE DYNAMIC SOLAR WIND

J. BIRN[1,*], M. HESSE[2] and K. SCHINDLER[3]
[1]*Los Alamos National Laboratory, Los Alamos, NM 87545, USA*
[2]*NASA/Goddard Space Flight Center, Greenbelt, MD 20771, USA*
[3]*Ruhr-Universität Bochum, Germany*
*(*Author for correspondence, E-mail: jbirn@lanl.gov)*

(Received 11 October 2005; Accepted in final form: 14 February 2006)

Abstract. We discuss quasi-static and dynamic models of the magnetotail response to perturbations imposed by the solar wind, focusing particularly on the formation of thin current sheets, their structure and breakup.

Keywords: magnetotail, thin current sheets, solar wind/magnetosphere interaction

1. Introduction

The Earth's magnetosphere with its elongated tail is formed, and deformed, by the impact of the solar wind onto the Earth's magnetic field. The solar wind hence controls, directly and indirectly, the structure and evolution of the magnetosphere. In this evolution one often distinguishes between effects "directly driven" by the solar wind and those caused by "unloading" of previously stored energy. In the latter case the energy released has also come primarily from the solar wind, but it is released in a manner not directly related to the solar wind driver. The unloading phase is primarily associated with magnetospheric substorms, although smaller-scale dissipation events apparently also contribute to magnetospheric activity. Alternatively, one might distinguish slow and fast magnetospheric evolution, where the major distinguishing factor is the deviation from instantaneous equilibrium. The distinctions are closely related but not identical. For instance, directly driven effects, such as the impact of an interplanetary shock, may lead to very rapid deformations and fast flows. Or quasi-static phases may be embedded in sequences of rapid deformation.

Traditionally, magnetohydrodynamic (MHD) simulations of the combined solar wind – magnetosphere system ("global simulations") have been the obvious, and most straightforward approach to model the magnetospheric response to various solar wind impact conditions (e.g., Fedder and Lyon, 1987; Walker *et al.*, 1993; Raeder, 1994). Here we use several approaches that complement global simulations and provide additional insight into the effects of the solar wind on magnetospheric structure and dynamics:

Space Science Reviews (2006) 124: 103–116
DOI: 10.1007/s11214-006-9100-3

– *Quasi-static MHD theory.* As mentioned above, the magnetospheric state prior to eruptions such as substorms can be considered as quasi-static, that is, governed by instantaneous equilibrium. The evolution under such conditions can be described as adiabatic, i.e. entropy conserving. We will demonstrate that this constraint, in combination with mass and flux conservation restricts the evolution, which may lead to the formation of a thin current sheet, embedded in the magnetotail plasma sheet, and ultimately to loss of MHD equilibrium.
– *Vlasov equilibrium theory.* MHD theory can be expected to break down when characteristic scales are reduced to typical ion, or even electron scales, such as the particle gyro radii or inertial lengths. In this regime, collisionless plasmas are better described by the Vlasov theory of the evolution of phase space distribution functions, rather than the moment equations of MHD (e.g., Krall and Trivelpiece, 1973). We show that thin current sheet equilibria persist at those scales, even though the sheet structure, and particularly the main current carriers may change from those at larger scales.
– *MHD and particle simulations of the magnetotail.* While the MHD approach is essentially the same as that of the global simulations, the restriction to the tail permits higher spatial resolution and the investigation of quasi-analytical initial equilibrium states, that do not require a history of solar wind input. Particularly, it permits highly stretched initial tail equilibria which are more difficult to obtain in global simulations because of the premature onset of reconnection. Particle simulations complement the MHD approach.

2. Adiabatic Tail Evolution

The quasi-static approach is based on the assumption that the evolution is slow compared to typical Alfvén times, so that inertia terms can be neglected and the instantaneous configurations are governed by force balance. In addition, a sequence of configurations must be consistent with the conservation of mass, magnetic flux, and entropy, if heat flux and Joule dissipation are neglected. General solutions of the magnetospheric equilibrium structure require iteration procedures or quasi-time-dependent approaches. However, the special structure of the magnetotail with a distinct difference between the characteristic spatial scales permits not only the derivation of general solutions through ordinary differential equations and ordinary integrals (Birn, 1987) but also the construction of sequences of solutions governed by the conservation laws (Birn, 1991, 2005) In the following we will outline some basics of the approach and present typical solutions.

We start out from magnetohydrostatic force balance, which can be written as

$$\frac{1}{\mu_o}\mathbf{B}\cdot\nabla\mathbf{B} = \nabla\left(p + \frac{B^2}{2\mu_o}\right) \tag{1}$$

assuming isotropic pressure. In the following we will use typical magnetospheric coordinates, with x pointing sunward, y from dawn to dusk, and z northward, perpendicular to the tail plasma/current sheet. Under typical tail conditions, excluding the near-Earth dipolar field and the magnetopause boundary regions, the characteristic scale length L_z for variations in z, equivalent to the characteristic current or plasma sheet half-thickness, is of the order of a few R_E or less, whereas the variations along x and y occur on scales of the order of tens of R_E. Under these conditions, the z component of the curvature term in the force balance (1) becomes small and one can derive the pressure balance (Siscoe, 1972; Schindler, 1972)

$$p + B^2/2\mu_0 = \hat{p}(x, y) \tag{2}$$

Consistent with the neglect of the curvature term in the z direction, the contribution of B_z^2 to the total pressure in (2) can also be neglected. Under these conditions, the partial differential Equation (1) can be solved explicitly (Birn, 1987), using Euler potentials α and β to represent the magnetic field, $\mathbf{B} = \nabla\alpha \times \nabla\beta$. The solution becomes particularly simple when the total pressure can be expressed as a function of a radial variable r, where $r^2 = (x - x_0)^2 + y^2$, using a cylindrical coordinate system r, ϕ, and z:

$$z(r, \alpha, \phi) = \frac{\beta'(\phi)}{r} \int_{\alpha_o(r)}^{\alpha} \frac{d\alpha}{\sqrt{2\mu_0(\hat{p}(r) - p(\alpha))}} \tag{3}$$

where $\beta = \beta(\phi)$ and $\alpha_0(r)$ is the value of the flux function α at the neutral sheet $z = 0$, defined by $p(\alpha_0) = \hat{p}(r)$, and we have assumed symmetry around $z = 0$, choosing $z \geq 0$ in (3) for simplicity. We have further chosen the Euler potentials such that p is a function of α only.

The solar wind influence on a magnetotail configuration is contained in the pressure distribution across field lines $p(\alpha)$, which involves the history of the configuration, and the total pressure $\hat{p}(r)$, which is more directly related to the solar wind via pressure balance at the magnetopause. Deriving a magnetotail configuration from the shape of the magnetopause involves an inversion of the solution (3) (Birn, 1991, 2005). This procedure and sample solutions will be presented in Section 2.1.

2.1. 3D Tail Equilibria with Prescribed Boundary Boundaries

The following approach is discussed in detail by Birn (1991) and Birn (2005). Here we provide only a short outline. Neglecting magnetic flux closure through the boundary, one can define the magnetopause (or some flux surface just inside the magnetopause) from (3) as field lines given by

$$a(r, \phi) = \frac{\beta'(\phi)}{r} \int_{\alpha_o(x)}^{\alpha_b(\phi)} \frac{d\alpha}{\sqrt{2\mu_0(\hat{p}(r) - p(\alpha))}} \tag{4}$$

where $\alpha_b(\phi)$ represents the value of α at the boundary $z = a(r, \phi)$.

We first note that the r dependence in (4) can be replaced by a dependence on the variable $\hat{p}$, assuming that $\hat{p}$ varies monotonically with r. Since the ϕ dependence

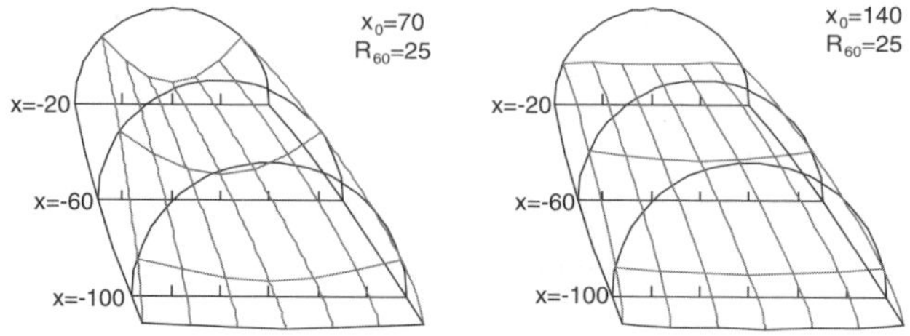

Figure 1. Magnetic flux surfaces $p = 0.1$ for two models with identical boundaries (indicated by the half-circles) but different amounts of field line flaring. The field lines lie in (r, z) planes crossing at the location $x = x_0$, $y = 0$. After Birn and Schindler (2005).

does not affect the integrand, the integral Equation (4) can be solved for the unknown function $\alpha(p, \phi)$ by replacing the integration over α by an integration over p. This yields

$$\alpha(p, \phi) = -\frac{\sqrt{2\mu_o}}{\pi\beta'(\phi)} \int_{p_b(\phi)}^{p} \frac{r(\hat{p})\,a(\hat{p})}{\sqrt{p - \hat{p}}} d\hat{p} \tag{5}$$

As discussed by Birn (2005), a solution for a given boundary surface $a(r, \phi)$ and a radial variation of the total pressure $\hat{p}(r)$, obtained from observations or approximate force balance with the solar wind, can now be obtained via (5). Figure 1 shows an example of two configurations obtained by Birn (2005) for the same boundary shape, the same x dependance of the total pressure but different y dependencies reflected in the different flaring of the field in the x, y projection. The solutions differ remarkably also in the variation of flux surfaces across the tail, which would be reflected in a variation of the plasma sheet thickness. A stronger flaring of the field in the y direction corresponds to a more pronounced increase of the plasma sheet thickness from midnight toward the flanks of the tail. These solutions therefore demonstrate that the instantaneous shape of the magnetotail depends not only on the boundary shape and the down-tail variation of the total pressure, both of which are determined closely from the instantaneous force balance with the solar wind, but also from, somewhat more obscure, parameters which depend on the history of the magnetosphere. In the following we will particularly look at this history and the constraints on the evolution of the tail configuration that result form the conservation laws for mass, entropy, and magnetic topology.

2.2. Adiabatic Boundary Deformation, Loss of Equilibrium

In addition to satisfying instantaneous force balance, sequences of quasi-equilibrium solutions are constrained by conservation of entropy. The entropy is related to the quantity

$$S = pV^\gamma \tag{6}$$

where γ is the ratio of specific heats and $V(\alpha)$ is the flux tube volume per unit magnetic flux, given by

$$V(\alpha) = \int_{r_n}^{r_o(\alpha)} \frac{dr}{B_r(r,\alpha)} \tag{7}$$

Here $B_r(r,\alpha)$ follows from the pressure balance (2) (with B_z^2 neglected)

$$B_r = \sqrt{2\mu_o[\hat{p}(r) - p(\alpha)]} \tag{8}$$

The integration boundaries in (7) are the near-Earth boundary, $r = r_n$, and the crossing point of a field line through the x, y plane, following from $\hat{p}(r_o) = p(\alpha)$. If we choose the flux value α to be frozen in the plasma fluid, i.e., ideal MHD, entropy conservation implies that $S(\alpha)$ is a fixed function, which follows from the initial state, while $p(\alpha)$ and $V(\alpha)$ might change when the configuration gets deformed.

As discussed by Birn *et al.* (2003a), an adiabatic sequence of tail equilibria can be constructed when the ϕ variation in (5) is neglected, setting $\beta = \phi$, so that

$$\alpha(p) = -\frac{\sqrt{2\mu_o}}{\pi} \int_{p_b}^{p} \frac{r(\hat{p})\,a(\hat{p})}{\sqrt{p - \hat{p}}} d\hat{p} \tag{9}$$

where p_b is now a constant. A sequence of equilibria that satisfy the imposed conservation laws can then be constructed by choosing a sequence of boundary functions $b(\hat{p}) \equiv r(\hat{p})\,a(\hat{p})$. This yields the instantaneous $\alpha(p)$ through (9), and the conservation of entropy (6) can then be used to derive

$$V(p) = \{S[\alpha(p)]/p\}^{1/\gamma} \tag{10}$$

Using (8), the flux tube volume (7) can be written as

$$V(p) = \int_{r_n}^{r_o(p)} \frac{dr}{\sqrt{2\mu_o(\hat{p}(r) - p)}}$$

$$= \int_{p}^{p_n} \left(-\frac{dr}{d\hat{p}}\right) \frac{d\hat{p}}{\sqrt{2\mu_o(\hat{p} - p)}} \tag{11}$$

where p_n is the value of $\hat{p}$ at the near-Earth boundary $r = r_n$. Inversion of the integral Equation (11) then leads to

$$r(\hat{p}) = \frac{\sqrt{2\mu_o}}{\pi} \int_{\hat{p}}^{p_n} V(p) \frac{dp}{\sqrt{p - \hat{p}}} \tag{12}$$

where $V(p)$ follows from (10). The solution (12) can then be inserted into (3) and (4), which yields explicit expressions of the magnetic field lines, including the boundary.

As discussed by Birn and Schindler (2002) and Birn *et al.* (2003a), the procedure of replacing the integration variables α with p and r with $\hat{p}$, respectively, implies that both $p(\alpha)$ and $\hat{p}(r)$ are monotonic functions. This excludes extrema (maxima,

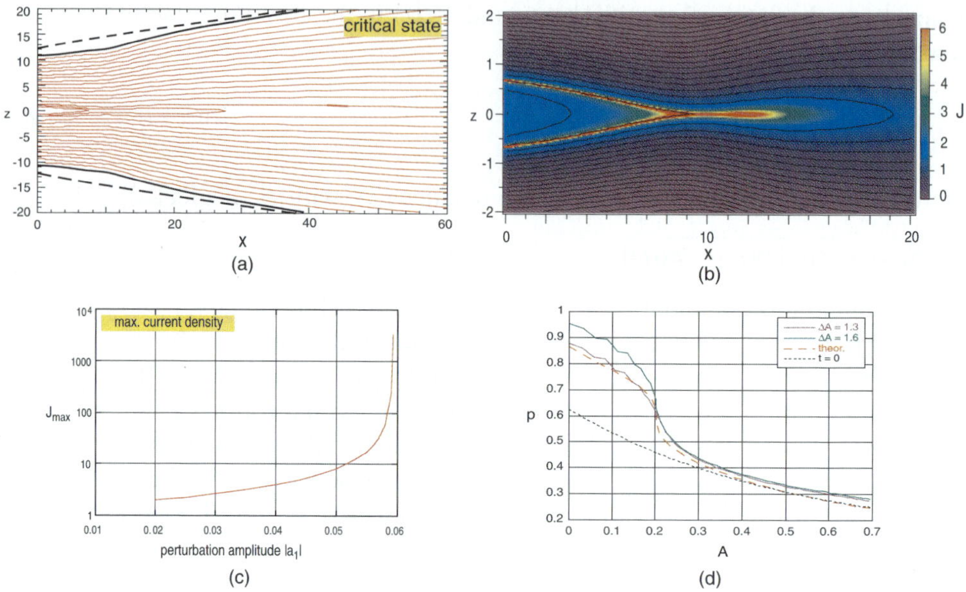

Figure 2. (a) Critical state resulting from an adiabatic boundary deformation (solid black lines). The dashed lines shows the original boundaries. (b) Enlarged interior structure, showing also the color-coded current density. (c) Maximum currrent density as function of the amplitude of the boundary deformation. (d) Pressure variation $p(A)$ for magnetotail configurations corresponding to boundary deformations as in Figure 2a–c. The black dotted lines show the initial variation and the red dashed line the variation of the critical state calculated from adiabatic theory. The green and purple solid lines show the end results of time-dependent MHD simulations for two amplitudes close to the theoretical limit. Adapted from Birn and Schindler (2002) and Birn *et al.* (2003).

minima, or saddle points) of the magnetic flux surfaces where the magnetic field would vanish. (Such null points are denoted as neutral points of x- and o-type, according to the field line structure in their vicinity, with x-points corresponding to saddle points and o-points corresponding to maxima or minima of a flux surface. Note that, in 2D, field lines correspond to lines of constant magnetic flux.). Therefore a sequence of solutions constructed in this fashion must also satisfy the topology conservation inherent to ideal MHD. An important result of these studies of adiabatic sequences of equilibria is that even modest boundary deformations may lead to a critical state at which neighboring equilibria that satisfy the imposed conservation laws cease to exist.

Figure 2 illustrates properties of such a critical state, obtained within 2D theory (Birn and Schindler, 2002). The top two panels (a) and (b) show the magnetic configuration of the critical state. Panel (b) is an enlargement of a central portion which also shows the current density in color. The heavy black line in (a) shows the deformation of the original boundary (dashed line). Panel (c) shows the maximum current density of the resulting equilibrium as a function of the amplitude of the boundary deformation for the chosen profile of indentation. These figures demonstrate that the critical state, reached for a finite deformation, is characterized by a thin current sheet, whose current density goes to infinity at the critical point. This current sheet is embedded within the wider plasma sheet and it bifurcates earthward into two current sheets.

Figure 2(d) further demonstrates that the critical state is characterized by the development of a steep local gradient of the pressure function $p(\alpha)$ (red dashed line), corresponding to the intensified current density in the embedded thin current sheet. This panel also contains results from ideal MHD simulations, modeling the dynamic response to a slow deformation of the boundary for two values of the deformation amplitude close to the predicted critical value (Birn *et al.*, 2003b). The resulting pressure functions also show the development of the local steep pressure gradient, indicating the increased current density.

While the examples of Figure 2 are obtained within 2D theory and simulations, Birn *et al.* (2003a) demonstrated that the basic results of a loss of equilibrium, coinciding with the formation of a thin embedded current sheet, also hold in three dimensions. They investigated further how properties of the underlying configuration and of the spatial variation of the boundary deformation affected the critical limit.

Figure 3 shows the effects of tail flaring on thin current sheet formation. The left panels show perspective views of four initial magnetic field configurations which differ by the amount of flaring in y and z. Configurations (b) and (c) have the same downtail variation of the lobe magnetic field strength $\hat{B}(x)$ but different flaring in y and z. Configurations (a) and (d) have nearly uniform flaring, approximately equal to the weakest or strongest flaring in (b) and (c), respectively. The right panel shows the maximum current density as function of the amplitude a_m of the boundary deformations, which had the same spatial profile for all cases. Obviously, the critical threshold is reached at larger amplitude when the flaring of the underlying configuration is larger (less stretched), whereas it makes little difference whether the initial flaring is predominantly in y or in z.

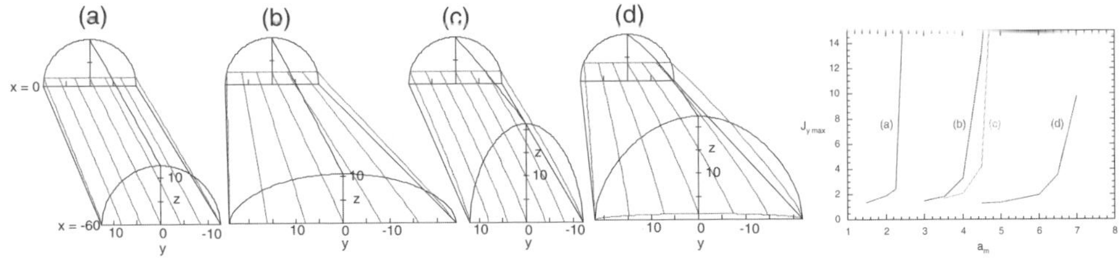

Figure 3. Effects of tail flaring on thin current sheet formation. The left panels show perspective views of four initial magnetic field configurations. Configurations (b) and (c) have the same downtail variation of the lobe magnetic field strength $\hat{B}(x)$ but different flaring in y and z. Configurations (a) and (d) have nearly uniform flaring, approximately equal to the weakest or strongest flaring in (b) and (c), respectively. The right panel shows the maximum current density as function of the amplitude a_m of the chosen boundary deformation for the four configurations. Adapted from Birn *et al.* (2003b).

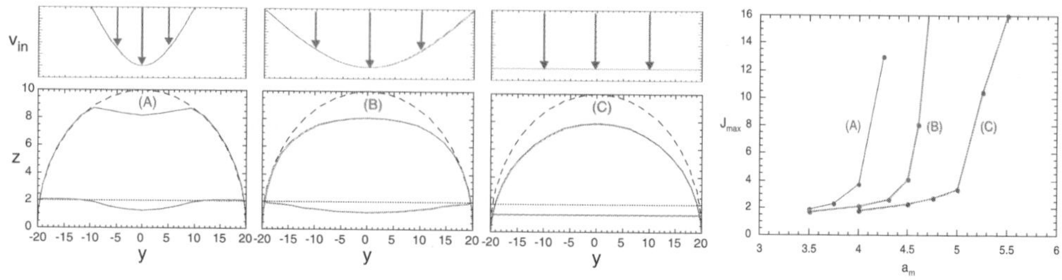

Figure 4. Effects of the cross-tail variation of the boundary deformation. The top panels on the left show three different inflow profiles leading to the deformations shown in the bottom panels. The right panel shows the maximum current densities as function of the amplitude a_m of the chosen boundary deformation for the three configurations. Adapted from Birn *et al.* (2003b).

Figure 4 shows the effects of the variation of the perturbation in the y direction, choosing three different profiles for the inflow, given in the top panels on the left. They result in deformations shown in the three panels below. The right panel shows the maximum current densities as function of the amplitude a_m of the boundary deformation for the three configurations. Obviously, the critical point is reached at lower amplitudes when the inflow profile is narrower. Also, Birn *et al.* (2003a) found, that for a given amplitude, the narrowest inflow profile (A) leads to the most significant enhancement of j_y near $y = 0$, despite the fact that the lobe field increase is weakest in this case and more spread out in y than the corresponding inflow.

3. Vlasov Models of Thin Current Sheets

The MHD approximation breaks down when the sheet thickness approaches typical particle scales, such as the ion gyroradius or ion inertial length. However, using two-dimensional Vlasov equilibrium theory (neglecting cross-tail, i.e., y variations), (Schindler and Birn, 2002) demonstrated that the thin current sheets predicted from MHD theory and simulations persist when the scale reaches ion or even electron scales. In this model the particle physics is contained in the phase space distribution function governed by the Vlasov equation (e.g., Krall and Trivelpiece, 1973). An example of such a thin embedded current sheet solution is illustrated in Figure 5, showing the current density (left), pressure contributions (center), and the electric potential (right) as functions of A, for a model with an ion gyroradius $\rho_i = 0.2$ (normalized to the half-width of the wide current sheet), and $T_i/T_e = 5$. This model is chosen to illustrate the transition from a relatively wide sheet to a thinner sheet at which the electron contribution to the current becomes more important. The left panel demonstrates that indeed the current within the thin sheet is carried by the electrons, although the ions still contribute significantly to the current in the wider sheet. However, the pressure is dominated by the ions (center). The right panel of Figure 5 shows the potentials for $\rho_i = 0.2$ and $\rho_i = 1.0$, respectively.

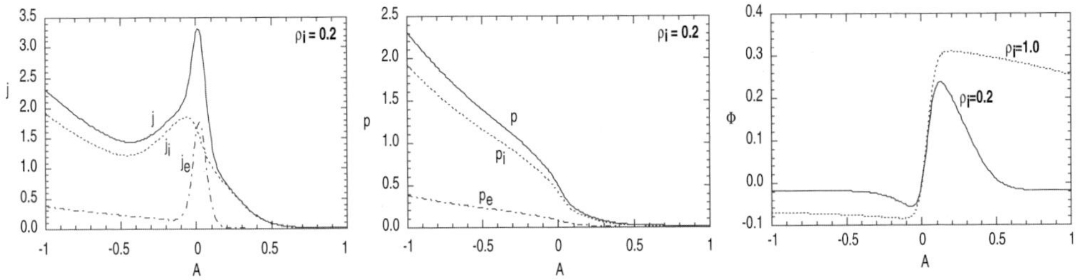

Figure 5. Current density (left), pressure contributions (center), and electric potential (right) as functions of the flux variable A for a thin current sheet model with $T_i/T_e = 5$. Adapted from Birn *et al.* (2004b).

In this case potential differences of 25–30% of the ion thermal potential are found. However, Birn *et al.* (2004a) demonstrated that for different parameters, potentials of more than 0.5 kT_i/e could be obtained. We note that the presence of such potentials follows from the requirement of charge neutrality in the model and that Φ vanishes only for special cases, such as the standard plane current sheet model by Harris (1962). Birn *et al.* (2004a) further explored the possible association of thin current sheets with the perpendicular electric fields that are part of the U-shaped potential structures above auroral arcs. For sufficiently small scales, the converging perpendicular electric field corresponds to two oppositely directed thin Hall current sheets, possibly associated with near-Earth consequences of bursty bulk flows in the magnetotail.

4. Dynamic Evolution

The studies discussed in the previous sections demonstrate that finite boundary deformations, resulting from the interaction of the solar wind with the magnetosphere, can lead to the loss of (ideal MHD) equilibrium and the formation of a thin current sheet in the magnetotail. The breakdown of MHD is then expected to lead to an unstable evolution, as observed in magnetospheric substorms. In the final part of this paper, we investigate the dynamic consequences of this current sheet formation and the initiation of reconnection in the magnetotail.

An earlier study (Birn *et al.*, 2001, and references therein) investigated the growth of magnetic islands by reconnection, using a broad variety of simulation codes. This study was initiated in the context of a "Geospace Environment Modeling (GEM)" workshop, and therefore dubbed the "GEM Reconnection Challenge." The simulations addressed the same initial state, consisting of a plane current sheet (Harris, 1962) with a total thickness of one ion inertia length. This state also included a finite magnetic-island perturbation to trigger the dynamic evolution. It was found that all simulations that included Hall electric fields led to essentially identical fast reconnection rates, regardless of the dissipation mechanism.

Using a similar multi-code approach, Birn *et al.* (2005) studied the initiation of fast reconnection by localized thinning of a wider current sheet, four times as thick as in the GEM challenge study. (This "Newton challenge" study was initiated in the context of a reconnection workshop held at the Isaac Newton Institute in Cambridge, England.) Current sheet thinning was forced by the finite deformation of the field above and below the current sheet, resulting from plasma inflow over a limited time, which caused local magnetic field enhancement of ~20–40%. Again the problem was studied by a variety of computer simulations codes, an MHD code, using spatially localized but fixed resistivity, a Hall MHD code without explicit dissipation term, three different electromagnetic particle-in-cell (PIC) codes, an implicit particle code, and a hybrid code, treating ions as particles but electrons as a fluid, with dissipation associated with electron anisotropy [for details, see Birn *et al.* (2005)]. The results demonstrated that fast magnetic reconnection can be initated by modest boundary deformations. Consistent with the GEM results, fast reconnection required inclusion of Hall effects or a resistivity model with highly localized resistivity. Again, the reconnection rate was found to be essentially independent of the dissipation mechanism when the whistler dynamics, associated with the Hall term, was included.

Figure 6 summarizes results from these simulations, showing, on the right, the reconnected flux, defined as the integral over B_z between the magnetic null points of x- and o-type, as function of time. All particle simulations results shown here were obtained using an ion/electron mass ratio of $m_i/m_e = 25$. Obviously, all particle studies, as well as the Hall-MHD study show very similar reconnection rates, although the onset times differ. All studies also show very similar final near-equilibrium stages; the result from a PIC simulation is shown in the left panel of Figure 6. This seems to indicate that entropy conservation operates similarly, despite the fact that kinetic approaches include anisotropy, a different dissipation

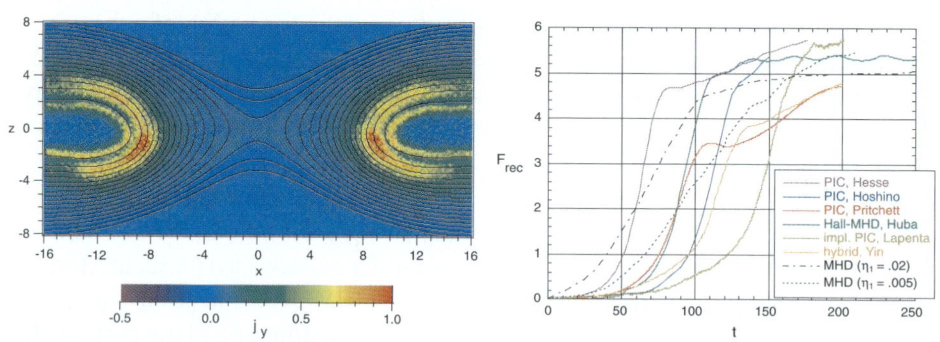

Figure 6. Results from the "Newton challenge," studying fast reconnection initiated by localized compression of a thick current sheet. Left: Late near-equilibrium stage in a PIC simulation; right: Time variation of the reconnected flux for various simulations with different codes. Adapted from Birn *et al.* (2005).

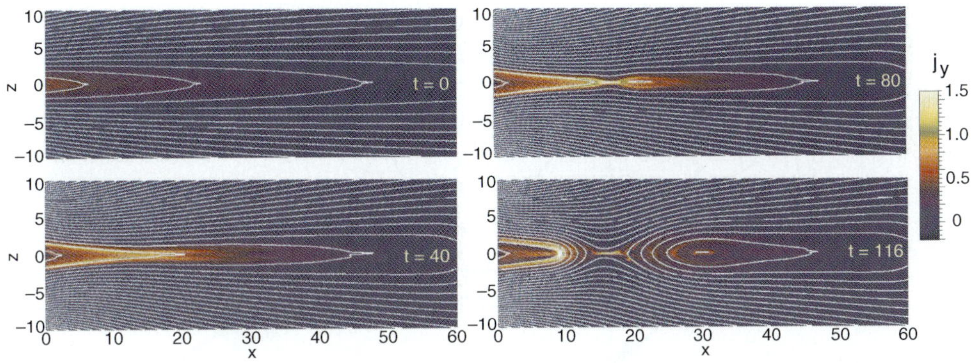

Figure 7. Current density (color) and magnetic field evolution in a PIC simulation of current sheet formation leading to magnetotail reconnection and plasmoid ejection. Adapted from Hesse and Schindler (2001).

mechanism, and different waves not included in MHD, and that Joule dissipation $\mathbf{E}' \cdot \mathbf{j}$ (where $\mathbf{E}'$ is the electric field in the plasma rest frame) is strongly localized and hence less significant than the adiabatic transport for the pressure distribution in the final state.

The validity of the concept of boundary deformation leading to current sheet formation and subsequent fast reconnection, which was studied in an idealized configuration discussed above, has been confirmed also by PIC simulations in a more realistic tail configuration by Hesse and Schindler (2001). This is illustrated by Figure 7, showing current density (color) and magnetic field lines in the x, z plane. A finite boundary deformation, similar to those discussed in Section 2 initially causes the formation of a (bifurcated) thin current sheet and subsequently leads to reconnection and the formation, and tailward ejection, of a plasmoid.

The near-Earth consequences of this process have been studied extensively by 3D resistive MHD simulations, again using boundary deformations to cause thin current sheet formation and, via imposed resistivity, the onset of reconnection and plasmoid ejection (Birn and Hesse, 1996). Figure 8 schematically shows major effects in the near tail. The current sheet thinning leads to the onset of reconnection and the formation of a magnetic neutral line. The fast earthward flow from the reconnection site gets braked and diverted away from midnight when the flow approaches the dipolar region. This diversion causes a distortion of field lines, corresponding to a twist or shear associated with the presence of field-aligned currents into the ionosphere on the dawnside and out of the ionosphere on the duskside, closing through the ionosphere by the auroral electrojet. This current pattern is called the substorm current wedge. It was inferred from observations as a major feature of magnetospheric substorms (McPherron *et al.*, 1973).

The collapse and earthward transport of field lines from the reconnection site leads to a dipolarization (Hesse and Birn, 1991), accompanied by an increase in

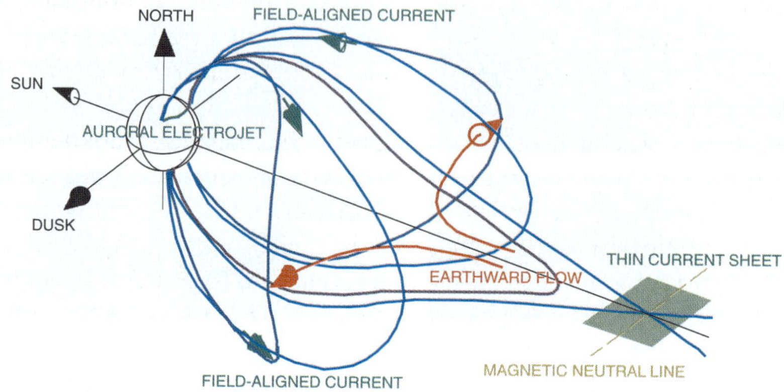

Figure 8. Schematic, summarizing substorm effects in the near tail, as found in MHD simulations: Earthward flows from the reconnection site cause a collapse (dipolarization) of the field in the near tail. The diversion of the flow away from midnight leads to a distortion of the field and the build-up of field-aligned currents, associated with the substorm current wedge. The strong cross-tail electric field in the collapsing region is also the source of particle acceleration.

the equatorial field strength. The induced electric field associated with this collapse was found to be the cause of energetic particle injections observed in the near tail, particularly by geosynchronous satellites (e.g., Birn *et al.*, 1997a). The acceleration may involve non-adiabatic orbits, particularly for ions (Birn *et al.*, 1997b), whereas electrons, which are trapped on collapsing field lines, experience betatron acceleration when they stay close to the equatorial plane and first-order Fermi acceleration at small pitch angles (Birn *et al.*, 2004b).

5. Summary and Conclusions

Using quasi-static theory and dynamic simulations, we have illustrated responses of the magnetotail to boundary deformations as resulting from its interaction with the solar wind. The results demonstrated the strong influence of the boundary deformation on the internal structure of the tail, which may lead to the formation of thin current sheets embedded in the near-tail plasma sheet and the possible loss of equilibrium and/or stability. Both MHD and particle simulations demonstrated that the localized current sheet thinning may lead to the onset of fast magnetic reconnection.

Quasi-static MHD models and time-dependent MHD simulations showed how properties of the boundary deformations affect current intensification and the critical limit for loss of equilibrium. Properties of thin current sheets were also investigated by Vlasov equilibrium theory. The self-consistent quasi-neutral structure of thin current sheets generally requires the presence of an electric potential, which is constant along field lines. These models confirm that the current in thin sheets is predominantly carried by the electrons, while the pressure is dominated by the ions. The models also show significant potentials, which are constant on field lines. These potentials might provide a link to auroral arc formation by closure through parallel potential drops.

The onset and consequences of fast magnetic reconnection were demonstrated by results from a multi-code approach. Thinning and reconnection in an initially wide plane current sheet was initiated by temporally limited, spatially varying, inflow of magnetic flux (Birn *et al.*, 2005). As in an earlier study (Birn *et al.*, 2001), full particle, hybrid, and Hall-MHD simulations lead to the same fast reconnection rates, apparently independent of the dissipation mechanism, whereas fast reconnection in MHD simulations required strongly localized resistivity. All simulations lead to surprisingly similar final states, despite differences in energy transfer and dissipation. This indicates that entropy conservation operates similarly, despite the fact that kinetic approaches include anisotropy, a different dissipation mechanism, and different waves not included in MHD, and that Joule dissipation is strongly localized and hence less significant than the adiabatic transport for the pressure distribution in the final state.

Large-scale effects of magnetic reconnection in the near tail were discussed on the basis of MHD simulations. They demonstrate the important role of earthward plasma flow and its diversion in distorting the magnetic field and building up the field-aligned currents of the substorm current wedge. The field collapse in this earthward flow region is also the main mechanism for accelerating charged particles and generating energetic particle injections.

Acknowledgements

This paper was supported by the US Department of Energy's Office of Basic Energy Sciences, by the National Science Foundation, and by NASA's Sun-Earth Connection Theory, and Living With a Star Programs. The authors also thank the Director and staff of the Isaac Newton Institute, Cambridge, England, for their hospitality and financial support concerning part of this work. JB further acknowledges gratefully the support and hospitality of the International Space Science Institute (ISSI) and of the organizers of the workshop on "Solar dynamics and its effects on the heliosphere and Earth."

References

Birn, J.: 1987, *J. Geophys. Res.* **92**, 11101.
Birn, J.: 1991, *J. Geophys. Res.* **96**, 19441.
Birn, J.: 2005, *J. Geophys. Res.* **110**, A07203, doi:10.1029/2004JA010869.
Birn, J., Dorelli, J. C., Schindler, K., and Hesse, M.: 2003a, *J. Geophys. Res.* **108**, A02215, doi:1029/2003JA010275.

Birn, J., Drake, J. F., Shay, M. A., Rogers, B. N., Denton, R. E., Hesse, M., *et al.*: 2001, *J. Geophys. Res.* **106**, 3715.

Birn, J., Galsgaard, K., Hesse, M., Hoshino, M., Huba, J. D., Lapenta, G., *et al.*: 2005, *Geophys. Res. Lett.* **32**, L06105, doi:10.1029/2004GL022058.

Birn, J., and Hesse, M.: 1996, *J. Geophys. Res.* **101**, 15345.

Birn, J., and Schindler, K.: 2002, *J. Geophys. Res.* **107**, SMP18, doi: 10.1029/2001JA0291.

Birn, J., Schindler, K., and Hesse, M.: 2003b, *J. Geophys. Res.* **108**, A01337, doi: 10.1029/2002JA009641.

Birn, J., Schindler, K., and Hesse, M.: 2004a, *J. Geophys. Res.* **109**, A02217, doi:1029/2003JA010303.

Birn, J., Thomsen, M. F., Borovsky, J. E., Reeves, G. D., McComas, D. J., and Belian, R. D.: 1997a, *J. Geophys. Res.* **102**, 2309.

Birn, J., Thomsen, M. F., Borovsky, J. E., Reeves, G. D., McComas, D. J., Belian, R. D., *et al.*: 1997b, *J. Geophys. Res.* **102**, 2325.

Birn, J., Thomsen, M. F., and Hesse, M.: 2004b, *Phys. Plasmas* **11**, 1825.

Fedder, J. A., and Lyon, J. G.: 1987, *Geophys. Res. Lett.* **14**, 14880.

Harris, E. G.: 1962, *Nuovo Cimento* **23**, 115.

Hesse, M., and Birn, J.: 1991, *J. Geophys. Res.* **96**, 19,417.

Hesse, M., and Schindler, K.: 2001, *Earth Planets Space* **53**, 645.

Krall, N. A., and Trivelpiece, A. W.: 1973, *Principles of Plasma Physics*. New York: McGraw-Hill.

McPherron, R. L., Russell, C. T., and Aubry, M. P.: 1973, *J. Geophys. Res.* **78**, 3131.

Raeder, J.: 1994, In: J. Kan, J. D. Craven, and S.-I. Akasofu (eds.), *Substorms 2, Proc. 2nd Int. Conf. on Substorms*. Fairbanks, AK, p. 561.

Schindler, K.: 1972, In: B. M. McCormac (ed.), *Earth's Magnetospheric Processes*. Dordrecht, Holland, p. 200.

Schindler, K., and Birn, J.: 2002, *J. Geophys. Res.* **107**, doi:10.1029/2001JA0304.

Siscoe, G. L.: 1972, *Planet. Space Sci.* **20**, 937.

Walker, R. J., Ogino, T., Raeder, J., and Ashour-Abdalla, M.: 1993, *J. Geophys. Res.* **98**, 17,235.

REVIEW OF IONOSPHERIC EFFECTS OF SOLAR WIND MAGNETOSPHERE COUPLING IN THE CONTEXT OF THE EXPANDING CONTRACTING POLAR CAP BOUNDARY MODEL

M. LESTER[1,*], S. E. MILAN[1], G. PROVAN[1] and J. A. WILD[2]
[1]*Department of Physics and Astronomy, University of Leicester, Leicester LE1 7RH, UK*
[2]*Department of Communication Systems, Lancaster University, Lancaster LA1 4WA, UK*
(**Author for correspondence: E-mail: mle@ion.le.ac.uk*)

(Received 19 September 2005; Accepted in final form 28 November 2006)

Abstract. This paper reviews the coupling between the solar wind, magnetosphere and ionosphere. The coupling between the solar wind and Earth's magnetosphere is controlled by the orientation of the Interplanetary Magnetic Field (IMF). When the IMF has a southward component, the coupling is strongest and the ionospheric convection pattern that is generated is a simple twin cell pattern with anti-sunward flow across the polar cap and return, sunward flow at lower latitudes. When the IMF is northward, the ionospheric convection pattern is more complex, involving flow driven by reconnection between the IMF and the tail lobe field, which is sunward in the polar cap near noon. Typically four cells are found when the IMF is northward, and the convection pattern is also more contracted under these conditions. The presence of a strong Y (dawn-dusk) component to the IMF leads to asymmetries in the flow pattern. Reconnection, however, is typically transient in nature both at the dayside magnetopause and in the geomagnetic tail. The transient events at the dayside are referred to as flux transfer events (FTEs), while the substorm process illustrates the transient nature of reconnection in the tail. The transient nature of reconnection lead to the proposal of an alternative model for flow stimulation which is termed the expanding/contracting polar cap boundary model. In this model, the addition to, or removal from, the polar cap of magnetic flux stimulates flow as the polar cap boundary seeks to return to an equilibrium position. The resulting average patterns of flow are therefore a summation of the addition of open flux to the polar cap at the dayside and the removal of flux from the polar cap in the nightside. This paper reviews progress over the last decade in our understanding of ionospheric convection that is driven by transient reconnection such as FTEs as well as by reconnection in the tail during substorms in the context of a simple model of the variation of open magnetic flux. In this model, the polar cap expands when the reconnection rate is higher at the dayside magnetopause than in the tail and contracts when the opposite is the case. By measuring the size of the polar cap, the dynamics of the open flux in the tail can be followed on a large scale.

Keywords: solar-terrestrial relations, plasmas

1. Introduction

In this paper we discuss the impact of solar wind magnetosphere coupling on the ionosphere during so called normal solar wind conditions. The ionosphere represents the lower boundary of the magnetosphere, but perhaps more importantly it represents one of the main sinks of energy which is transmitted from the solar wind to the magnetosphere. Since the ionosphere is only weakly ionised and is

Space Science Reviews (2006) 124: 117–130
DOI: 10.1007/s11214-006-9132-8

coupled strongly to the neutral atmosphere, it also plays a role in coupling that energy through to the atmosphere. There is not sufficient space to discuss the detail of the ionosphere and this can be found in text books such as Rishbeth and Garriott (1969), Schunk and Nagy (2000). The main physical process which mediates the coupling between the solar wind and magnetosphere is reconnection between the interplanetary magnetic field (IMF) and the geomagnetic field as originally proposed by Dungey (1961). Thus, in this paper we concentrate on the flow excitation in the ionosphere by reconnection at the dayside and in the tail.

2. Ionospheric Convection

The coupling between the solar wind and magnetosphere is dominated mainly by magnetic reconnection at the dayside magnetopause between the IMF and the geomagnetic field, and was first described for southward IMF by Dungey (1961) and for northward IMF by Dungey (1963). The scenario for southward IMF is illustrated in Figure 1 in which a cross section in the noon-midnight plane of the magnetosphere is presented. Reconnection occurs at the sub-solar point of the magnetopause, resulting in newly opened magnetic field lines which are connected to the IMF at one end and the geomagnetic field at the other. Newly reconnected magnetic flux tubes are transported to the nightside due to the motion of the solar wind and then reconnection of the tail field lines causes closed magnetic field lines to return earthward. This process is often referred to as the Dungey cycle.

Under northward IMF conditions the situation is more complex and involves reconnection at locations which are at much higher latitudes well away from the sub-solar point. Typically it is thought that this occurs at points in the tail lobes where the geomagnetic field would be anti-parallel to the draped northward IMF. The

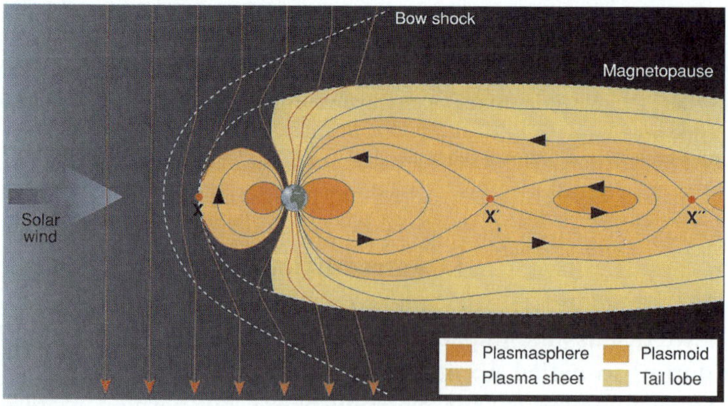

Figure 1. Schematic representation of reconnection at the dayside magnetosphere and in the tail which drives magnetospheric dynamics (after Dungey, 1961).

convection process that is driven under such conditions is clearly more complex. For example, reconnection can occur between an individual IMF field line and one lobe only, or when the IMF clock angle is low reconnection can occur simultaneously with both lobes (e.g. Imber *et al.*, 2006).

The magnetospheric convection driven by the Dungey cycle also results in ionospheric convection. When the Z component of the IMF is negative, and there is no Y component, ionospheric convection is a simple two cell convection pattern with antisunward flow across the polar cap and sunward flow at lower latitudes. The addition of a Y component adds an asymmetry to the flows (e.g. Cowley *et al.*, 1991; see Cowley, 1998 for a review) such that one cell becomes more dominant than the other. Under purely northward IMF conditions, two lobe cells are present with sunward convection in the centre of the polar cap. The addition of a Y component to the Z component causes further asymmetry in the flow, and if the Y component is larger than the Z component then a distorted two cell pattern with strong azimuthal flow in the dayside region is often observed.

Average patterns of ionospheric convection were produced primarily by averaging data sets from ground based radars or from low earth orbiting satellites (e.g. Heppner and Maynard, 1987). However, the interpretation of Flux Transfer Events (Haerendal *et al.*, 1978; Russell and Elphic, 1978, 1979) as transient and patchy reconnection and subsequent observations by other authors of this transient process has lead to an alternative consideration of how ionospheric convection is excited. This is referred to as the expanding/contracting polar cap model and follows an initial discussion by Russell (1972), theoretical work by Siscoe and Huang (1985) and Freeman and Southwood (1988), before being finally formalised by Cowley and Lockwood (1992). In this model as flux is added to the polar cap, a twin cell convection pattern is stimulated in the dayside ionosphere, as illustrated by Figure 2a, where the flow is stimulated by the motion of the boundary as the polar cap seeks to return to equilibrium after the addition of the newly created flux. In Figure 2a, the polar cap boundary, i.e. the boundary between open and closed magnetic field lines, is indicated by the solid circle, with the portion of the boundary at which reconnection occurs indicated by the dashed line. The motion of the boundary is indicated by the open arrows. If flux is removed from the polar cap by reconnection in the geomagnetic tail, then a similar twin cell convection pattern is stimulated in the nightside ionosphere (Figure 2b). A standard two cell convection pattern consists of the summation of the two processes. Cowley and Lockwood (1992) proposed that the flow would be stimulated in order to move the newly reconnected flux at the dayside magnetopause into the polar cap, thereby moving the polar cap boundary equatorward. Likewise, a region of newly closed flux on the nightside would result in the same pattern but the polar cap would contract. The timescale for this process in the absence of other processes would be of order 15–20 minutes (Cowley and Lockwood, 1992). Thus the difference between reconnection rates at the dayside and in the nightside would determine whether the polar cap was expanding or contracting.

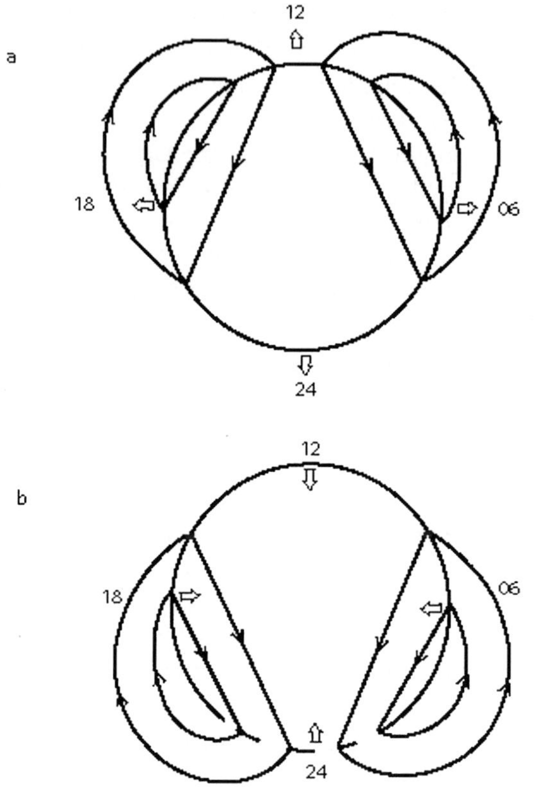

Figure 2. Sketch of the basic form of the time dependent flow excited by reconnection at the dayside magnetopause and in the geomagnetic tail (based on Cowley and Lockwood, 1992).

The study of ionospheric flows in the last decade has been improved with the expansion of the Super Dual Auroral Radar Network (SuperDARN) which consists of networks of radars in both northern and southern hemispheres (Greenwald *et al.*, 1995). Doppler velocity measurements made by these radars can be combined to provide maps of the large scale convection pattern at intervals of 2 minutes in general, and, at times, 1 minute (Ruohoniemi and Baker, 1998). In the remainder of this paper we discuss some results that have been attained with SuperDARN over the last 10 years which address flow excitation by the model proposed by Siscoe and Huang (1985) and Cowley and Lockwood (1992).

3. FTE Flow Signatures

There have been many reports of flux transfer events, with the first model described by Russell and Elphic (1978, 1979) based on typical signatures of the magnetic field

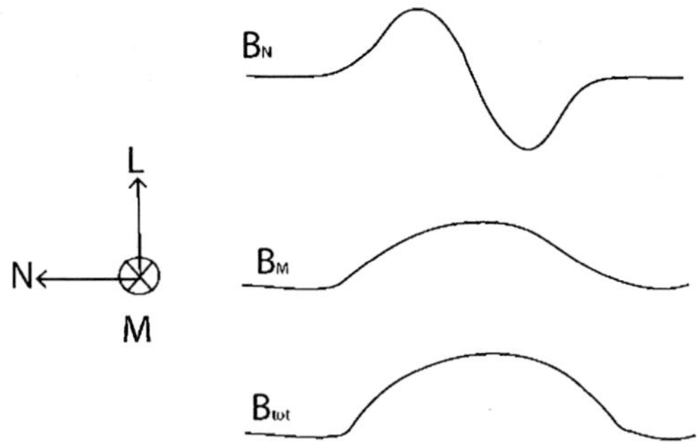

Figure 3. Sketch of the typical signatures of a FTE seen by a space borne magnetometer in boundary normal co-ordinates where the N component is normal to the local magnetopause boundary while the L and M components are in the plane of the local magnetopause boundary.

in a boundary normal co-ordinate system, where the N component is normal to the local magnetopause boundary while the L and M components are in the plane of the magnetopause. Figure 3 presents sketches of the typical signatures of an FTE which are the bipolar signature in the B_N component and an increase in the total field, B_{tot}. The variation in the other two components depends on the relative motion of the FTE and the spacecraft, although in Figure 3, the M component is shown as increasing. The first unambiguous observation of the ionospheric flow stimulated by FTEs was that by Elphic *et al.* (1990). Thereafter, a number of observations were made by HF radars of signatures which were assumed to be the response to transient reconnection (e.g. Pinnock *et al.*, 1993; Provan *et al.*, 1998; Milan *et al.*, 2000; McWilliams *et al.*, 2000). The first simultaneous observations of a FTE at the magnetopause and flow enhancement in the ionosphere measured by a SuperDARN radar were made by Neudegg *et al.* (1999). In this study a clear magnetospheric FTE was observed by Equator-S and this was accompanied by a near simultaneous enhancement of the ionospheric flow by the CUTLASS Hankasalmi radar. Furthermore, it was subsequently shown that this FTE excited strong UV aurora equatorward of the footprint of the newly reconnected field lines (Neudegg *et al.*, 2001). This event also demonstrated that the signatures in HF radar data termed pulsed ionospheric flows (Provan *et al.*, 1998; McWilliams *et al.*, 2000), which had been discussed as likely signatures of FTEs, were indeed ionospheric signatures of transient reconnection.

A subsequent statistical study (Neudegg *et al.*, 2000) demonstrated that if the repetition rate of FTEs was greater than 5 minutes, then the tendency was to identify a clear one-to-one response between FTE and ionospheric flow response. If on the other hand the repetition rate was faster, then the flows tended less to the pulsed behaviour and more to flow behaviour associated with continuous excitation.

More recent observations of FTEs and ground signatures have been made utilising the Cluster spacecraft (Wild *et al.*, 2001, 2003). In the first of these a series of FTEs were observed as the Cluster spacecraft moved through the outer post noon

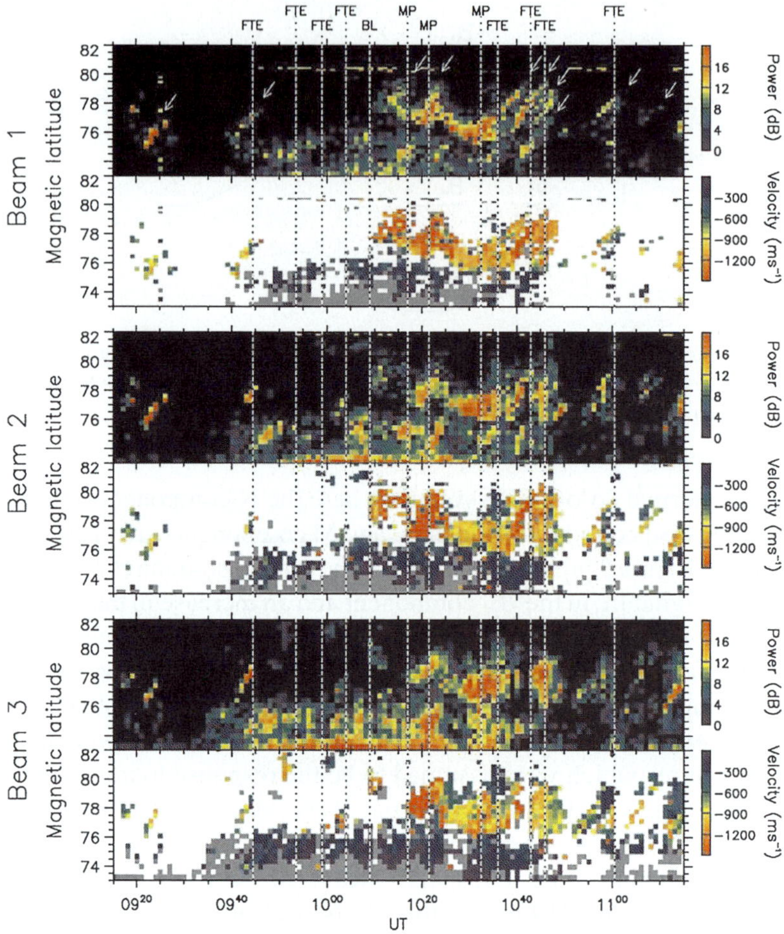

Figure 4. Radar backscatter power and Doppler velocity as a function of magnetic latitude and UT from 3 beams of the CUTLASS Finland radar (after Wild *et al.*, 2001). FTEs and excursions into a boundary layer (BL) and across the magnetopause (MP) are identified by vertical dashed lines.

magnetosphere to the magnetosheath in a 2 hour interval. The FTEs were seen as enhancements in the total magnetic field, a bipolar variation in the component of the field normal to the magnetopause and also as small bursts of mixed magnetosphere and magnetosheath plasma. Figure 4 presents the radar observations made during the interval in which Cluster observed the FTEs. In this figure the radar backscatter power and Doppler velocity are presented for each of beams 1, 2 and 3. The dashed vertical lines represent the times at which FTEs were observed by Cluster, in addition to an observation of a boundary layer (BL) and three magnetopause crossings (MP). Although the estimated spacecraft footprint was some 2h in local time to the east of the data presented in Figure 4, there were clear pulsations in both Doppler velocity and backscatter power which moved polewards. These are classic pulsed

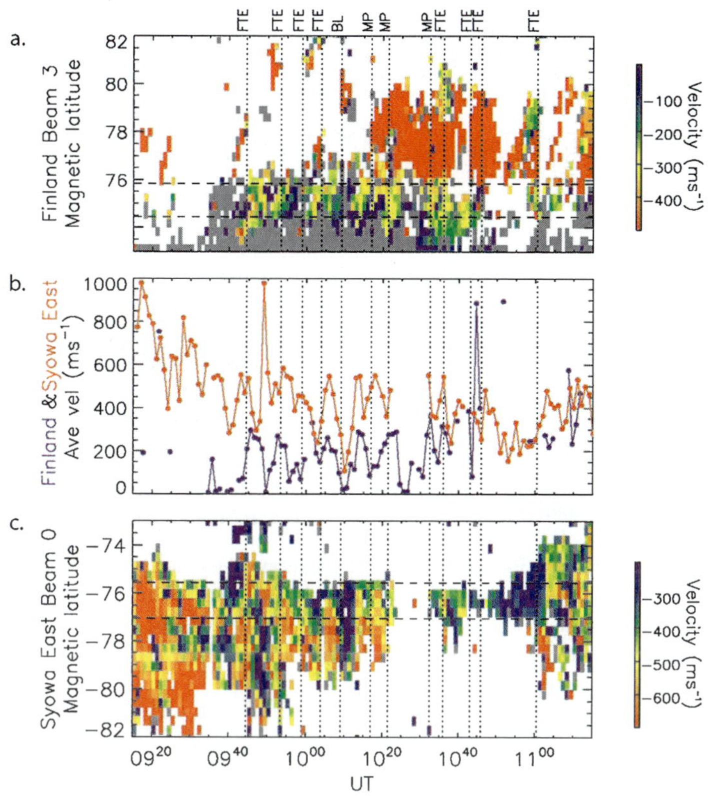

Figure 5. Doppler velocity from Beam 2 of CUTLASS Finland (panel a) and Syowa (panel c). Panel b shows the averaged velocity between the horizontal dashed lines in panels a and c (after Wild *et al.*, 2003).

ionospheric flows (PIFs) and poleward moving radar auroral forms (PMRAFs), the latter being identified by arrows in the panels presenting backscatter power. The time variation in the velocity is less clear and so Figure 5 presents the Doppler velocity in beam 3 (top panel) together with, as a time series, the average Doppler velocity between 75 and 76° magnetic latitude in the middle panel (blue line). It is clear that with each of the first 4 FTEs there is an enhancement in the Doppler velocity. It should be noted that where there are such enhancements at these lower latitudes, they precede a PMRAF at higher latitudes. Wild *et al.* (2001) concluded that the PIF/PMRAF originated at the lower latitudes (close to the footprint of Cluster) and then propagated poleward to higher latitudes where they resemble the more classic signatures. This further suggests that the PMRAFs are fossils of ionospheric structuring which takes place at lower latitudes at the footprint of the reconnection site, or merging gap, as also suggested by Davies *et al.* (2002).

The flows at this lower latitude region were consistent with westward flow. Since the satellite observations were in the post noon sector and the IMF was

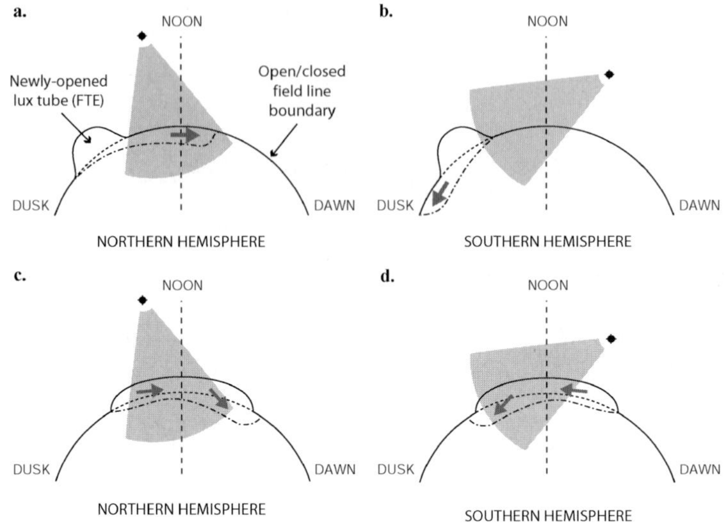

Figure 6. Schematic representations of the possible locations of the newly reconnected magnetic flux at the polar cap boundary and the flows that would be generated for the events shown in Figures 3 and 4. In panels a and b the newly reconnected flux is limited to the post noon sector, while in panels c and d the newly reconnected flux extends across noon.

directed dawnward and southward, Wild *et al.* (2001) proposed that this region of westward flow corresponded to the newly opened flux tubes. The extent of this westward flow region is difficult to judge from the northern hemisphere alone. Figure 5 (lower panel) illustrates that there were also pulsed flows in the southern hemisphere in a range nearly magnetically conjugate to the northern hemisphere flows presented in this figure. Furthermore, by averaging the Doppler velocity over the same latitude region as the northern hemisphere observations (Figure 5, middle panel, red curve) we see that in association with the first 4 FTEs the flows in the southern hemisphere are also enhanced in near synchronisation with the northern hemisphere observations. Furthermore, a map potential analysis (Ruohoniemi and Baker, 1998) also demonstrates that the flows in the southern hemisphere occur in the dawn LT sector and are clearly eastward directed (see Wild *et al.*, 2003).

This latter observation is important to help assess the location and potential extent of the reconnection site for these events. The two general scenarios are presented in Figure 6 (based on Wild *et al.*, 2003). The two top panels represent the scenario where the reconnection region is localised in the post noon sector and to the east of the radar field of view. The lower two panels represent a case where the reconnection site extends across the noon local time sector. The flow response is summarised by the arrow and while the northern hemisphere radar data alone would not be able to determine which of the two scenarios is the most likely, the addition of the southern hemisphere data demonstrates that the latter case of an extended region across noon is the most likely case.

4. Flow Excitation During Magnetospheric Substorms

The expanding/contracting polar cap model of flow excitation indicates that reconnection in the geomagnetic tail should stimulate flow in much the same way that reconnection at the dayside magnetopause does. There seem to be two distinct categories of reconnection in the tail, azimuthally localised features, termed bursty bulk flows (BBFs) which have a variety of signatures in the ionosphere (e.g. Grocott *et al.*, 2004; Boralv *et al.*, 2005) and the larger scale reconnection which occurs at some stage during the expansion phase of a magnetospheric substorm. A general discussion of substorms is dealt with elsewhere in more detail (Nakamura, 2006), but here we consider the flow excitation associated with substorms. In one such study, Grocott *et al.* (2002) used the SuperDARN data to demonstrate the excitation of a twin vortex flow on the nightside associated with a modest substorm which occurred during a period of weak northward IMF but with a strong B_y component. The flows were found to occur in the region of the substorm auroral bulge as observed by Polar VIS. Furthermore, the transpolar voltage increased from 40 kV in the pre expansion phase onset interval to 80 kV after onset, before decaying to about 35 kV some 10 minutes into the recovery phase.

This result was confirmed by a statistical study of flows measured by SuperDARN using the IMAGE FUV auroral imager (Mende *et al.*, 2000) to identify the location of the auroral break-up for 67 substorms (Provan *et al.*, 2004). In this study, the ionospheric flows were ordered by the magnetic latitude and magnetic local time of the auroral break-up to remove any spatial averaging caused by differences in the break-up region. The flows were averaged at 2 minute intervals from 30 minutes before onset to 30 minutes after onset. There is a clear development of a two cell convection pattern during the growth phase. Following expansion phase onset the flows in the nightside region near the auroral break-up region become very weak, and the flow appears to be diverted around this region. The estimated cross polar cap potential appears to peak some 10 minutes after the onset although the variability in this parameter is quite large. The enhancement in the cross polar cap potential from just before onset is of order 30 kV, in agreement with Grocott *et al.* (2002). This work demonstrated for the first time on a statistical basis that the dayside and nightside reconnection both drive flows independently of each other.

5. Large-Scale Response of the Polar Ionosphere

The last decade has seen an improvement in the ability to image the global auroral and polar regions from space with good (seconds to minutes) time resolution. The

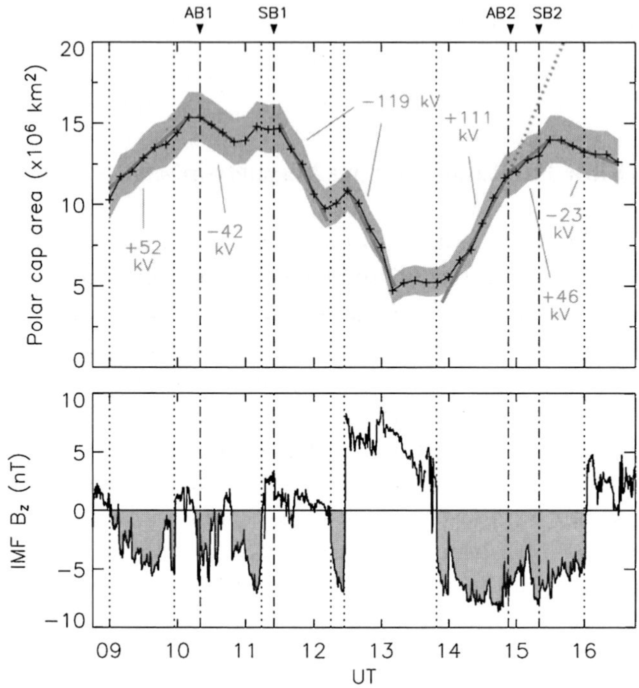

Figure 7. The variation of the polar cap area for the time interval 09–16 UT and the IMF B_z component during the same time interval (after Milan *et al.*, 2003).

VIS (Frank *et al.*, 1995), and UVI (Torr *et al.*, 1995) imagers on board Polar and the FUV imager on IMAGE (Mende *et al.*, 2000) have all contributed to the observations of the dynamics of the auroral oval and polar cap for intervals of 10–12 hours at a time.

By comparing observations of the UV aurora with other observations such as the spectral width parameter measured by the SuperDARN radars and low earth orbiting particle measurements, Milan *et al.* (2003) have demonstrated that the size of the polar cap, i.e. the region of open flux, can be estimated from global auroral images. Since the variation in the amount of magnetic flux can be used to calculate the transpolar voltage, this is an alternative global method of investigating the expanding contracting polar cap model. Figure 7 illustrates how the polar cap area varies during a period when there were a number of changes in the polar cap area as well as 2 magnetospheric substorms. The polar cap area is plotted in the top panel while the lower panel presents the IMF B_z component for this interval. The grey shading of the polar cap area line represents the uncertainty assuming that the estimate of the polar cap boundary is incorrect by ± 1 degree of latitude at all magnetic local times.

There are several points to make about this figure. Firstly, it is clear that during the intervals of southward IMF, the polar cap area does increase. This is most easily seen following 0900 UT and after 1400 UT. In the first case the voltage associated with the expansion of the polar cap area is +52 kV and in the second it is +111 kV, with the higher value being associated with the more negative B_z. The two substorms (SB1 and SB2) result in decreases in area of the polar cap. This is particularly clear in the first case when the polar cap area decreases by a factor of 3 following the expansion phase onset, resulting in a voltage of −119 kV. The second substorm occurs during a period of negative B_z and so the decrease is not as marked as reconnection at the dayside still continues. Furthermore, psuedobreakups or small auroral break ups also result in the decline of the polar cap area in the first case (AB1) and a slowing of the increase in the second case (AB2).

In addition, during the second interval of southward IMF, Milan *et al.* (2003) were able to calculate the reconnection voltage in the frame of the polar cap boundary by using the flows measured by the SuperDARN radars across the boundary. This technique is described in detail by Baker *et al.* (1997) and has subsequently been used in a number of studies (e.g. Chisham *et al.*, 2004). Using this technique Milan et al. (2003) calculated the integrated reconnection electric field between 05 and 19 MLT to be 104 kV during the second substorm growth phase, in good agreement with the value measured from the change in the polar cap area. There was also evidence during this interval for pulsed reconnection at the dayside magnetopause. The good agreement between the two different estimates of the reconnection voltage during this interval is strong evidence in support of the convection excitation discussed by Siscoe and Huang (1985) and Cowley and Lockwood (1998). Finally, when the IMF was northward and reconnection in the tail had stopped, the polar cap area remained constant. This indicates that, during this interval at least, there was no closure of open flux by lobe reconnection in both hemispheres simultaneously. Observation of several intervals allowed Milan *et al.* (2006) to estimate the rate and duration of flux closure during magnetospheric substorms. These authors found that during a typical substorm ~0.25 GWb of open magnetic flux in the tail was closed.

6. Outstanding Questions

Despite the success of the expanding-contracting polar cap model in predicting many of the signatures of ionospheric flow associated with reconnection at the dayside and in the tail, there remain a number of outstanding questions. In terms of the coupling between the solar wind, magnetosphere and ionosphere, we still have a poor understanding of what happens during northward IMF conditions. Suggestions have been made about the nature of flow excitation but there remain unresolved questions concerning the location of reconnection, whether it occurs in one hemisphere only or both hemispheres simultaneously (e.g. Imber *et al.*, 2006), and the consequences for the ionosphere. There also remain a number of questions about the flow generated during substorms. For example, what are the flows and how much flux is typically reconnected during a pseudobreakup, how quickly after expansion phase onset does flow become excited, how long into the recovery phase does flow continue to be excited? Furthermore, there remain questions about the flow excitation in both hemispheres. Also the high latitude SuperDARN radars have been less successful at measuring the ionospheric flow during very disturbed conditions (e.g. magnetic storms), due mainly to ionospheric effects on the propagation of the HF signal. Lower latitude radars are now being planned and deployed to investigate the ionospheric flow during storms.

This paper has concentrated on flow excitation and the dynamics of the polar cap and to a lesser extent the auroral oval. Of course there are other questions that are important in understanding the response of the ionosphere to the solar wind-magnetosphere coupling that have not been addressed due to space constraints. These include the different spatial and temporal scales of auroral activity, the coupling between the ionosphere and thermosphere and how energy can be transported into the lower atmosphere.

7. Summary

The paper presents an overview of the coupling between the solar wind and magnetosphere and the consequences in the ionosphere. It concentrates on reconnection as the main mechanism for coupling and the flow that is then excited in the ionosphere as a marker of magnetospheric convection. The flow excitation model of Cowley and Lockwood is shown to be supported by a wide range of observations, including dayside reconnection, ionospheric flow stimulation during magnetospheric substorms, and the variation in the polar cap area.

Acknowledgements

The authors are grateful to the PIs of all the instruments and missions from which data are used in this study. ML acknowledges funding from the International Space Science Institute. ML, SEM and GP acknowledge funding from PPARC via grant PPA/G/O/2003/00013.

References

Baker, K. B., Rodger, A. S., and Lu, G.: 1997, *J. Geophys. Res.* **102**, 9603.

Borälv, E., Opgenoorth, H. J., Kauristie, K., Lester, M., Bosqued, J.-M., Dewhurst, J. P., *et al.*: 2005, *Ann. Geophysicae* **23**, 997.

Chisham, G., Freeman, M. P., Coleman, I. J., Pinnock, M., Hairston, M. R., Lester, M., *et al.*: 2004, *Ann. Geophysicae* **22**, 4243.

Cowley, S. W. H.: 1998, in J. Moen, *et al.* (eds.), *Polar Cap Boundary Phenomena*, pp. 127–140. Kluwer Academic Publishers, Netherlands.

Cowley, S. W. H., and Lockwood, M.: 1992, *Ann. Geophysicae* **10**, 103.

Cowley, S. W. H., Morelli, J. P., and Lockwood, M.: 1991, *J. Geophys. Res.* **96**, 5557.

Davies, J. A., Yeoman, T. K., Rae, I. J., Milan, S. E., Lester, M., Lockwood, M., *et al.*: 2002, *Ann. Geophysicae* **20**, 781.

Dungey, J. W.: 1961, *Phys. Res. Lett.* **6**, 47.

Dungey, J. W.: 1963, in C. DeWitt, J. Hiebolt, and A. Lebeau (eds.), *Geophysics: The Earths Environment*, pp. 526–535. Gordon and Breach, Newark, NJ.

Elphic, R. C., Lockwood, M., Cowley, S. W. H., and Sandholt, P. E.: 1990, *Geophys. Res. Lett.* **17**, 2241.

Frank, L. A., Sigwarth, J. B., Craven, J. D., Cravens, J. P., Dolan, J. S., Dvorsky, M. R., *et al.*: 1995, *Space Sci. Rev.* **71**, 297.

Freeman, M. P., and Southwood, D. J.: 1988, *Planet. Space Sci.* **36**, 509.

Greenwald, R. A., Baker, K. B., Dudeney, J. R., Pinnock, M., Jones, T. B., Thomas, E. C., *et al.*: 1995, *Space Sci. Rev.* **71**, 761.

Grocott, A., Cowley, S. W. H., Sigwarth, J. B., Watermann, J. F., and Yeoman, T. K.: 2002, *Ann. Geophysicae* **20**, 1577.

Grocott, A., Yeoman, T. K., Nakamura, R., Cowley, S. W. H., Frey, H. U., Reme, H., *et al.*: 2004, *Ann. Geophysicae* **22**, 1061.

Haerendel, G., Paschmann, G., Sckopke, N., Rossenbauer, H., and Hedgecock, P. C.: 1978, *J. Geophys. Res.* **83**, 3195.

Heppner, J. P., and Maynard, N. C.: 1987, *J. Geophys. Res.* **92**, 4467.

Imber, S. M., Milan, S. E., and Hubert, B.: 2006, *Ann. Geophys.* **24**, 3115.

Lockwood, M.: 1998, in J. Moen, A. Egeland, and M. Lockwood (eds.), *Identifying the Open-Closed Field Line Boundary, Polar Cap Boundary Phenomena*, pp. 73–90. Kluwer Academic Publishing, Netherlands.

McWilliams, K. A., Yeoman, T. K., and Provan, G.: 2000, *Ann. Geophysicae* **18**, 445.

Mende, S. B., Heetderks, H., Frey, H. U., *et al.*: 2000, *Space Sci. Rev.* **91**, 243.

Mende, S. B., Carlson, C. W., Frey, H. U., Peticolas, L. M., and Østgaard, N.: 2003, *J. Geophys. Res.* **108**(A9), 1344, doi: 10.1029/2002JA009787.

Milan, S. E., Lester, M., Cowley, S. W. H., and Brittnacher, M.: 2000, *J. Geophys. Res.* **105**, 15741.

Milan, S. E., Lester, M., Cowley, S. W. H., Oksavik, K., Brittnacher, M., Greenwald, R. A., *et al.*: 2003, *Ann. Geophysicae* **21**, 1121.

Milan, S. E., Wild, J. A., Grocott, A., and Draper, N. C.: 2006, *Adv. Space Res.* **38**, 1671.

Nakamura, R.: 2006, Substorms and their solar wind causes. *Space Sci. Rev.*, this volume, doi: 10.1007/s11214-006-9131-9.

Neudegg, D. A., Yeoman, T. K., Cowley, S. W. H., Provan, G., Haerendel, G., Baumjohann, W., *et al.*: 1999, *Ann. Geophysicae* **17**, 707.

Neudegg, D. A., Cowley, S. W. H., Milan, S. E., Yeoman, T. K., Lester, M., Provan, G., *et al.*: 2000, *Ann. Geophysicae* **18**, 416.

Neudegg, D. A., Cowley, S. W. H., McWilliams, K. A., Lester, M., Yeoman, T. K., Sigwarth, J., *et al.*: 2001, *Ann. Geophysicae* **19**, 179.

Pinnock, M., Rodger, A. S., Dudeney, J. R., Baker, K. B., Newell, P. T., Greenwald, R. A., *et al.*: 1993, *J. Geophys. Res.* **98**, 3767.

Provan, G., Yeoman, T. K., and Milan, S. E.: 1998, *Ann. Geophysicae* **16**, 1411.

Provan, G., Lester, M., Mende, S. B., and Milan, S. E.: 2004, *Ann. Geophysicae* **22**, 3607.

Rishbeth, H., and Garriott, O.: 1969, *Introduction to Ionospheric Physics*. Academic Press, London.

Ruohoniemi, J. M., and Baker, K. B.: 1998, *J. Geophys. Res.* **103**, 20797.

Russell, C. T.: 1972, in E. R. Dyer (ed.), *Critical Problems of Magnetospheric Physics*, pp. 1–16. Inter Union Committee on STP, National Academy of Sciences, Washington, DC.

Russell, C. T., and Elphic, R. C.: 1978, *Space Sci. Rev.* **22**, 681.

Russell, C. T., and Elphic, R. C.: 1979, *Geophys. Res. Lett.* **6**, 33.

Schunk, R. W., and Nagy, A. F.: 2000, *Ionosphere – Physics, Plasma Physics and Chemistry*. CUP, Cambridge.

Siscoe, G., and Huang, T. S.: 1985, *J. Geophys. Res.* **90**, 543.

Torr, M. R., Torr, D. G., Zukic, M., Johnson, R. B., Ajello, J., Banks, P., *et al.*: 1995, *Space Sci. Rev.* **71**, 329.

Wild, J. A., Cowley, S. W. H., Davies, J. A., Khan, H., Lester, M., Milan, S. E., *et al.*: 2001, *Ann. Geophysicae* **19**, 1491.

Wild, J. A., Milan, S. E., Cowley, S. W. H., Dunlop, M. W., Owen, C. J., Bosqued, J. M., *et al.*: 2003, *Ann. Geophysicae* **21**, 1807.

THE EVOLVING SIGMOID: EVIDENCE FOR MAGNETIC FLUX ROPES IN THE CORONA BEFORE, DURING, AND AFTER CMES

S. E. GIBSON[1,*], Y. FAN[1], T. TÖRÖK[2] and B. KLIEM[3]
[1]*NCAR/HAO, P.O. Box 3000, Boulder, CO 80301, U.S.A.*
[2]*Mullard Space Science Laboratory, University College London, Holmbury St. Mary, Dorking, Surrey RH5 6NT, U.K.*
[3]*Astrophysical Institute Potsdam, An der Sternwarte 16, 14482 Potsdam, Germany*
*(*Author for correspondence: E-mail: sgibson@ucar.edu)*

(Received 7 July 2005; Accepted in final form 6 February 2006)

Abstract. It is generally accepted that the energy that drives coronal mass ejections (CMEs) is magnetic in origin. Sheared and twisted coronal fields can store free magnetic energy which ultimately is released in the CME. We explore the possibility of the specific magnetic configuration of a magnetic flux rope of field lines that twist about an axial field line. The flux rope model predicts coronal observables, including heating along forward or inverse S-shaped, or sigmoid, topological surfaces. Therefore, studying the observed evolution of such sigmoids prior to, during, and after the CME gives us crucial insight into the physics of coronal storage and release of magnetic energy. In particular, we consider (1) soft-X-ray sigmoids, both transient and persistent; (2) The formation of a current sheet and cusp-shaped post-flare loops below the CME; (3) Reappearance of sigmoids after CMEs; (4) Partially erupting filaments; (5) Magnetic cloud observations of filament material.

Keywords: coronal mass ejections, flares, sigmoids

1. Introduction: The Magnetic Flux Rope Paradigm

We define a magnetic flux rope as a set of magnetic field lines that wind more than once about some common axial field line. Figure 1 shows a picture of the so-called "Granddaddy" prominence which exhibits apparently twisted structure, next to a cartoon showing two views of a flux rope magnetic field line twisting about a straight axis (note that the axial field line does not in general have to be straight). Magnetic flux rope models have been employed to explain a wide range of solar and heliospheric physics phenomena, from the solar interior out into interplanetary space. In particular, a range of CME and CME-associated phenomena have been modeled with magnetic flux ropes. This is not surprising, because the energy source for CMEs is widely agreed to lie in their twisted or sheared magnetic fields. Although coronal magnetic fields are not yet commonly observed, observations of photospheric vector magnetic fields have long shown that non-potential magnetic fields are common (Hagyard, 1984; Tanaka, 1991; Leka *et al.*, 1996; Lites *et al.*, 1995), and observations such as the twisted prominence of Figure 1 indicate that the coronal field tied to the plasma is likewise significantly non-potential. Twisted magnetic flux ropes are good candidates for metastable coronal MHD equilibria

Space Science Reviews (2006) 124: 131–144
DOI: 10.1007/s11214-006-9101-2

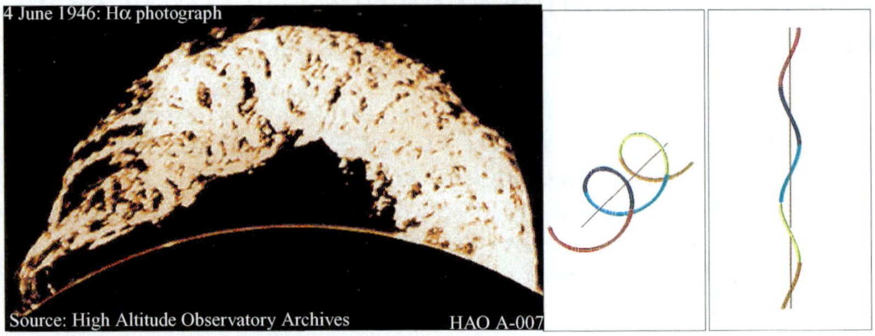

Figure 1. (Left) The "Granddaddy" prominence (HAO H-alpha), (right) two views of flux rope cartoon, demonstrating that a single twisted field line (in this case, left-handed) includes both forward and inverse S shapes.

capable of storing free magnetic energy which may be tapped to drive coronal dynamic phenomena such as CMEs (Low, 1996; Rust, 2003).

CMEs have been related to interplanetary counterparts, such as magnetic clouds, which are well-modeled as magnetic flux ropes (Burlaga *et al.*, 1982). The idea of the CME as a magnetic flux rope has gained acceptance over the years, and a brief survey of recent CME model publications shows that they are united in describing the erupting CME as a flux rope (Amari *et al.*, 2003a,b; Chen and Krall, 2003; Roussev *et al.*, 2003; MacNeice *et al.*, 2004; Manchester *et al.*, 2004a). However, the question of whether the flux rope is formed during the eruption, or whether the flux rope existed prior to the eruption, remains controversial. In this paper we will argue for the existence of a stable, pre-CME magnetic flux rope, by showing that models of this type can explain the observed evolution of soft-X-ray sigmoids in relation to CMEs and filament eruptions.

2. Sigmoids

2.1. Sigmoid Observations

Sigmoids are forward or inverse S-shaped structures observed in the solar corona. Active regions possessing sigmoids have been shown to be significantly more likely to produce flares or CMEs than non-sigmoid active regions (Canfield *et al.*, 2000). Sigmoids can be classified as transient or persistent. Persistent sigmoids can be a collection of sheared loops that together indicate an S or inverse-S shape for days or weeks, while transient sigmoids tend to be more sharply focussed into apparently a single, sigmoid loop, and can appear and disappear multiple times during their disk passage (Pevtsov, 2002b). Transient sigmoids are often associated with a CME, and in such cases may transition into post-flare cusped loops. Figure 2 gives examples of a variety of sigmoids. Sigmoids are generally observed in soft-X-ray emission, but can be visibile in UV or EUV, particularly transient sigmoids

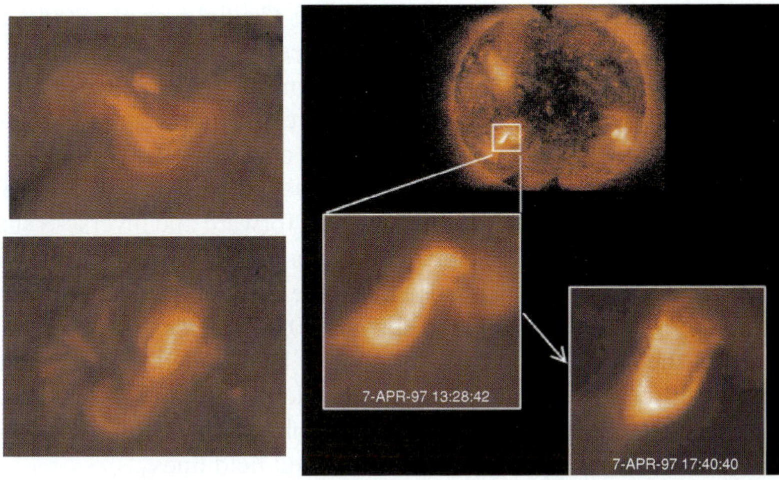

Figure 2. Examples of soft-X-ray sigmoids, as observed by Yohkoh SXT. (left top) Persistent sigmoid; (left bottom) transient sigmoid; (right) sigmoid associated with CME, transitioning to cusp.

(Sterling and Hudson, 1997; Gibson *et al.*, 2002). Forward or inverse S-shaped filaments also can be visible in H-alpha along a neutral line in sigmoid active regions (Rust and Kumar, 1994; Pevtsov *et al.*, 1996; Lites and Low, 1997; Gibson *et al.*, 2002). Even after CMEs and sigmoid-to-cusp transitions, active regions can exhibit sigmoid structures again within a matter of hours. Likewise, a sigmoid filament can also reform after an eruption, or even sometimes appear unaffected altogether by the flare/CME/sigmoid-to-cusp transition occuring apparently just above it (Tang, 1986; Pevtsov, 2002a; Gibson *et al.*, 2002). We will discuss the implications of this observed phenomenon further below.

Both S and inverse-S morphology are observed, but more S-shaped sigmoids are observed in the southern hemisphere, and more inverse-S shapes are observed in the northern hemisphere (Rust and Kumar, 1996; Pevtsov *et al.*, 2001). This is consistent with other observed patterns, such as a predominantly negative (left-handed) current helicity observed in the northern hemisphere, and positive (right-handed) in the southern hemisphere (Seehafer, 1990; Pevtsov *et al.*, 1995), as well as patterns of chirality in H-alpha filaments (Martin *et al.*, 1992; 1994). Thus, sigmoids are of interest because of their association with dynamic phenomena such as flares and CMEs, but also because they provide clues to the global organization of magnetic helicity, which is believed to be very nearly conserved as a global quantity in the highly conducting corona (Berger and Field, 1984).

2.2. Sigmoid as Flux Rope

With the advent of the Yohkoh soft-X-ray Telescope, the first comprehensive sigmoid studies were obtained (Manoharan *et al.*, 1996; Hudson *et al.*, 1998; Sterling and Hudson, 1997; Pevtsov and Canfield, 1999; Canfield *et al.*, 2000) and immediately it was suggested that they were manifestations of a magnetic flux rope topology (Rust and Kumar, 1996). Sheared magnetic field lines are intrinsically S-shaped, and the hemisphere rules of sigmoid direction are plausibly connected to the direction of magnetic helicity implied by other observations. Thus, a left (right)-handed magnetic flux rope should yield an inverse (forward) S-shaped soft-X-ray sigmoid. However, as Figure 1 demonstrates, both forward and inverse-S shapes are contained in a twisting field line, and simulations of flux ropes emerging into the corona generally contain both forward and inverse S-shaped field lines (Fan, 2001; Magara and Longcope, 2001; Fan and Gibson, 2003; Abbett *et al.*, 2003; Archontis *et al.*, 2004). In the case of Figure 1, the rope is left-handed, and the bottoms of the winding field lines are inverse-S shaped, while the tops are forward-S shaped. Thus, to be consistent with the observed hemispheric rules, sigmoids should be showing the bottoms, or dipped portions of sheared field lines.

It has been proposed that soft-X-ray sigmoids are the manifestations of flux ropes undergoing the kink instability. For example, the sigmoid could indicate the kinked axis of the flux rope (Rust and Kumar, 1996). Usually, however, if the axial field line is arched upwards when a left-handed flux rope kinks, the rope axis behaves as the tops of flux rope field lines and forms a forward-S shape (Fan and Gibson, 2003; 2004; Kliem *et al.*, 2004). Cases have been found where a left-handed rope's axis is dipped downwards and possesses an inverse-S shape (Magara and Longcope, 2001; 2003) or where an originally upward-arched axial field line kinks downward as it undergoes the kink instability (Kliem *et al.*, 2004). However, if we wish to relate these or any other flux rope field lines to the soft-X-ray sigmoid, it is necessary to provide a physical reason why these particular dipped field lines are heated.

One potential heat source could be Joule heating in regions of enhanced currents. Török and Kliem (2003) and Aulanier *et al.* (2005a) found that as a line-tied flux rope is subjected to photospheric twisting motions, e.g. as might arise from sunspot rotation, the current density peaks in a sigmoid flux bundle in the bottom part of the twisting flux rope or in a current layer of sigmoid projected shape (of correct direction) below the flux rope. Since these structures exist both in quasi-steady and in erupting twisting flux ropes and since sunspot rotations may persist over periods of days (Brown *et al.*, 2003), such ropes or current layers might give rise to either persistent or transient sigmoids. However, it remains to be shown that the current density steepens sufficiently in this system so that the energy requirement of the sigmoid soft-X-ray emission would be met.

2.3. Sigmoid Due to Heating at Current Sheets

Electric current sheets are regions where the magnetic field is discontinuous across very thin spatial scales, and, generally speaking, the thinner the sheet, the stronger the current density. Such regions are thus good candidates for providing sufficient heating to raise sigmoid structures to soft-X-ray temperatures. The discontinuous magnetic fields at the current sheets can reconnect, and the thermal energy released by reconnecting magnetic fields is widely invoked to explain the required heating rate for soft-X-ray solar flares (Yokoyama and Shibata, 1998). In general, reconnections arising in numerical simulations such as will be discussed below result from numerical diffusion in regions of large gradients, and may not model realistic reconnection rates. 3D numerical simulations also tend not to explicitly model the dissipation processes which would heat the soft-X-ray loops. However, the locations of numerically-driven reconnections can have clear physical origins, e.g., current sheets. It is reasonable to consider where in a given magnetic topology current sheets would be likely to form. [In addition to the cases discussed below, see also Amari *et al.* (2000) and Kusano (2005) for examples of how reconnecting sigmoid field lines might arise during the (noneruptive) formation of flux ropes].

2.3.1. *Current Sheets at Interface of Rope and Ambient Field*

Current sheets are known to form at the boundary between a straight, cylindrically symmetric flux rope and its surrounding magnetic fields when it undergoes the kink instability (Rosenbluth *et al.*, 1973; Arber *et al.*, 1999; Gerrard *et al.*, 2001). An analogous helical current sheet can form around an arched flux rope that undergoes the kink instability. However, as discussed above, the forwards-S shape of this helical current sheet for a left-handed rope kinking upwards is inconsistent with sigmoid hemisphere rules. On the other hand, if such a left-handed rope kinks downwards instead of erupting upwards, the helical current sheet forms an inverse-S shape. Thus, one possible explanation for sigmoids could be heating along helical current sheets of downward kinking ropes (Kliem *et al.*, 2004).

2.3.2. *Current Sheets at Bald-Patch-Associated Separatrix Surface*

An alternative location for current sheet formation is a separatrix surface that arises because of line-tying at a rigid boundary, such as the photosphere (Parker, 1994; Titov and Demoulin, 1999, hereafter T&D; Low and Berger, 2003). Figure 3 shows a set of dipped field lines which form such a surface within a left-handed flux rope, and which possess the correct sigmoid direction (inverse-S). The bald patch (BP) of a coronal magnetic field structure is defined as the locus of points where dipped field just touches the photosphere (i.e. at the centers of the purple field lines shown in Figure 3). The bald-patch-associated separatrix surface (BPSS) is made up of the field lines that contain the BP points. Magnetic connectivity is discontinuous across the BPSS, because the field lines are line-tied to the (assumed) rigid photosphere. Dynamic evolution of the flux rope field lying above and within this separatrix surface relative to the shorter, arcade-type field below and external to it, could result in tangential discontinuities, leading to the the formation of electric current sheets along this sigmoid separatrix surface.

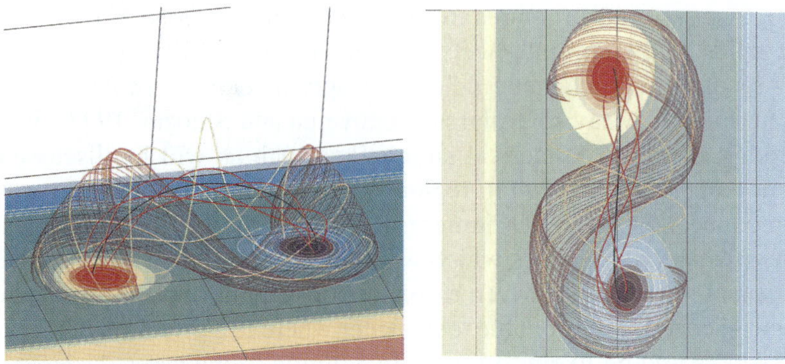

Figure 3. Bald-patch-associated separatrix surface (BPSS) (group of purple field lines) overlaid on sample flux-rope field lines for Fan and Gibson (2003; 2004) simulation time step 39. Color contours at lower boundary represent normal magnetic field at the photosphere. From Gibson *et al.* (2004).

2.3.3. *Current Sheets at Magnetic X-Line*

A well-studied location for current sheets to occur is in the region of magnetic X-points, and in three dimensions we can generalize this to a magnetic X-line along which the poloidal field comes to an X-point (the axial component of the field is not necessarily zero) (see e.g. Gorbachev and Somov, 1988). Such an X-type magnetic topology can exist below the flux rope instead of or in addition to a BPSS (see Section 4). In the classical picture of an eruptive flare, a vertical current sheet is formed behind an erupting flux rope where oppositely directed field lines reconnect to form the detached rope plasmoid with arcade field below (e.g. Anzer and Pneuman, 1982; Yokoyama and Shibata, 1998). More generally, this X-line could pinch into a current sheet under very general, possibly small perturbations (Titov *et al.*, 2003; Galsgaard *et al.*, 2003; Aulanier *et al.*, 2005b). Thus, as in the case of a BPSS, any type of perturbation of such configurations might be expected to light up sigmoid field lines reconnecting at the current sheets forming in the vicinity of the X-line.

3. Dynamic Perturbations

The question we are faced with is, what could cause dynamic perturbations of a flux rope and lead to current sheet formation and heating along sigmoid field lines? In this section we will briefly discuss some possibilities, which together can explain the range of observed sigmoids, from transient to persistent.

3.1. Eruptive Perturbations of Flux Ropes

In two related simulations, Fan and Gibson (2003; 2004) and Török *et al.* (2004) demonstrated that a flux rope in a coronal atmosphere underwent the kink instability

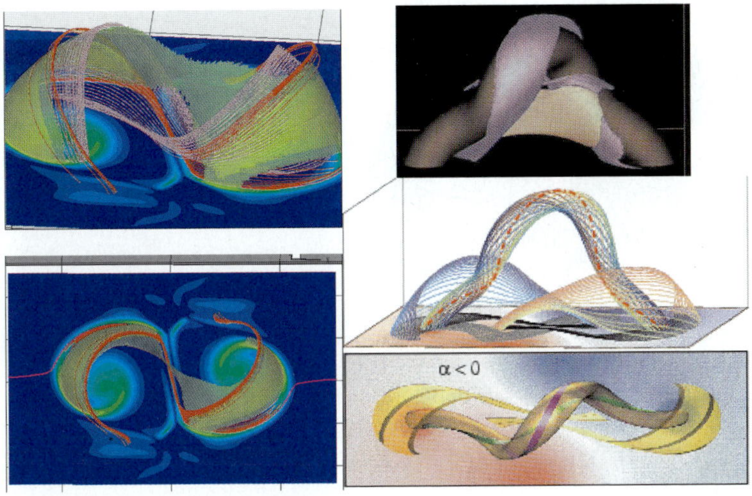

Figure 4. (Left bottom) Comparison of Fan and Gibson (2003; 2004) $t = 56$ BPSS (red field lines) to current sheets (yellowish-green isosurfaces), and (left top) same, with $t = 39$ BP-associated separatrix surface also shown (purple field lines) (both figures from Gibson *et al.*, 2004). (Right top) Vertical current sheet forming in the vicinity of magnetic X-line (yellow isosurface) during flux rope eruption (Török *et al.*, 2004). (Right middle) Field lines associated with this eruption, including two sets (red and blue) that intersect vertical current sheet, and (right bottom) same field lines seen from top (from Kliem *et al.*, 2004).

and erupted as it crossed a threshold for magnetic twist. Note that by "eruption" we mean a significant sudden upward motion, not necessarily implying an ejection of material from the corona. In both of these simulations, the left-handed flux rope axis kinked into a forwards S, inconsistent with observed hemispheric trends.

In the simulation of Fan and Gibson (2003; 2004), current sheets formed in the correct shape of an inverse S, and Gibson *et al.* (2004) demonstrated that these current sheets indeed formed along the BPSS as predicted (Figure 4, left). Thus the kink-instability-triggered eruption acted as a strong dynamic perturbation of the BPSS, creating a sharply defined, transient soft-X-ray sigmoid. Gibson and Fan (2006) analyzed the end-state of a related simulation (in spherical coordinates, see e.g. Fan, 2005), which resulted in the ejection of the upper part of the flux rope. They found that a vertical current sheet formed behind the ejecting portion of the flux rope (Figure 5 left) and cusped field lines closed down via reconnection along this current sheet (Figure 5 middle), as in the classical model of eruptive flares.

In the Török *et al.* (2004) simulation, the magnetic topology differed in that there was an X-line rather than a BPSS. We will discuss the reasons for and implications of this difference in greater detail below. In this analysis, a vertical current sheet formed beneath the erupting flux rope in the vicinity of the X-line. (Figure 4 right, top). Kliem *et al.* (2004) demonstrated that if one traced field lines passing close to this current sheet, they formed a sigmoid shape in the correct direction for

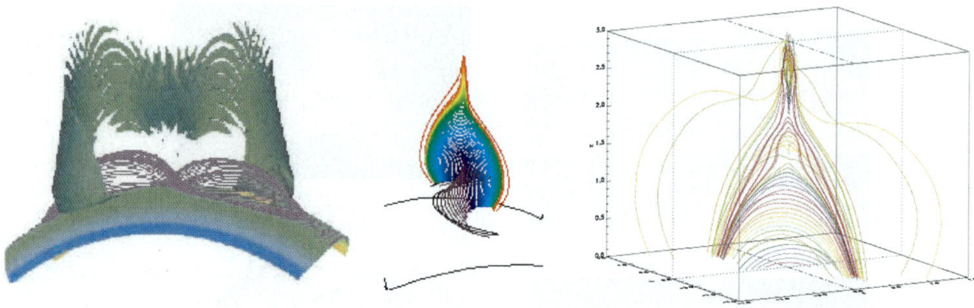

Figure 5. End-states: (left) Vertical current sheet forming behind portion of flux rope that escapes and (middle) cusp-shaped field lines reconnecting below erupting rope (Gibson and Fan, 2006). (Right) Cusped field lines reconnecting below erupting rope (Török and Kliem, 2005). Not all of the twisted rope has erupted in the Gibson and Fan (2006) case: a sigmoid BPSS is still present beneath the reconnecting field (left and middle frames).

the direction of magnetic twist (Figure 4 right, middle and bottom). As in the Fan (2005) simulation, cusped field lines were found beneath another, related simulated erupting flux rope in which the flux rope was ejected (Török and Kliem, 2005) (Figure 5, right).

Both the Gibson and Fan (2006) and the Török and Kliem (2005) results are consistent with a transient sigmoid brightening transitioning to a soft-X-ray cusp as in Figure 2. One difference between these two simulations, however, is that Gibson and Fan (2006) demonstrated the continued presence of a flux rope lying below the post-flare loops, as evidenced by the BPSS present in the left and middle frames of Figure 5. We will discuss the consequences of such a partly-expelled flux rope in more detail below.

3.2. Noneruptive Perturbations of Flux Ropes

As discussed in Section 2, any dynamic perturbation could cause current sheet formation along the BPSS or in the vicinity of the X-line. Indeed, Török *et al.* (2004) found that the X-line pinched into a current sheet even during the relaxation of the approximate analytical equilibrium of Titov and Demoulin (1999, hereafter T&D) to a nearby stable numerical equilibrium. Similarly, Fan and Gibson (2006) found that, even during the quasistatic evolution of the a confined flux rope before its eruption, sigmoid current sheets formed along the BPSS. These topologies present a "fault line" in the coronal magnetic field across which field lines behave very differently when driven dynamically. Thus, many different perturbations, ranging from flux emergence to photospheric motions at the footpoints of the field lines, may constantly cause the development of magnetic tangential discontinuities (or current sheets) where reconnections heat persistent, or long-lived sigmoids.

For example, the dynamic emergence of the flux rope from below the photosphere into the corona is a possible physical driver of current sheet formation at the BPSS. Dense photospheric material weighing down the dipped field prevents easy emergence when this is simulated (Fan, 2001; Magara and Longcope, 2001, 2003). However, as more axial flux is transported upward by the Lorentz-force induced shear, a sigmoid current sheet forms and reconnection allows the dipped field to lose some of its anchoring mass and emerge into the corona (Manchester *et al.*, 2004b). Depending upon whether this is a continuous process or one that occurs in fits and starts, it could result in either the continuous heating of a persistent sigmoid, or a series of transient sigmoids.

Cases where the flux rope kinks but does not erupt might also drive sigmoid heating. Fan and Gibson (2006) demonstrated such a case where a less twisted flux rope reached a stable equilibrium with a somewhat kinked axis, and indeed current sheets formed along the BPSS during this writhing motion. In another example, as mentioned above, Kliem *et al.* (2004) described a case where a downward kinking left-handed rope formed an inverse-S-shaped, helical current sheet at its interface with surrounding magnetic fields. These scenarios are then consistent with reconnection heating forming a transient, but non-CME related sigmoid.

4. Partly vs. Fully Expelled Flux Ropes

The degree to which the flux rope is expelled may depend upon whether reconnections occur behind or within the rope, and this in turn may depend upon whether or not the flux rope lies down low enough in the corona to possess a BPSS. Figures 6 and 7 illustrate this point in 2D and 3D respectively. Figure 6 shows 2D cartoons of a flux rope viewed along its axis. If reconnection occurs at an X-point below the flux rope (left), it can be completely expelled. If reconnection occurs within the flux rope (right), some of the rope remains behind and it is partly expelled.

Figure 7 demonstrates this in 3D. The flux rope shown in this image is based on the T&D model, an analytic model of a flux rope within an arcade field. When the T&D model flux rope is only partly "emerged" above the photosphere so that there is a single, continuous BP of dipped field grazing the central portion of the neutral line (left image), there is a corresponding single BPSS. This configuration is very similar to that of Fan and Gibson (2003; 2004), and is the 3D analogue to the right-hand images of Figure 6. Such a configuration is shown in Gibson and Fan (2006) to lead to partial rope expulsion, with a BPSS left behind. In this case there is no magnetic X-line before the eruption, but reconnection occurs as opposing upper portions of BPSS field lines are squeezed together when the flux rope axis kinks. The lower portions of these same field lines reconnect to form the lower, remaining flux rope BPSS shown in Figure 5.

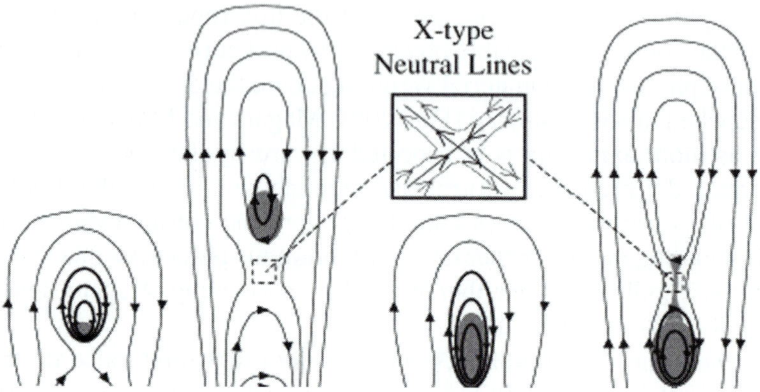

Figure 6. Cartoons of erupting flux ropes, viewed along their axes. (Left) Reconnection occurring at the X point below the rope, leading to the total expulsion of the flux rope and prominence. (Right) Reconnection occurring within the flux rope and prominence, leading to the partial expulsion of the flux rope and prominence. Adapted from Gilbert *et al.* (2001).

Figure 7. T&D-type flux rope at various stages of "emergence". (Left) Low enough in the coronal atmosphere so that there is no X-line, but the T&D rope would have a single BPSS. (Middle) Higher in the atmosphere, so that the X-line (blue line) is present, and the BPSS would be bifurcated. (Right) High enough to have an X-line but no BPSS in the T&D rope. Note that the images shown are adapted from Roussev *et al.* (2003), and show a version of the T&D model with purely poloidal field outside the rope, so the comments regarding BPSSs made here should be taken with regards to the analogous stage of "emergence" of the true T&D model, which does have finite twist outside the rope.

If the axis of the flux rope lies high enough in the atmosphere, the magnetic X-line will be present and the BPs will either bifurcate into two (middle image), or disappear altogether if the curvature of the rope legs is such that no concave-up fields intersect the photospheric neutral line (right image: see Titov and Demoulin (1999) for discussion). The Török and Kliem (2005) simulation uses the T&D model with no BP as its starting point, and demonstrates how a flux rope can be expelled essentially in entirety during eruption, reconnecting in the vicinity of the magnetic X-line and leaving behind cusped, post-flare fields.

4.1. Observational Consequences of Partly-Expelled Flux Rope

4.1.1. *Partial Filament Eruptions*

Since the partly-expelled flux rope leaves behind a BPSS, a sigmoid could reform soon after an eruption. It also explains how a filament might partially erupt, or even be unperturbed by the flux rope eruption as has been observed (Tang, 1986; Pevtsov, 2002a; Gibson *et al.*, 2002). The filament is expected to lie within the dips of the magnetic flux rope, and so a BPSS-sigmoid would wrap around it, with loop apexes lying above it (see Figure 3). When modeled this way observations of related quiescent filaments and persistent sigmoids are well-matched (Gibson *et al.*, 2004). Referring to Figure 6 again, we see that depending on where the reconnection happens, the filament will partially erupt, or not erupt at all (Gilbert *et al.*, 2001). As the eruption begins, the BPSS is dynamically forced, causing a transient sigmoid brightening. Gibson and Fan (2006) demonstrated that the kinking rope's legs then are squeezed together, creating a vertical current sheet where sigmoid field lines reconnect, breaking the rope in two. As the eruption continues, the field-lines reconnecting at the vertical current sheet become less sigmoid-shaped, and more cusp-shaped, closing down above the surviving dipped, possibly filament-containing field which is essentially unaffected by the eruption.

4.1.2. *Relevance for Magnetic Cloud Observations*

Magnetic clouds, known to be associated with CMEs and filament eruptions, can be examined for evidence of entrained, cool, filament material, by examining the charge states of solar wind ions within them which are "frozen in" at coronal temperatures. Such analyses of magnetic clouds tend to indicate only relatively hot coronal material. However, some cases have been found (He^+ events) that imply cool material coexisting with hot material (Skoug *et al.*, 1999; Gloeckler *et al.*, 1999). One recent case (Zurbuchen *et al.*, 2005) (Jan 9–10, 2005) demonstrates purely cold material, and it is worth noting that this event had no associated flare.

It has been argued that the coronal temperature diagnostics provided by magnetic clouds may be able to help distinguish between different eruption models (Lynch *et al.*, 2004). For example, if the filament contained in a pre-CME flux rope were to erupt without reconnections, it should result in the presence of cool magnetic cloud material, whereas if a flux rope were formed via reconnections during eruption, any entrained filament material would be heated. However, we have demonstrated that in a partly-expelled flux rope, filament-entrained field lines can also experience reconnections and heating. Indeed, Figure 6 demonstrates that all, some, or none of the filament-carrying field lines might undergo reconnections during a flux rope eruption, making it tempting to speculate that all three types of magnetic clouds (hot, mixed, or cool) might arise. We hesitate to go this far, since we feel that equating the heated magnetic cloud material with localized reconnections during eruption is likely to be an oversimplification. However, we do assert that the internal reconnections during the partial eruption of a pre-existing flux rope (or for that matter, the reconnections between an erupting flux rope and external field (Amari *et al.*, 2003b)), are as likely to heat entrained filament material as reconnections that occur when a rope is formed during eruption.

5. Conclusions

The combined results presented here demonstrate that sigmoid evolution, before, during, and after a CME, is consistent with the presence of a long-lived coronal magnetic flux rope. Before any CME occurs, the sigmoid can appear as a general brightening possessing an S or inverse-S pattern, for example during the more-or-less continuous dynamic perturbation of field lines at the BPSS or in the vicinity of an X-line. During the CME, transient sigmoids might occur as the rope loses equilibrium and erupts, causing current sheet formation along the BPSS or again at the X-line. During and after the CME, field lines reconnect into cusped field lines below the erupting portion of the rope in accordance with the observed transition of the transient sigmoid into a cusp. If the flux rope is only partly expelled, the BPSS may remain, allowing a quick return to persistent or transient soft-X-ray sigmoids, or indeed eventually to more eruptions. A partly-expelled flux rope could also explain partially or non-erupting filaments, as well as the presence of hot or mixed temperature charge states within magnetic clouds.

Acknowledgements

S.E.G. wishes to thank all the ISSI workshop participants for an excellent tutorial on the integrated solar-heliosphere-magnetosphere system, and T. Zurbuchen in particular for discussions on magnetic cloud charge states. She also thanks J. Goldstein and B. T. Klöun for their animated debates at the workshop. We thank H. Gilbert and I. Roussev for permitting us to adapt their figures, and T. Holzer and an anonymous referee for helpful comments. T.T. and B. K. thank the John von Neumann-Institut for Computing, Jülich for computer time. S. E. G. and Y. F. were supported in part by AFOSR grant F49620-02-0191, and B. K. was supported by DFG grant MA 1376/16-2. The Yohkoh SXT project is a collaborative project of LMSAL, the National Astronomical Observatory of Japan, and the University of Tokyo, supported by NASA and ISAS.

References

Abbett, W. P., and Fisher, G. H.: 2003, *Astrophys. Journ.* **582**, 475.
Anzer, U., and Pneuman, G. W.: 1982, *Solar Phys.* **79**, 129.
Amari, T. *et al.*: 2000, *Astrophys. Journ.* **529**, L49.

Amari, T. *et al.*: 2003a, *Astrophys. Journ.* **585**, 1073.
Amari, T. *et al.*: 2003b, *Astrophys. Journ.* **95**, 1231.
Arber, T. D., Longbottom, A. W., and van der Linden, R. A. M.: 1999, *Astrophys. Journ.* **517**, 990.
Archontis, V. *et al.*: 2004, *Astron. and Astrophys.* **426**, 1047.
Aulanier, G., Démoulin, P., and Grappin, R.: 2005a, *A&A* **430**, 1067.
Aulanier, G., Pariat, E., and Démoulin, P.: 2005b, *A&A* **444**, 961.
Berger, M. A., and Field, G. B.: 1984, *J. Fluid Mech.* **147**, 133.
Brown, D. S. *et al.*: 2003, *Solar Phys.* **216**, 79.
Burlaga, L. *et al.*: 1982, *Geophys. Res. Lett.* **9**, 1317.
Canfield, R. C., Hudson, H. S., and Pevtsov, A. A.: 2000, *IEEE Trans. Plasma Sci.* **28**, 1786.
Chen, J., and Krall, J.: 2003, *Journ. Geophys. Res.* **108**, CiteID 14010.
Fan, Y.: 2005, *Astrophys. Journ.* **630**, 543.
Fan, Y.: 2001, *Astrophys. Journ.* **554**, L111.
Fan, Y., and Gibson, S. E.: 2006, *ApJL* **641**, L149.
Fan, Y., and Gibson, S. E.: 2004, *Astrophys. Journ.* **609**, 1123.
Fan, Y., and Gibson, S. E.: 2003, *Astrophys. Journ. Lett.* **589**, L505.
Galsgaard, K., Titov, V. S., and Neukirch, T.: 2003, *ApJ* **595**, 506.
Gerrard, C. L., Arber, T. D., Hood, A. W., and van der Linden, R. A. M.: 2001, *Astron. and Astrophys.* **373**, 1089.
Gibson, S. E., and Fan, Y.: 2006, *ApJL* **637**, L65.
Gibson, S. E. *et al.*: 2004, *Astrophys. Journ.* **617**, 600.
Gibson, S. E. *et al.*, 2002, *Astrophys. Journ.* **574**, 1021.
Gilbert, H. R., Holzer, T. E., Low, B. C., and Burkepile, J. T.: 2000, *Astrophys. Journ.* **549**, 1221.
Gorbachev, V. S., and Somov, B. V.: 1988, *Solar Phys.* **117**, 77.
Gloeckler, G. *et al.*: 1999, *Geophys. Res. Lett.* **26**, 157.
Hagyard, M. J., Teuber, D., West, E. A., and Smith, J. B.: 1984, *Solar Phys.* **91**, 115.
Hudson, H. S. *et al.*: 1998, *Geophys. Res. Lett.* **25**, 248.
Kliem, B., Titov, V. S., and Török, T.: 2004, *Astron. and Astrophys.* **413**, L23.
Kusano, K.: 2005, *Astrophys. Journ.* **631**, 1260.
Leka, K. D., Canfield, R. C., McClymont, A. N., and van Driel-Gesztelyi, L.: 1996, *Astrophys. Journ.* **462**, 547.
Lites, B. W. *et al.*: 1995, *Astrophys. Journ.* **446**, 877.
Lites, B. W., and Low, B. C.: 1997, *Solar Phys.* **174**, 91.
Low, B. C.: 1996, *Solar Physics* **167**, 217.
Low, B. C., and Berger, M.: 2003 *Astrophys. Journ.* **589**, 644.
Lynch, B. J. *et al.*: 2004, *Astrophys. Journ.* **617**, 589.
MacNeice, P. *et al.*: 2004, *Astrophys. Journ.* **614**, 1028.
Magara, T., and Longcope, D.: 2003, *Astrophys. Journ.* **586**, 630.
Magara, T., and Longcope, D.: 2001, *Astrophys. Journ. Lett.* **559**, L55.
Manchester, W. *et al.*: 2004a, *Journ. Geophys. Res.* **109**, CiteID A01102.
Manchester, W., Gombosi, T., DeZeeuw, D., and Fan, Y.: 2004b, *Astrophys. Journ.* **610**, 588.
Manoharan, P. K., van Driel-Gesztelyi, L., Pick, M., and Demoulin, P.: 1996, *Astrophys. Journ. Lett.* **468**, L73.
Martin, S. F., Marquette, W. H., and Bilimoria, R.,1992, in K. Harvey (ed.), *ASP Conf. Proc. 27, The Solar Cycle*, (San Francisco: ASP), 53.
Martin, S. F., Bilimoria, R., and Tracadas, P. W.: 1994, in R. J. Rutten and C. J. Schrijver (eds.), (*NATO ASI Ser. C, 433*) Solar Surface Magnetism, (Dordrecht: Kluwer), 303.
Parker, E. N.: 1994, Spontaneous Current Sheets in Magnetic Fields, Oxford University Press, New York.
Pevtsov, A. A., Canfield, R. C., and Metcalf, T. R.: 1995, *ApJ* **440**, L109.

Pevtsov, A. A., Canfield, R. C., and Zirin, H.: 1996, *Astrophys. Journ.* **473**, 533.
Pevtsov, A. A., and Canfield, R. C.: 1999, Magnetic Helicity in Space and Laboratory Plasmas, in M. R., Brown, R. C. Canfield, and A. A. Pevtsov, (eds.), *Geophys. Monogr. Ser., AGU*, Washington D. C., vol. 111, 103.
Pevtsov, A. A., Canfield, R. C., and Latushko, S. M.: 2001, *Astrophys. Journ. Lett.* **549**, L261.
Pevtsov, A. A.: 2002, *Sol. Phys.* **207**, 111.
Pevtsov, A. A.: 2002, Yohkoh 10th Anniversary Meeting Proceedings, eds. Martens and Cauffman, COSPAR Colloquia Series, Elsevier, 125.
Rosenbluth, M. N., Dagazian, R. Y., and Rutherford, H. P.: 1973, *Phys. Fluids* **16**, 1894.
Roussev, I. I. *et al.*: 2003, *Astrophys. Journ.* **588**, L45.
Rust, D. M., and Kumar, A.: 1994, *Solar Phys.* **155**, 69.
Rust, D. M., and Kumar, A.: 1996, *Astrophys. Journ. Lett.* **464**, L199.
Rust, D. M.: 2003, *Adv. in Space Res.* **32**, 1895.
Seehafer, N.: 1990, *Sol. Phys.* **125**, 219.
Skoug, R. *et al.*: 1999, *Geophys. Res. Lett.* **26**, 161.
Sterling, A. C., and Hudson H. S.: 1997, *Astrophys. Journ. Lett.* **491**, L55.
Tanaka, K.: 1991, *Sol. Phys.* **136**, 133.
Tang, F.: 1986, *Sol. Phys.* **105**, 399.
Titov, V. S., and Demoulin, P.: 1999, *Astron. Astrophys.* **351**, 70.
Titov, V. S., Galsgaard, K., and Neukirch, T.: 2003, *ApJ* **582**, 1172.
Török, T., and Kliem, B. 2003, *A&A* **406**, 1043.
Török, T., Kliem B., and Titov, V. S.: 2004, *Astron. and Astrophys.* **413**, L27.
Török, T., and Kliem, B.: 2005, *Astrophys. Journ. Lett.* **630**, L97.
Yokoyama, T., and Shibata, K.: 1998, *Astrophys. Journ. Lett.* **494**, L113.
Zurbuchen, T. H., *et al.*: 2005, AGU Spring meeting 2005, abstract SH54B-04.

PROPERTIES OF INTERPLANETARY CORONAL MASS EJECTIONS

NAT GOPALSWAMY
NASA Goddard Space Flight Center, Greenbelt, MD 20771, USA
(E-mail: gopals@ssedmail.gsfc.nasa.gov)

(Received 6 June 2006; Accepted in final form 12 September 2006)

Abstract. Interplanetary coronal mass ejections (ICMEs) originating from closed field regions on the Sun are the most energetic phenomenon in the heliosphere. They cause intense geomagnetic storms and drive fast mode shocks that accelerate charged particles. ICMEs are the interplanetary manifestations of CMEs typically remote-sensed by coronagraphs. This paper summarizes the observational properties of ICMEs with reference to the ordinary solar wind and the progenitor CMEs.

Keywords: coronal mass ejections, interplanetary CMEs, solar wind, magnetic clouds, shocks, solar flares, geomagnetic storms

1. Introduction

There are two classes of large-scale interplanetary (IP) structures related to the two types magnetic field topology on the Sun: interplanetary coronal mass ejections (ICMEs) originating from closed field regions and corotating interaction regions (CIRs) due to high speed streams originating from open field regions (see, e.g., Gosling, 1996, for a review). Two earlier ISSI volumes have extensively dealt with CIRs (Balogh *et al.*, 1999) and CMEs (Kunow *et al.*, 2006). Both CIRs and ICMEs are capable of driving shocks, which in turn, accelerate charged particles. The CIR shocks generally form far beyond 1 AU, although they are occasionally observed near 1 AU. ICME are the IP manifestations of CMEs. CMEs drive shocks from close to the Sun to far into the IP medium, the shocks being the strongest near the Sun. CME-driven shocks accelerate charged particles from close to the Sun and in the IP medium. ICMEs are responsible for the severest of geomagnetic storms when they impinge upon Earth's magnetosphere. ICMEs also provide an enormous plasma laboratory to study physical processes in space. The white light CMEs observed near the Sun are typically 10 times more abundant than the ICMEs observed *in situ* (Gopalswamy, 2004). ICMEs are therefore a special population that makes significant impact on the heliosphere, in particular on Earth's space environment. In fact Earth spends in the flows related to ICMEs anywhere from 10% of the time during solar minimum to 35% of the time during solar maximum (Cliver *et al.*, 2003). ICMEs are also observed throughout the heliosphere (see, e.g., Burlaga, 1995; Balogh, 2002).

Space Science Reviews (2006) 124: 145–168
DOI: 10.1007/s11214-006-9102-1

Early ideas on transient plasma ejections from the Sun can be found in the reviews by Burlaga (1995) and Gosling (1997). The first IP shock was identified from the Mariner 2 plasma and magnetic field measurements (Sonnet *et al.*, 1964). Since solar flares were known for a much longer time, early studies focused on the connection between solar flares and IP shocks (see, e.g., Dryer, 1994, for a review). White-light CMEs were first detected by NASA's seventh Orbiting Solar Observatory (OSO-7) on December 14, 1971 (Tousey, 1973). Although the connection between coronal shocks (inferred from metric type II radio bursts) and IP shocks (detected *in situ* and inferred from IP type II bursts) was recognized early on (Pinter, 1973), the correspondence between CMEs and their IP counterparts (ICMEs) became clear when Helios 1 observed an ICME (June 20, 1980) that left the Sun two days before as a CME observed by the Solwind coronagraph on board P78-1 (Burlaga *et al.*, 1982).

Helios 1 detected a magnetic loop behind an IP shock, which Burlaga *et al.* (1981) defined as a magnetic cloud (MC), a name given to those ICMEs that have a high magnetic field, smooth field rotation, and low proton temperature (Lepping *et al.*, 1990). The term "magnetic cloud" was originally used by Parker (1957) in a much broader sense in his theoretical study of the dynamics of hydromagnetic gas clouds ejected from the Sun into the IP space. ICMEs are identified using plasma, magnetic, compositional and energetic particle signatures (see, e.g., Gosling *et al.*, 1990). These signatures include bidirectional streaming of superthermal electrons and ions, unusual abundances and charge states, low electron and proton temperatures, strong magnetic fields with flux rope structures, and Forbush decrease. It must be noted that not all of the signatures are present in all events (see Neugebauer and Goldstein, 1997). *In situ* observations can be used to infer the magnetic field topology of the ICMEs and the physical conditions of their birthplace near the Sun (see, e.g., Henke, 1998; Lepri *et al.*, 2001).

This paper summarizes the properties of ICMEs in relation to the white light CMEs. After an observational description of ICMEs in Section 2, their statistical properties are summarized in Section 3 in comparison with the solar wind. The solar origin of ICMEs, the source locations, and their variation with the solar activity cycle are discussed in Section 4. The charge state composition of ICMEs and its solar origin is discussed in Section 5. In Section 6, the connection between ICMEs and geomagnetic storms is discussed. Section 7 gives the summary and conclusions.

2. An Observational Description of ICMEs

ICMEs are transient disturbances in the solar wind referred to by various names such as driver gas, ejecta, and plasma cloud. ICMEs are recognized as large-scale magnetic structures with magnetic field enhanced with respect to the solar wind and having plasma and composition signatures distinct from the solar wind in which

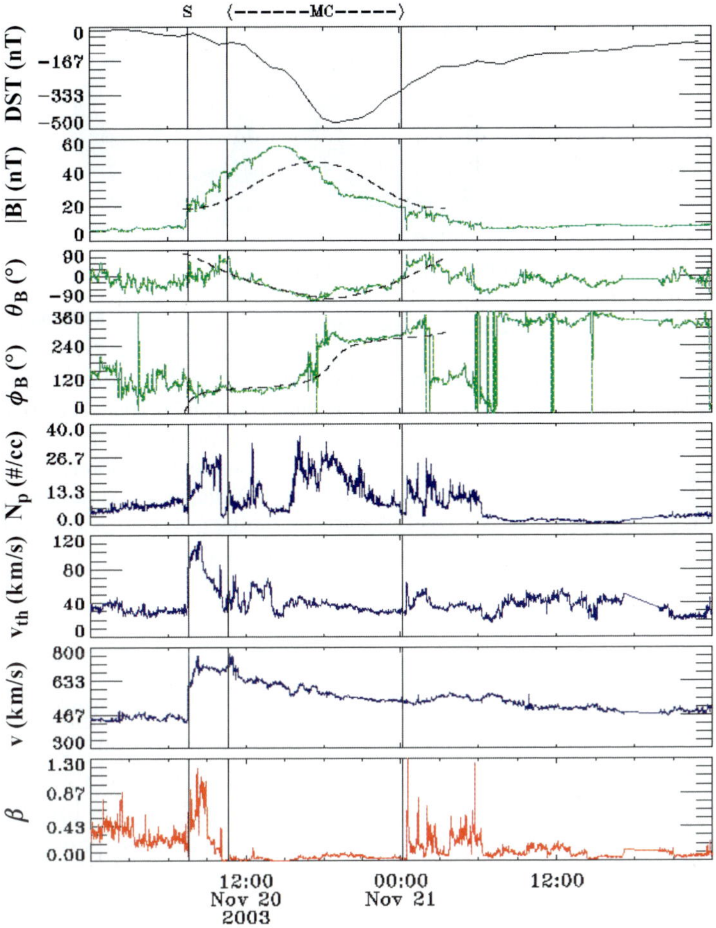

Figure 1. The magnetic cloud event of 2003 November 20 with the associated geomagnetic storm index (Dst), the magnetic field strength ($|\mathbf{B}|$), the latitude (θ_B), longitude (ϕ_B) of the cloud, and the solar wind parameters (proton density N_p, proton thermal speed V_{th}, flow speed V and the plasma beta β). The dashed curves in the $|\mathbf{B}|$, θ_B, and ϕ_B are model fits. The boundaries of the cloud (MC) and the shock (S) are marked at the top from Goplswamy *et al.* (2005a).

they are embedded. When the ICME has a flux rope structure, it is called an MC, as defined by Burlaga *et al.* (1981). Observations of an ICME (MC in this case) are shown in Figure 1. The MC interval is marked by the two vertical lines. The MC was driving a shock, which was located about 2.5 h ahead of the MC. The MC was associated with an intense geomagnetic storm ($Dst = -472$ nT) shown at the top (Gopalswamy *et al.*, 2005a). The magnetic field strength is very high, with the peak value of ~ 56 nT. The polar and azimuthal angles of the magnetic field are also smooth during the cloud interval. The polar angle shows that the cloud is highly inclined to the ecliptic plane (–73.4 degrees) and south-pointing throughout

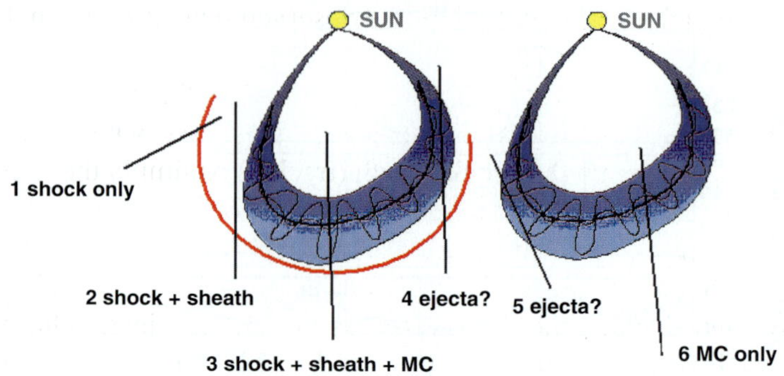

Figure 2. Six possible tracks of an observing spacecraft through an MC with a leading shock (left) and another without (right). Tracks 1 and 2 never encounter the MC proper. They pass through the shock and the compressed ambient medium in one of the flanks of the MC. Track 4 passes through the nose of the MC. This situation arises when the observing spacecraft is along the Sun-Earth line and a fast and wide CME erupts from close to the Sun center. Trajectory 4 passes through the shock, sheath, and through the edge of the MC. Tracks 5 and 6 are similar to 4 and 3, respectively, except that the MC is slow and hence it does not drive a shock. Trajectories 4 and 5 are not expected to observe an MC structure.

the cloud. In addition, the azimuthal angle shows a smooth rotation from east to west, so such clouds are called ESW cloud. The solar wind density is enhanced in the region between the cloud and the shock (a region known as the shock sheath). The proton thermal speed also has a jump in the sheath and is low in the cloud. The MC was expanding, evidenced by the smooth decrease in speed from the front (738 km/s) to the back (531 km/s), with an average value of 625 km/s. There was a slight increase in speed after the MC, probably due to a high speed stream originating from a coronal hole. The plasma β is extremely low during the cloud interval, showing the dominance of the magnetic field. The duration of the cloud is $\sim$14 h. Since the cloud was moving with an average speed of $\sim$625 km/s, this duration corresponds to a cloud thickness of $\sim$0.21 AU. This thickness is consistent with the cloud diameter inferred from multi-spacecraft observations. For example, Burlaga *et al.* (1990) used Helios 1, Helios 2, IMP-8 and Voyager-2 data ICMEs to infer a loop-like structure (see Figure 2) with a radius of curvature of $\sim$0.35 AU and a cloud diameter of 0.25 AU. They also inferred that the legs of the loop must be connected to the Sun at both ends.

As the observing spacecraft passes from the outer boundary of the loop structure in Figure 2 to the axis and then to the inner boundary, the azimuthal field changes sign at the axis, indicating the rotation of the field direction. The magnitude of the azimuthal component also changes, peaking at the axis and falling off on either side. The cloud can drive a shock if it is super-Alfvenic. The shock standoff distance corresponds to a lead time of $\sim$0.5 day. i.e., a spacecraft in the solar wind first encounters the shock and then the cloud several hours later. The trajectory of the observing spacecraft essentially decides the observed structure. When the trajectory goes through the nose of the magnetic cloud (tracks 3 and 6), one observes an MC as in Figure 1. Tracks 4 and 5 will not observe a cloud structure because of the skimming trajectory. The extreme case is when the trajectory never passes through the cloud, but only through the shock sheath (track 2) or simply the shock (track 1). The "shock only" and "shock + sheath" configurations are readily observed. The interpretation of the general ejecta corresponding to tracks 4 and 5 is not clear. According the picture presented in Figure 2, all ICMEs have cloud structure, but the vantage point decides the appearance as an MC or ejecta. One of the other possible interpretations is that the ICME contains just untwisted loops ejected from active regions and hence do not show any MC structure (Gosling, 1990). It is really an open question whether all the ICMEs have flux rope structures.

3. Statistical Properties

3.1. Solar Wind Properties

To see the ICME as disturbances superposed on the solar wind, we first show the distributions of magnetic field strength, speed, density and temperature of the solar wind obtained from Omniweb for a time period 1996–2003 (see Figure 3). Note that the solar wind parameters obtained from Omniweb correspond to a heliocentric distance of 1 AU and may have different values at other distances. The number of hours with missing data was only a small fraction ($<8\%$). The distributions with and without ICMEs are similar because the duration over which Earth was immersed in CME-related solar wind is only $\sim 28\%$ on the average. When the ICME intervals (see, e.g., Cane and Richardson, 2003; Lepping *et al.*, 2005) are removed, the solar wind magnetic field and speed show slightly lower values, the density did not show much change and the temperature was slightly higher. This is consistent with the general property that ICMEs have enhanced magnetic field and reduced temperature.

3.2. Properties of ICMEs

To get the properties of ICMEs, we need an accurate list. The identification of ICMEs, especially their boundaries, has been highly subjective and controversial (see, e.g., Russell and Shinde, 2005 and references therein), but the overall numbers obtained by various authors is similar. Here we use the following subsets of ICMEs: MCs identified manually using the Burlaga criteria, the cloud-like events identified by automatic detection (Lepping *et al.*, 2005), and ICMEs following IP shocks (Manoharan *et al.*, 2004). The automatic detection of cloud-like events employed the following criteria (Lepping *et al.*, 2005): (1) low plasma beta (≤ 0.3),

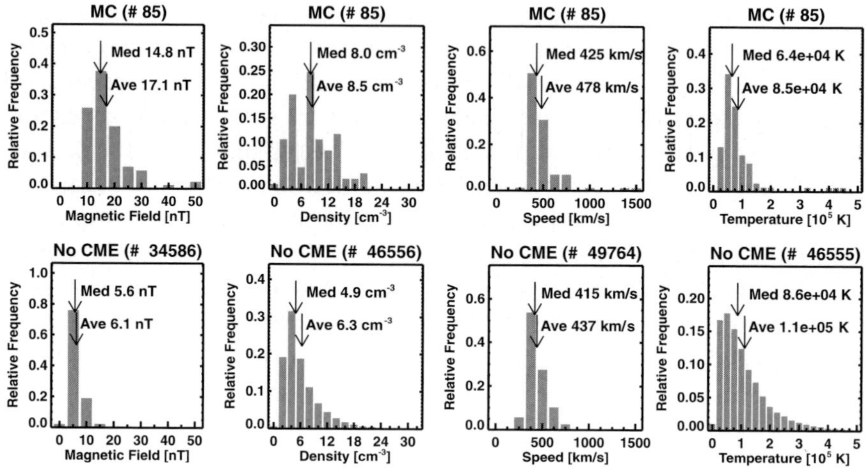

Figure 3. Distributions of magnetic field strength, density, speed, and temperature of the solar wind over the interval 1996–2003 (inclusive) with (top) and without (bottom) ICMEs. Data from Omniweb (http://omniweb.gsfc.nasa.gov) with one-hour time resolution were used. The number of data points (i.e., the number of hours) is indicated at the top. The total number of hours during the study interval is 70128, which means a relatively small data gaps for magnetic field (1.7%), density (7.8%), speed (2.2%), and temperature (7.8%). The mean and median values of the distributions are marked. The plots excluding ICMEs ("No CME") were created by removing the data points corresponding to ICME intervals.

(2) magnetic field with slowly changing direction, (3) enhanced magnetic field over an interval of $\geq$8 h, (4) average magnetic field $\geq$8 nT, (5) low proton thermal speed ($\leq$30 km/s), and (6) large latitudinal difference angle ($\geq$45 deg). Lepping *et al.* (2005) could identify most of the MCs also by this method. From the list generated by automatic detection, the manually identified MCs were eliminated to get the list of cloud-like events. Figure 4 summarizes the properties of MCs, cloud-like events, and shock-driving ICMEs. The maximum magnetic field strength in the three populations of ICMEs ranged from a few nT to several tens of nT, generally much larger than the average solar wind value (6.1 nT, see Figure 3). The mean speeds of MCs (478 km/s) and shock-driving ICMEs (484 km/s) were generally larger than those of the cloud-like events (412 km/s) and the slow solar wind (437 km/s). During the occasional extreme events, MCs had speeds exceeding 1000 km/s. Overall, MCs and shock-driving ICMEs have higher maximum field strength, speed, and duration compared to the cloud-like events. The proton thermal speed of the cloud-like events was higher than that of the MCs but lower than that of the shock-driving ICMEs.

Figure 5 shows the scatter plots between maximum magnetic field strength and speed of ICMEs measured *in situ*. There is certainly a weak correlation between the maximum field strength and speed for the MCs, but there is no such correlation for the cloud-like events. The shock-driving ICMEs, on the other hand, have a

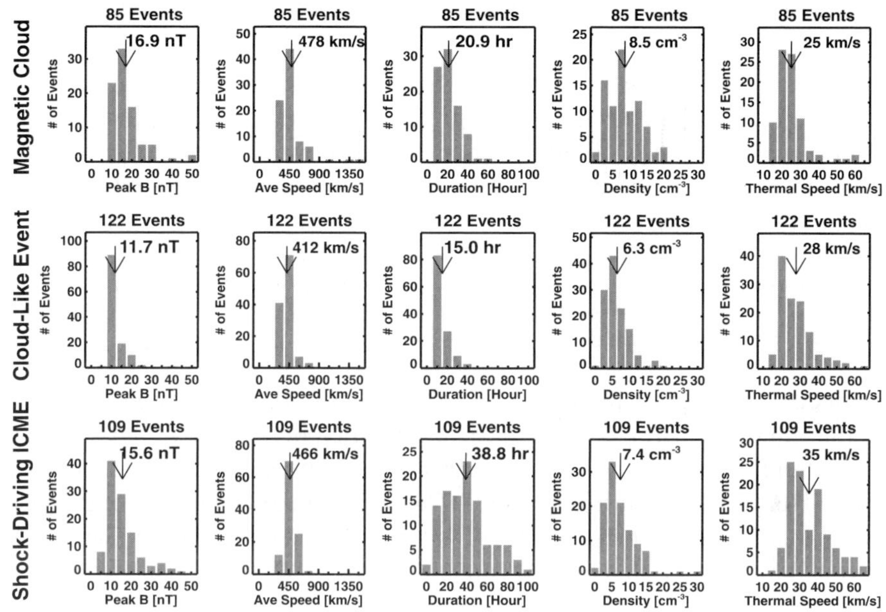

Figure 4. Properties of ICMEs grouped into magnetic cloud (top), cloud-like (middle) events, and shock driving ICMEs (bottom) that are not MCs. Distributions of the magnetic field strength, speed, duration, density, and thermal speed within the cloud are shown. The magnetic field strength is the peak value within the ICME interval while all others are average values within the same interval. The averages of the distributions are indicated on the plots. The cloud-like events were identified using an automatic program. The sharp cutoffs of some of the distributions at the low end are due to the selection criteria employed by the automatic feature recognition routine.

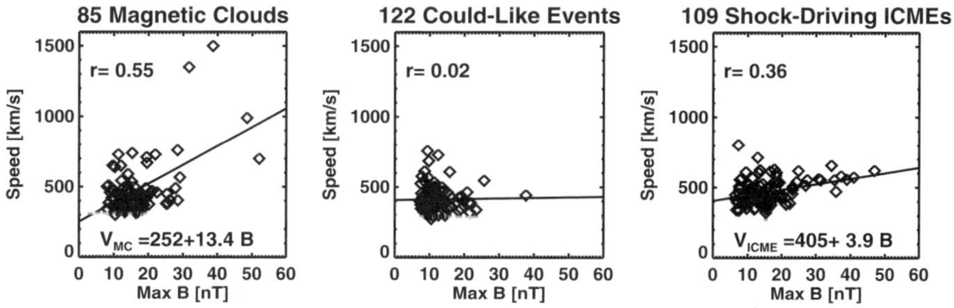

Figure 5. Correlation between the speed and magnetic field strength of MCs (left), cloud-like events (middle) and shock-driving ICMEs (right). The correlation coefficient (r) is marked in each case.

correlation much weaker than that for the MC events, but definitely positive compared to the cloud-like events. Gonzalez *et al.* (1998) had found a better correlation ($r = 0.75$) using a sample of 30 MCs. A bigger sample of 85 MCs (see Figure 5) yields a correlation coefficient of only 0.55. It must be pointed out that the MCs in Figure 5 have a wider range of speeds (300 to 1500 km/s) compared to that in Gonzalez *et al.* (1998) (350 to 750 km/s). The corresponding regression lines are: $V_{MC} = 23.4 + 21.28B_{MC}$ from Gonzalez *et al.* (1998) compared to MC speed $V_{MC} = 252 + 13.4$ magnetic field B_{MC} for the 85 MCs in Figure 5 (V_{MC} in km/s and B_{MC} in nT). Gonzalez *et al.* (1998) suggested that the poor correlation between speed and magnetic field strength in non-MC driver gas may be due to the geometrical effect that the observing spacecraft does not pass through the center of the cloud and hence measures lower field strength (corresponding to tracks 4 and 5 in Figure 2).

3.3. ICMEs and Shocks

Interplanetary shocks are important players in Sun-Earth connection because they accelerate energetic particles near the Sun and when they arrive at 1 AU (energetic storm particle events, see Rao *et al.*, 1967), mark the sudden commencement of magnetic storms (Chao and Lepping, 1974), and signal the impending arrival of ICMEs. Most of the IP shocks observed within 1 AU can be identified with ICMEs. Occasionally, shocks driven by CIRs are observed, but these generally form beyond 1 AU (see, e.g., Burlaga *et al.*, 1995). Lindsay *et al.* (1994) surveyed IP shocks within 0.72 AU from the Sun and found that most of them (80%) were associated with ICMEs (the other 20% being CIR-driven). However, starting from MCs, Klein and Burlaga (1982) found that only a third of them were preceded by IP shocks. Much higher rate of association was also reported in two later studies (70% by Marsden *et al.*, 1987 and 80% by Zhang and Burlaga, 1988). Gopalswamy *et al.* (2000) found that 25 of the 28 ICMEs (89%) during the period 1996–1998 were driving shocks. They found similar results (92–93% association) for an expanded data set (Gopalswamy *et al.*, 2001a). The number of shocks varies in phase with the solar activity cycle (Lindsay *et al.*, 1994). Since the shocks need faster and wider CMEs to drive them the solar-cycle variation of the number of shocks actually tracks the fast and wide CMEs better (Gopalswamy *et al.*, 2003). Figure 6 shows a scatter plot between the shock and MC speeds. The correlation coefficient is $r = 0.95$. Such a tight relationship suggests that the shocks are driven even at 1 AU (Gopalswamy *et al.*, 2005b). Occasionally, shocks are observed at 1 AU, which do not seem to have a driver behind them. However, this is due to an observation limitation: the spacecraft passes through the shock, but not through the driver as illustrated in Figure 2 (tracks 1 and 2). In these cases, the associated CME at the Sun is ejected roughly at right angles to the Sun-Earth line.

4. The Solar Connection

The close connection between CMEs and ICMEs was recognized in 1982 when a Helios-1 MC was identified with a white-light CME that left the Sun two days before (Burlaga *et al.*, 1982). The high speed plasmas with helium abundance

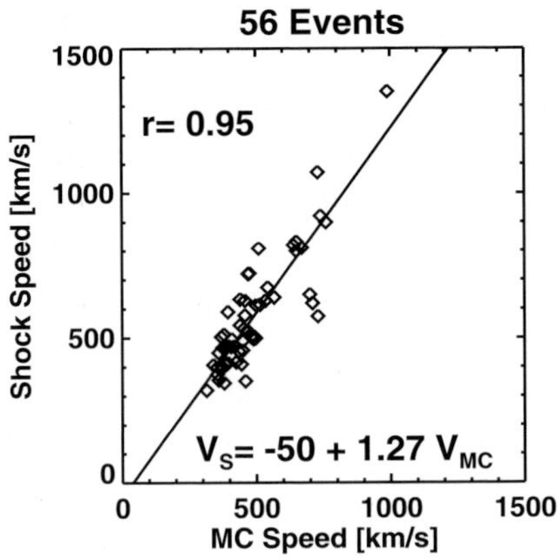

Figure 6. Scatter plot between MC speeds (V_{MC}) and the speeds of the associated shocks (V_S) for the period 1996–2003 from Gopalswamy *et al.* (2006).

enhancement behind IP shocks was recognized as the IP counterparts of CMEs (Borrini *et al.*, 1982). Howard *et al.* (1982) identified a halo CME observed by the Solwind coronagraph with an IP shock observed by ISEE-3 three days later at 1 AU. Sheeley *et al.* (1985) used quadrature observations to show that most of the IP shocks had associated white light CMEs. They also found that nearly half of these IP shocks had driver gas (ICME) behind them. Magnetic clouds have also been found to be related to prominence eruptions (e.g. Burlaga *et al.*, 1982; Wilson and Hildner, 1986; Marubashi, 1986; Bothmer and Schwenn, 1994; Rust, 1999) and flares (see, e.g., Bravo *et al.*, 1998). However, prominence eruptions form the inner core of white-light CMEs (Hori and Culhane, 2002; Gopalswamy *et al.*, 2003b) and eruptive flares are indicative of CMEs (Munro *et al.*, 1979). Thus these surface eruptive signatures are proxies to CMEs. With the simultaneous availability of white light images of CMEs and *in situ* observations, it has become possible to routinely identify the CME corresponding to each ICME. Similarly, when bright halo CMEs appear from close to the disk center, they are highly likely to result in an ICME 1–4 days later.

Unfortunately, there is no simple way to continuously observe CMEs evolving into ICMEs at present. The situation will change soon since the STEREO mission has been launched. Tracking type II radio bursts from the corona into the IP medium is one way of tracking the shocks ahead of CMEs, but not the CMEs. Interplanetary scintillation observations is another possible way, but what is tracked may not be the CME proper, but the disturbed region in the front (see, e.g., Manoharan *et al.*, 2006). MHD modeling can also bridge the gap between coronal and solar wind observations (see, e.g., Roussev and Sokolov, 2006). As the standoff distance of the shock from the CME corresponds to a separation of ~0.5 day, it is important to identify the substructures of an ICME for a proper understanding of the CME-ICME relationship. Gopalswamy (2003) proposed the following scenario for the relation between CMEs and ICMEs. Shocks in the corona associated with white light CMEs are well established, so associating them with the leading shocks in ICMEs is straightforward. The sheath behind ICMEs can be associated with the bright front of CMEs. There are some difficulties with this regarding the topology of the field lines in the sheath (closed or open). The magnetic cloud structure in ICMEs can be thought of as the evolved form of the coronal cavity seen as dark void in coronagraphic images. Recent observations clearly show the existence of flux rope structure in the void region (see, e.g., Chen *et al.*, 1997). Finally, as we noted above, the eruptive prominences are the slowest and most sunward feature of CMEs. Such features are rarely observed inside ICMEs, but occasionally show up at the bottom of MCs as cool material (Burlaga *et al.*, 1998; Gopalswamy *et al.*, 1998). Post-flare loops and arcades observed in H-alpha, X-ray and EUV are features anchored to the Sun and are indicative of a CME eruption. Thus it is possible that the four-part structure (shock, sheath, flux rope, and prominence) is maintained throughout the inner heliosphere. Of course one or more of the substructures may not be present in all CMEs. For example, if the CME is sub-Alfvenic, it will not drive a fast-mode shock.

4.1. CME Acceleration

While most of the early works concentrated on relating white-light CMEs to IP shocks, recent studies have focused on the connection between white light CMEs and ICMEs (Lindsay *et al.*, 1999; Gopalswamy *et al.*, 2000, 2001b, 2005b; Schwenn *et al.*, 2005). Lindsay *et al.* (1999) found a linear relation between CME and ICME speeds and confirmed that slow CMEs accelerate and fast CMEs decelerate as a result of interaction with the solar wind. Gopalswamy *et al.* (2000) quantified this acceleration using SOHO and Wind observations and improved it further using quadrature data from Solwind coronagraph and Pioneer Venus Orbiter (PVO) data (Gopalswamy *et al.*, 2001b). The interplanetary acceleration a (m/s^2) was found to be related to the CME speed V (km/s) in the corona as $a = -0.0054(V - 406)$. Thus CMEs with $V < 406$ km/s accelerate, while those with $V > 406$ km/s decelerate. CMEs have constant speed when $V = 406$ km/s. The critical speed of 406 km/s was identified with the solar wind speed. Figure 7 shows the effect of this IP acceleration for a set of 59 CME-MC pairs. These MCs form a subset of those in Figure 2 for which the corresponding white light CMEs were identified. Note that the white light CMEs have a much wider distribution of speeds than the MCs. The average MC speed is also considerably lower than the average white light CME speed. The IP acceleration was found to be the same whether they are MCs or ejecta. Figure 7 also shows that CMEs resulting in MCs are much faster than the solar wind on the average.

The acceleration can be used to determine the ICME speed and arrival time based on the white light CME speed. Figure 8 (left) shows a scatter plot of observed

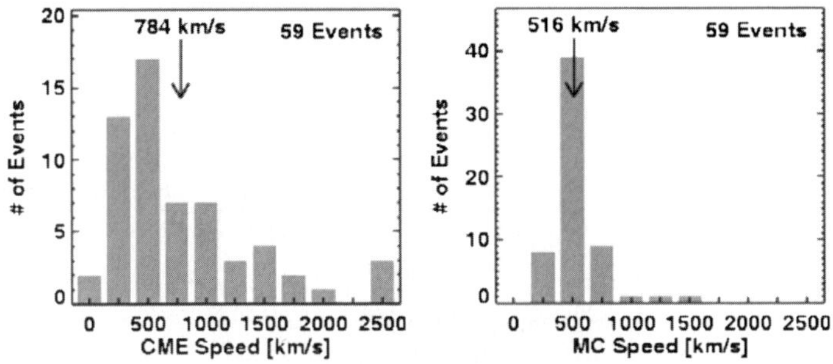

Figure 7. The speed distribution of CMEs (left) and ICMEs (right) for a set of 59 CME-ICME pairs. This is an updated version of the original figure in Gopalswamy *et al.* (2000).

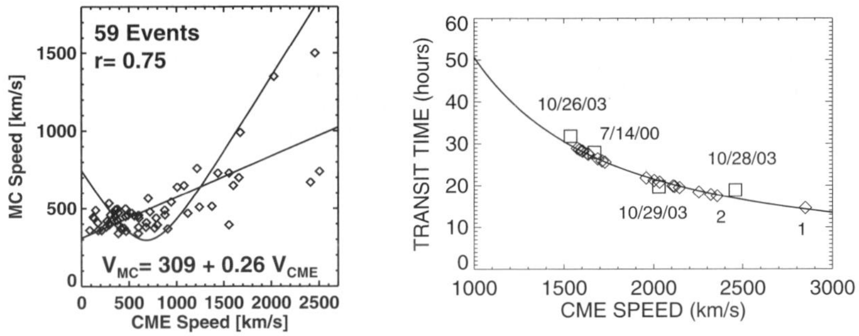

Figure 8. (left) CME and MC speeds for the 59 CME-MC pairs. The straight line is Lindsay *et al.* (1999) formula. The parabolic curve is from Gopalswamy *et al.* (2001b). (right) the empirical shock arrival model for (solid curve) with some very fast events (diamonds) including the Carrington events of 1859 September 1 (marked 1) and August 4, 1972 (marked 2). The squares are actual measurements of CME speeds from SOHO and the observed shock transit times.

MC and white light CME speeds, along with the final-speed curves according to Lindsay *et al.* (1999) and Gopalswamy *et al.* (2001b). The regression line ($V_{MC} = 309 + 0.26V_{CME}$) in Figure 8 is almost the same as the one obtained by Lindsay *et al.* ($V_{MC} = 360 + 0.25V_{CME}$). White light CMEs have been observed to have speeds exceeding 2500 km/s. For these CMEs, Lindsay *et al.* (1999) formula gives a final speed of only ~985 km/s. During the October November 2003 period, one ICME exceeded a speed of ~1500 km/s, while the progenitor CME had a speed of ~2500 km/s (Gopalswamy *et al.*, 2005c). For such high speed ICMEs, the parabolic curve suits better. It must be pointed out that the IP acceleration was obtained from quadrature observations, so the projection effects are minimal. The CME speeds used in Figure 8 are in the plane of the sky. Using space speeds should yield a better comparison. The IP acceleration can also be used to estimate the arrival time of ICMEs based on the speed and onset time at the Sun. The tight relationship between IP shocks and the driving ICMEs shown in Figure 6 has been used to extend the CME arrival model to shock arrival model shown in Figure 8 (right).

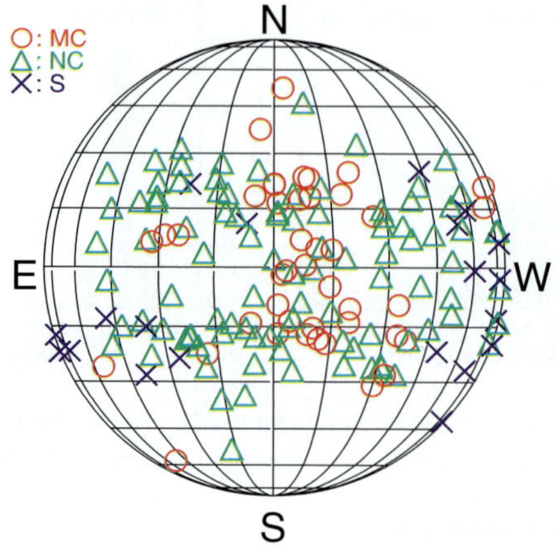

Figure 9. The solar source locations of CMEs associated with MCs (circles) and non-cloud ICMEs (NC – triangles) for the period 1996–2002 (inclusive). The non-cloud ICMEs were all driving shocks. The crosses indicate shocks (S) which have their driving CMEs propagating orthogonal to the Sun-Earth line, so the driver is not encountered by the observing spacecraft located along the Sun-Earth line. Note that the magnetic clouds generally originate from close to the disk center, shocks without drivers are due CMEs originating from close to the limb. The events were chosen based on the detection of *in situ* shocks at 1 AU.

Schwenn *et al.* (2005) have suggested a way to avoid the projection effects. They find that the radial speed ($V_{\rm rad}$) is related to the expansion speed ($V_{\rm exp}$) in the sky plane according to: $V_{\rm rad} = 0.88 V_{\rm exp}$.

Propagation of CMEs can also be affected by interaction with other CMEs (Gopalswamy *et al.*, 2001c) near the Sun (Manoharan *et al.*, 2004) and in the IP medium (Burlaga, 1995; Burlaga *et al.*, 2002). It was shown that the Sun-Earth transit time of ICMEs interacting with preceding CMEs is generally larger than those of isolated CMEs. At 1 AU, complex ejecta have been observed when successive CMEs interact and produce extended structures at 1AU, whose thickness may fill the entire Sun-Earth space.

4.2. Solar Sources

Figure 9 shows the solar sources of CMEs associated with MCs, shocks, and non-cloud ICMEs. The identification of non-cloud ICMEs is different from the CLs used above although there is significant overlap. The non-cloud ICMEs were compiled by starting with a list of IP shocks detected *in situ*, identifying the driver behind them (see Manoharan *et al.*, 2004), and eliminating known MCs from the list. The selection criterion excludes those non-cloud ICMEs that do not drive a shock, resulting in a slight underestimation of the number of non-cloud ICMEs. As for the MCs, we have included all the events that had corresponding white-light CMEs. The solar sources were identified from the location of the associated flare, filament

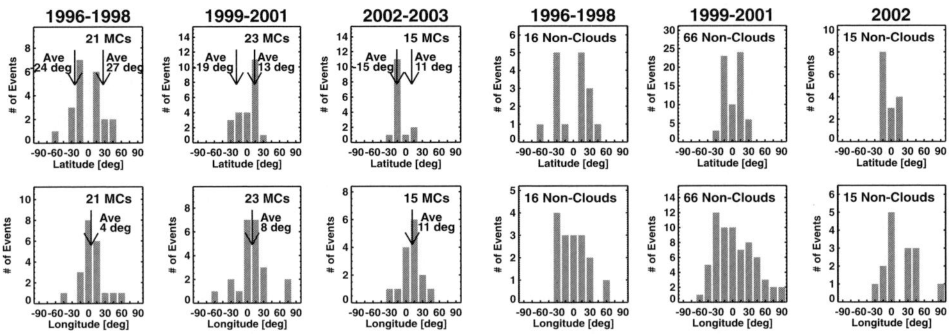

Figure 10. Latitude and longitude distributions of solar sources of MCs (left) and non-cloud events (right) for three phases of the solar cycle: minimum to rise phase (1996–1998), maximum phase (1999–2001) and the declining phase (2002–2003). Data from Gopalswamy *et al.* (2006).

eruption, post eruption arcades (X-rays and EUV), or EUV dimming signature. The non-cloud events also include a few shocks with no associated ICMEs, but did have associated white-light CMEs. These CMEs are heading in a direction predominantly orthogonal to the Sun-Earth line, so the ICMEs are not encountered by the observing spacecraft. One can see the following pattern in the source location of CMEs associated with IP events: (i) CMEs associated with MCs occur very close to the disk center with a slight westward bias, (ii) CMEs associated with non-cloud ejecta are scattered over a wide range of longitudes and have intermediate central meridian distances, and (iii) CMEs associated with just IP shocks (no driver gas) occur at large central median distances. Such a pattern supports the idea that all ICMEs are MCs, but the trajectory of the observing spacecraft can make an MC appear as ejecta without flux-rope structure, as illustrated in Figure 2. It must be pointed out that there are quite a few non-cloud MCs that originate close to the disk center. It is not clear whether these ICMEs have an inherent non-cloud structure (Gosling *et al.*, 1990) or they have cloud structure but the propagation results in sampling of the flux rope edges.

4.3. Solar Cycle Variation of CME Latitudes

The latitude and longitude distributions of the MC and non-cloud events are shown grouped into three phases of the solar activity cycle in Figure 10: the minimum phase (1996–1998), the maximum phase (1999–2001) and the declining phase (2002–2003). The longitude distributions generally peak close to the central meridian during all the three phases. The latitudinal distribution is dramatically different during different phases. During 1996–1998, the latitude distribution has a clear bimodal structure, corresponding to the active region belt in the north and south. The MCs originated predominantly in the northern hemisphere during the maximum phase, and in the southern hemisphere during the declining phase (2002–2003). It is not clear if this is due to the level of activity in the hemisphere or the global field has some influence. During the minimum phase, there is certainly the influence of the global solar dipole field which tends to move CMEs originating from higher latitudes towards the equatorial plane. This may be one of the reasons for a higher abundance of the MCs during the minimum phase and an under-abundance during the maximum phase, while the total number of ICMEs had a solar cycle variation similar to that of the general population of white light CMEs (see Wu *et al.*, 2006; Riley *et al.*, 2006).

It is well known that MCs originating from the northern (southern) hemisphere of the Sun have negative (positive) chirality (Bothmer and Schwenn, 1994). This pattern generally matches with the active region helicity and filament chirality (see, e.g., Pevtsov *et al.*, 2001). Since CMEs are expected to remove the accumulated helicity in the corona (see, e.g., Low, 1996), the hemispheric dependence of occurrence rate is expected to result in a preferred chirality of the MCs.

The latitude and longitude distributions of the non-cloud ICMEs have roughly the same solar cycle variation as the MCs, but less pronounced. One important difference is the number and distribution of non-cloud ICMEs during the maximum phase: the longitude distribution is much wider and shows pronounced western bias. The number of non-cloud events during the maximum phase is about 3 times larger than the number of MCs during the same phase. One possibility is that during solar maximum there is a large number of CMEs, but they are not channeled towards the equator (due to the weaker dipole field).

5. Composition Signatures

The charge state distributions of heavy ions observed in the interplanetary plasmas reflect the conditions in the coronal environment where the plasmas originate. This applies to both the ordinary solar wind and the ICMEs. The charge state composition can often distinguish the fast wind from slow wind (von Steiger and Zurbuchen, 2003) because they originate from different coronal regions (fast wind from coronal holes and slow wind from streamers). In fact the difference in charge state distributions is one of the important ways of identifying ICMEs against the background solar wind (see, e.g., Galvin, 1997; von Steiger *et al.*, 1995). Since the time scale for solar wind expansion is typically much shorter than that of ionization and recombination, the relative ionization states get frozen, maintaining the source values (Hundhausen *et al.*, 1968). Thus high ionization states are indicative of a hot source region, and one can construct the thermal history of the interplanetary plasmas by comparing the freezing-in temperatures of different charge state pairs.

5.1. High Charge States

Henke *et al.* (1998, 2001) studied the compositional signatures of more than 50 ICMEs observed by Ulysses and found that the O^{7+}/O^{6+} ratio was significantly

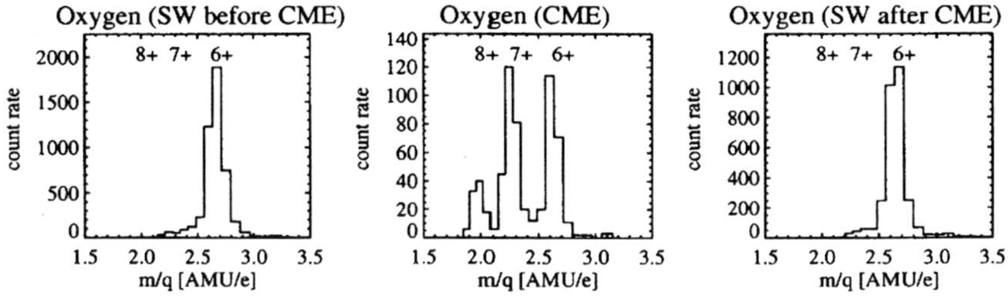

Figure 11. Ulysses observations of Oxygen charge states before, during, and after an ICME from Henke *et al.* (2001).

higher inside magnetic clouds (Xn + denotes the number density of atoms of an element X ionized n times). The correlation between the magnetic cloud structure and enhanced O^{7+}/O^{6+} ratio was confirmed for a much larger sample of ICMEs from the ecliptic (Aguilar-Rodriguez *et al.*, 2006). They found that the O^{7+}/O^{6+} and Mg^{10+}/O^{6+} ratios in magnetic clouds show a clear increase with respect to the ambient solar wind, whereas non-cloud ICMEs do not show such an increase. However, all types of ICMEs (MCs, non-cloud ejecta or complex ejecta) generally have a higher ionization states suggesting that ICMEs do originate from hot regions near the Sun. Figure 11 illustrates the dramatic increase in O^{7+} and O^{8+} ions within an ICME compared to the solar wind before and after the passage of the ICME. Similar effects could be seen in the charge state composition of other heavy elements such as Fe. In fact, Henke *et al.* (2001) found that almost all events with enhanced O^{7+}/O^{6+} ratio displayed an enhancement of the Fe^{12+}/Fe^{11+} ratio as well. Lepri *et al.* (2001) found that more than 90% of long-duration enhancements in Fe charge state in the solar wind were associated with ICMEs. However, the enhancements only occur in ~50% of ICMEs during 1998–2000. This again may be related to the requirement of magnetic cloud structure in ICMEs. Figure 12 shows a magnetic cloud event with distinct enhancement of O^{7+}/O^{6+} ratio and high Fe charge states only during the magnetic cloud interval. The magnetic cloud was driving a shock, but the sheath region has no charge state enhancement.

Reinard (2005) found that both O^{7+}/O^{6+} and Fe charge state (Q_{Fe}) are more likely to be enhanced for CMEs associated flares occurring close to the disk center. For a set of 67 ICMEs she studied, the average O^{7+}/O^{6+} ratio was 0.94, compared to the solar wind value (0.33) reported by Henke *et al.* (2001). When the central meridian events alone were considered, the average O^{7+}/O^{6+} ratio increased to 1.10. On the other hand, the average value for CMEs originating outside the disk-center region was only 0.45, close to the solar wind value. The average Q_{Fe} showed a similar behavior: the average values for central and non-central events were 12.7 and 10.5, respectively. The non-central value is very close to the solar wind values (10) obtained by Lepri and Zurbuchen (2004). It is well known that the CME plasma is observed at Earth only when the solar source is close to the disk center, typically within a central meridian distance of 30° (see, e.g., Gopalswamy *et al.*,

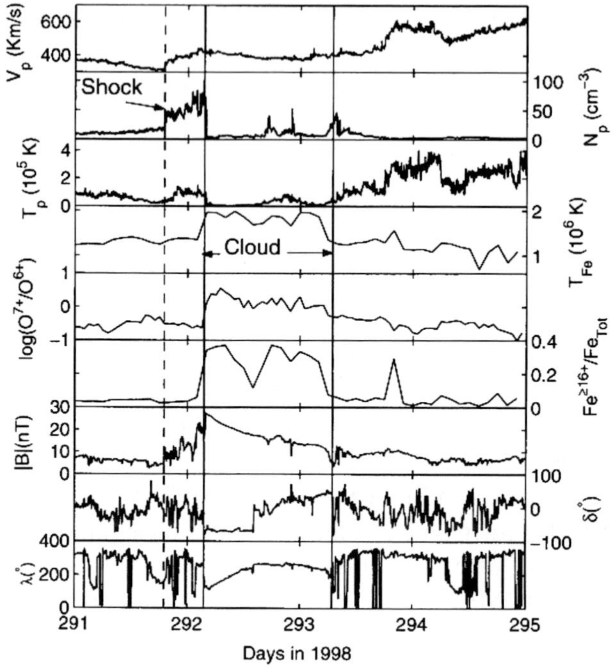

Figure 12. Enhancement of Fe charge state ($Fe^{\geq 16}/Fe_{tot}$), and O^{7+}/O^{6+} ratios at the time of a magnetic cloud during 1998 October 18–21. Also plotted are the proton velocity (V_p), density (N_p), temperature (T_p), Fe freeze-in temperature (T_{Fe}), magnetic field strength, latitude (δ), and longitude (λ). Note that the charge state enhancement occurs only during the MC interval. The Fe charge state enhancement is defined as the density ions with charge states Fe^{16+} and above relative to the density of all Fe ions from Lepri *et al.* (2001).

2000; Bravo and Blanco-Cano, 1998). This may be taken to indicate that only a spacecraft crossing along the center of the ICME will intersect the enhanced charge state ratios. Reinard (2005) also found a moderate correlation between flare size and the charge state ratio, further suggesting that flare heated material is trapped in the magnetic cloud structure.

5.2. Low Charge States

Magnetic clouds occasionally contain low charge states (Zwickl *et al.*, 1982; Galvin, 1997; Burlaga *et al.*, 1998; Gopalswamy *et al.*, 1998). Fe, O, and C charge states down to 5+, 2+, and 4+ , respectively were reported by Zwickl *et al.* (1982), but these events are extremely rare. Such low charge states are indicative of cool material. The inner cores of CMEs contain eruptive prominence, which is the coolest component (~8000 K) of the CME plasma at least initially. Gopalswamy *et al.* (1998) reported Fe charge states ranging from 5 to 11 in structures associated with the 1997 February 9 ICME. Towards the end of the ICME interval, a cold (proton thermal speed <20 km/s and dense (proton density ∼40 cm^{-3}) solar wind was observed by the WIND/MASS instrument. The presence of the low Fe^{5+} and Fe^{6+} in relatively high abundance ratio indicated that this plasma had originated from a relatively cold region in the corona, viz. the prominence. There were two other weaker pulses prior to the largest one and they are consistent with other prominence fragments ejected from under the same CME envelope.

5.3. Connection to Flares

It has been shown that only ∼10% of CMEs observed near the Sun are detected in the heliosphere, irrespective of their latitude of origin (Gopalswamy, 2004). These CMEs are generally faster and wider on the average and hence more energetic. There is also a reasonable correlation between the CME kinetic energy and the X-ray flare size (see, e.g., Hundhausen, 1997). Therefore, one can see that ICMEs observed in the heliosphere are generally associated with large flares. Such flares are eruptive in nature, meaning they are associated with mass motion or CMEs. Flare sites are the only locations in the solar atmosphere were temperatures exceeding 10 MK are observed. The flare sites are connected to the CMEs so one can expect hot plasmas within CMEs. In the standard model of eruptive events (the so-called CSHKP model), the reconnection beneath the prominence causes injection of the prominence and hot plasma into the CME flux rope. The prominence structure is typically much smaller than the overall CME volume, which may be the reason for the rare occurrence of the prominence-related solar wind at 1 AU.

The correlation between O^{7+}/O^{6+} and the flare size reported by Reinard (2005) can easily be explained by the CSHKP model. The high charge states require a plasma near the Sun, hotter than the quiet corona. Flare plasma can have temperatures exceeding 10 MK. However, there is a small problem: the temperature of the flare material inferred from the charge state ratios (both O^{7+}/O^{6+} and Q_{Fe}) is much lower (<6 MK) than the flare temperature (>10 MK) inferred from soft X-ray observations. One possible reason is that the hot plasma entering the MC from the reconnection region (see Figure 12) might cool due to the expansion of the MC before reaching a height of $\sim 4R_s$ where freezing in takes place. Another point to note is that the flare temperature is determined by the hot plasma detected in soft X-rays. This plasma is likely to have formed due to chromospheric evaporation due to energetic electron bombardment from the reconnection region. Similar situation does not exist in the upward direction.

One other place where higher temperature might exist is the shock down stream region (between the shock and the CME). However, no charge state enhancement if found in this region, again confirming the requirement of the magnetic topology. How about ICMEs originating close to the disk center, but not having magnetic cloud structure? These also seem to have enhanced charge state ratio. It is possible that these ICMEs also have MC structure, but the single point observation some how does not reveal it.

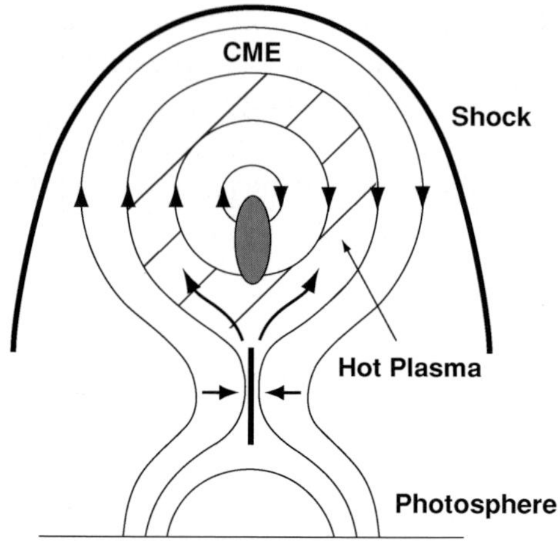

Figure 13. Model of a solar eruption adapted from Martens and Kuin (1989). The reconnection takes place at the current sheets (small vertical line above the closed field lines). The hot plasma trapped in the CME structure is indicated by the hashed region. The gray ellipse represents the prominence material. The solid curve at the top is the bow shock driven by the CME. The closed field region above the prominence is supposed to become the flux rope in the IP medium, to appear as in Figure 2.

6. ICMEs and Geomagnetic Storms

It has been established that ICMEs are the main source of major ($Dst < -100$ nT) geomagnetic storms (Gosling *et al.*, 1991; Bothmer and Schwenn, 1998; Tsurutani and Gonzalez, 1997; Webb *et al.*, 2000, 2002; Zhang *et al.*, 2003; Srivastava and Venkatakrishnan, 2004). CMEs on the Sun are intrinsically magnetic entities with large magnetic field strengths; they also compress any IMF at their leading regions when they travel through the interplanetary (IP) medium and interact with other IP CMEs (ICMEs) and/or the ambient solar wind driving IP shocks. Southward magnetic field (Bs) in shock sheaths and ICMEs contribute to the generation of the geomagnetic storms. As we noted before, the CMEs have to originate close to the disk center of the Sun to arrive at Earth before causing geomagnetic storms. Most of such CMEs are observed as halo CMEs. In fact, the properties of halo CMEs and geoeffective CMEs are nearly identical (Gopalswamy, 2006). This is because most of the front-side halos cause geomagnetic storms. While after-the-fact analysis shows the close connection between halo CMEs and geomagnetic storms, we are still far from predicting the occurrence and intensity of geomagnetic storms based on halo CME observations.

In order to illustrate the effect of the source location of CMEs on their geoeffectiveness, we consider a set of white-light CMEs from active region 0486 and the associated geomagnetic storms (see Figure 14). The first two CMEs originated from close to the disk center (and hence appear as full halos) and resulted in intense geomagnetic storms (–363 and –401 nT). The corresponding ICMEs were

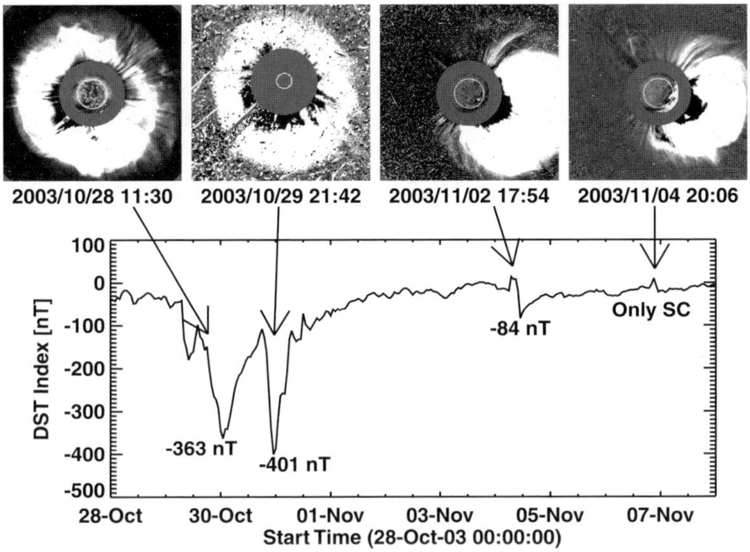

Figure 14. (top) Large CMEs from active region 0486 during October–November 2003 observed by SOHO/LASCO. The first two are symmetric full halos because the CMEs originated from close to the disk center. The last two CMEs occurred when the active region was close to the west limb, so they appear as partial halos. (bottom) A plot of the Dst index for the interval 2003 October 28 to November 7. The CMEs responsible for significant Dst events are shown by arrows from Gopalswamy *et al.* (2005c).

preceded by shocks and had magnetic cloud structure. On the other hand, the third CME is clearly a glancing blow to Earth as it is heading in the southwest direction. The CME was an asymmetric halo and only the eastern flank of the CME and the shock were encountered by Earth. The resulting geomagnetic storm had a sudden commencement due to shock impact followed by a weak storm due to the sheath. The last CME originated very close to the west limb and is unlikely to be encountered by Earth. The CME was one of the fastest of the cycle and was driving shock, yet the only thing happened was a sudden commencement. The first two CMEs correspond to track 3 in Figure 2 for a spacecraft along the Sun-Earth line. The second CME corresponds to track 2 while the last CME corresponds to track 1. Thus both the geoimpact and the identification of the ICME as an MC depend on the source location of the corresponding CME at the Sun.

An isolated magnetic storm typically lasts for ~1 day with a main phase of 3–12 hours and a recovery phase of ~14 hours (e.g., Tsurutani and Gonzalez, 1997). But geomagnetic storms are often complex, with a multiple-step decrease in Dst. These storms often have a longer duration and higher intensity. These complex geomagnetic storms (CGS) are sometimes referred to as multi-step storms or long-lived geomagnetic storms (LLGMS). High-speed streams (HSS) in corotating interaction regions (CIRs) cause only moderate to weak storms ($Dst > -100$ nT). CMEs on the Sun are intrinsically magnetic entities with large magnetic field strengths;

they also compress any IMF at their leading regions when they travel through the interplanetary (IP) medium and interact with other IP CMEs (ICMEs) and/or the ambient solar wind driving IP shocks. Southward magnetic field (Bs) in shock sheaths and ICMEs or magnetic clouds (MCs) contribute to the generation of the geomagnetic storms. While typical MCs have a south-to-north or a north-to-south structure, which ensures that the MCs are geoeffective either at the front or back, some MCs have high inclinations (see Mulligan *et al.*, 2001). The example given in Figure 1 has an axial field which is fully southward and produced the largest geomagnetic storm of cycle 23. In the opposite case of northward axial field, the MC is not geoeffective at all (Yurchyshyn *et al.*, 2001).

Burlaga *et al.* (2002) studied a set of fast ejecta observed at 1 AU from February 5, 1998 to November 29, 1999 and found that all MC events and two complex ejecta resulting from the interaction of multiple CMEs, produced geomagnetic storms. When HSS encounter and interact with CMEs, they can further compress Bz and enhance geoeffectiveness (e.g., Burlaga *et al.*, 1987; Burlaga, 1995; Gopalswamy *et al.*, 2005a). Thus, there are many sources of complexity in the IP medium, a systematic investigation of which will provide refinement to the existing knowledge on geoeffective solar events. Recently, Xie *et al.* (2006) studied a set of 37 complex storms with $Dst < -100$ nT and identified the associated white light CMEs. They found that ~65% of such storms are caused by successive CMEs. They also found that ~22% of the complex storms were caused by isolated CMEs while CIRs accounted for ~13%. The storm duration is well correlated with the number of participating CMEs (correlation coefficient $r = 0.78$). This study also found that more intense storms resulted when there is interaction among CMEs and/or with HSS events.

7. Summary

Interplanetary coronal mass ejections represent the most energetic events in the heliosphere originating from the closed field regions on the Sun. White-light CMEs observed by coronagraphs evolve into ICMEs interacting with the solar wind and other CMEs. Most ICMEs tend to attain the speed of the background solar wind although the very fast ones continue to have high speeds at 1 AU and beyond. ICMEs are responsible for the intense geomagnetic storms and large solar energetic particle events. The interplanetary magnetic field enhances significantly during the passage of ICMEs, typically increasing higher by a factor of 4 above the background value at 1 AU. Many CMEs drive shocks and continue to do so at 1 AU. Typically more shocks (than ICMEs) are observed by spacecraft along the Sun-Earth line, because the shocks are more extended than the driving ICMEs. Ions with high charge are observed in ICMEs, especially those with magnetic cloud structure. These ions are indicative of the higher temperature in the source region near the Sun. Such high temperatures prevail in the flaring regions, generally located beneath the erupting CMEs. ICMEs have to originate close to the disk center to have direct impact on Earth's magnetosphere. Most of the ICMEs originating close to the disk center seem to have flux rope structure. ICMEs that cannot be classified as magnetic clouds generally originate at larger central meridian distances than MCs. This suggests that all ICMEs may have cloud structure when viewed appropriately. During solar minimum, ICMEs have wide range of latitudes, but almost all ICMEs seem to be magnetic clouds. This may be due to the strong influence of the global dipolar field of the Sun, which forces the CMEs to move close to the ecliptic lane. Although the rate of magnetic clouds do not follow the CME rate as a function of the solar cycle, all ICMEs taken together have a rate that agrees with the solar cycle variation of white-light CMEs.

The current knowledge on ICMEs and their solar origin has been primarily obtained from two-point observations: near the Sun from coronagraphs near Earth using in-situ measurements. The coronagraphic observations give information on the mass distribution in CMEs. One has to use non-coronagraphic observations such as in EUV and X-rays to get additional information such as temperature. On the other hand *in situ* observations provide a lot more information, but only along the trajectory of the observing spacecraft through the ICME. Radio observations at long wavelengths and interplanetary scintillation observations are the only way to get information on CMEs throughout the inner heliosphere, but they provide information on the shocked plasma ahead of CMEs. Future observations from the Heliospheric Imager on board the STEREO observations will enable direct comparison of CMEs and ICMEs using Thomson scattering technique. One of the major hindrances to the understanding of CMEs is the lack of direct observation of magnetic properties of CMEs near the Sun. Most investigations use magnetograms to get information on the CME magnetic field. Microwave and infrared techniques show some promise, but they need further development before routine information becomes available.

Acknowledgements

I thank I. Schweizer for help with formatting of the manuscript. The author is indebted to S. Yashiro, S. Akiyama, and H. Xie in preparing several of the figures. This work was supported by NASA's SR&T and LWS TR&T programs.

References

Aguilar-Rodriguez, E., Blanco-Cano, X., and Gopalswamy, N.: 2004, *Adv. Space Res.* **38**(3), 522.

Balogh, A.: 2002, *ASP Conf. Ser.* **269**, 37.

Balogh, A., Gosling, J. T., Jokipii, J. R., Kallenbach, R., and Kunow, H. (eds.): 1999, *Space Sci. Rev.* **89**, 1.

Borrini, G., Gosling, J. T., Bame, S. J., and Feldman, W. C.: 1982, *J. Geophys. Res.* **87**, 7370.

Bothmer, V., and Schwenn, R.: 1994, *Space Sci. Rev.* **70**, 215.

Bothmer, V., and Schwenn, R.: 1998, *Ann. Geophys.* **16**, 1.

Bravo, S., and Blanco-Cano, X.: 1998, *Ann. Geophys.* **16**, 359.

Burlaga, L. F.: 1995, *Interplanetary Magnetohydrodynamics*, Oxford University Press.

Burlaga, L., Sittler, E., Mariani, F., and Schwenn, R.: 1981, *J. Geophys. Res.* **86**, 6673.

Burlaga, L. F.: 1982, *Geophys. Res. Lett.* **9**, 1317.

Burlaga, L. F., Lepping, R. P., and Jones, J. A.: 1990, *Geophys. Monogr. Ser.* **58**, 373.

Burlaga, L. F., Behannon, K. W., and Klein, L. W.: 1987, *J. Geophys. Res.* **92**, 5725.

Burlaga, L. F., *et al.*: 1998, *J. Geophys. Res.* **103**, 227.

Burlaga, L. F., Plunkett, S. P., and St. Cyr, O. C.: 2002, *J. Geophys. Res.* **107**, A10, 1266, doi:10.1029/2001JA000255.

Cane, H. V., and Richardson, I. G.: 2003, *J.Geophys. Res. A* **108**, 6.

Chen, J., *et al.*: 1997, *Astrophys. J.* **490**, L191.

Chao, J. K., and Lepping, R. L.: 1974, *J. Geophys. Res.* **79**, 1799.

Cliver, E. W., Ling, A. G., and Richardson, I. G. : 2003, *Astrophys. J.* **592**, 574.

Dryer, M.: 1994, *Space Sci. Rev.* **67**, 363.

Galvin, A. B.: 1997, in N. Crooker, J. A. Joselyn, and J. Feynman (eds.), *Coronal Mass Ejections*, AGU, Washington D.C., 253.

Gonzalez, W. D., Chía de Gonzalez, A. L., Dal Lago, A., Tsurutani, B. T., Arballo, J. K., Lakhina, G. K., *et al.*: 1998, *Geophys. Res. Lett.* **25**, 963.

Gopalswamy, N.: 2002, *COSPAR Colloq. Ser.* **14**, 157.

Gopalswamy, N.: 2003, *Adv. Space Res.* **31**, 869.

Gopalswamy, N.: 2004, in G. Poletto and S. T. Suess (eds.), *The Sun and the Heliosphere as an Integrated System*, Kluwer, Boston, p. 201.

Gopalswamy, N.: 2006, *J. Astrophys. Astron.* **27**, 243.

Gopalswamy, N., Hanaoka, Y., Kosugi, T., Lepping, R. P., Steinberg, J. T., Plunkett, S., *et al.*: 1998, *Geophys. Res. Lett.* **25**, 2485.

Gopalswamy, N., Lara, A., Lepping, R. P., Kaiser, M. L., Berdichevsky, D., and St. Cyr, O. C.: 2000, *Geophys. Res. Lett.* **27**, 145.

Gopalswamy, N., Lara, A., Kaiser, M. L., and Bougeret, J.-L.: 2001a, *J. Geophys. Res.* **106**, 25,261.

Gopalswamy, N., Lara, A., Yashiro, S., Kaiser, M. L., and Howard, R. A.: 2001b, *J. Geophys. Res.* **106**, 29,207.

Gopalswamy, N., Yashiro, S., Kaiser, M. L., Howard, R. A., and Bougeret, J.-L.: 2001c, *Astrophys. J.* **548**, L91.

Gopalswamy, N., Shimojo, M., Lu, W., Yashiro, S., Shibasaki, K., and Howard, R. A.: 2003a, *Astrophys. J.* **586**, 562.

Gopalswamy, N., Yashiro, S., Lara, A., Kaiser, M. L., Thompson, B. J., Gallagher, P. T., *et al.*: 2003b, *Geophys. Res. Lett.* **30**, CiteID 8015.

Gopalswamy, N., Yashiro, S., Michalek, G., Xie, H., Lepping, R. P., and R. A.: 2005a, *Geophys. Res. Lett.* **32**, 12.

Gopalswamy, N., Lara, A., Manoharan, P. K., and Howard, R. A.: 2005b, *Adv. Space Res.* **36**, 2289.

Gopalswamy, N., Yashiro, S., Liu, Y., Michalek, G., Vourlidas, A., Kaiser, M. L., *et al.*: 2005c, *J. Geophys. Res.* **110**, 9.

Gopalswamy, N., Fleck, B., and Gurman, J. B.: 2005d, in *Proc. Asia Pacific Regional Conference of IAA*, Bangalore, India.

Gopalswamy, N., Akiyama, S., Yashiro, S., Michalek, G., Xie, H., Petty, S., *et al.*: 2006, *J. Atmos. Sol. Terr. Phys.*, submitted.

Gosling, J. T.: 1990, *Geophys. Monogr. Ser.* **58**, 343.

Gosling, J. T.: 1997, in N. Crooker, J. A. Joselyn, and J. Feynman (eds.), *Coronal Mass Ejections*, AGU, Washington D.C., 9.

Gosling, J. T.: 1996, *Ann. Rev. Astron. Astrophys.* **34**, 35.
Gosling, J. T., McComas, D. J., Phillips, J. L., and Bame, S. J.: 1991, *J. Geophys. Res.* **96**, 731.
Henke, T., *et al.*: 1998, *Geophys. Res. Lett.* **25**, 3465.
Henke, T., Woch, J., Schwenn, R., Mall, U., Gloeckler, G., von Steiger, R., *et al.*: 2001, *J. Geophys. Res.* **106**, 10,597.
Hirshberg, J., Bame, S. J., and Robbins, D. E.: 1972, *Sol. Phys.* **23**, 467.
Hori, K., and Culhane, J. L.: 2002, *Astron. Astrophys.* **382**, 666.
Howard, R. A., Michels, D. J., Sheeley, Jr. N. R., and Koomen, M. J.: 1982, *Astrophys. J.* **263**, L101.
Hundhausen, A. J.: 1997, in N. Crooker, J. A. Joselyn, and J. Feynman (eds.), *Coronal Mass Ejections*, AGU, Washington D.C., 1.
Hundhausen, A. J., Gilbert, H. E., and Bame, S. J.: 1968, *Astrophys. J.* **157**, L3.
Klein, L. W., and Burlaga, L. F: 1982, *J. Geophys. Res.* **87**, 613.
Kunow, H., Crooker, N., Linker, J., Schwenn, R., and von Steiger, R. (eds.): 2006, *Space Sci. Rev.* **123**, 1, doi:10.1007/s11214-006-9007-z.
Lepping, R. P., Burlaga, L. F., and Jones, J. A.: 1990, *J. Geophys. Res.* **95**, 11957.
Lepping, R. P., Wu, C.-C., and Berdichevsky, D. B.: 2005, *Ann. Geophys.* **23**, 2687.
Lepri, S. T., Zurbuchen, T. H., Fisk, L. A., Richardson, I. G., Cane, H. V., and Gloeckler, G.: 2001, *J. Geophys. Res.* **106**, 29,23,2001.
Lepri, S. T., and Zurbuchen, T. H.: 2004, *J. Geophys. Res.* **109**, A01112.
Lindsay, G. M., Russell, C. T., Luhmann, J. G., and Gazis, P.: 1994, *J. Geophys. Res.* **99**, 11.
Lindsay, G. M., Luhmann, J. G., Russell, C. T., and Gosling, J. T.: 1999, *J. Geophys. Res.* **104**, 12,515.
Low, B. C.: 1996, *Sol. Phys.* **167**, 217.
Manoharan, P. K., Gopalswamy, N., Yashiro, S., Lara, A., Michalek, G., and Howard, R. A.: 2004, *J. Geophys. Res.* **109**, 6109.
Marsden, R. G., Sanderson, T. R., Tranquille, C., Wenzel, K.-P., and Smith, E. J.: 1987, *J. Geophys. Res.* **92**, 11,009.
Martens, P. C. H., and Kuin, N. P. M.: 1989, *Sol. Phys.* **122**, 263.
Marubashi, K.: 1986, *Adv. Space Res.* **6**, 335.
Mulligan, T., Russell, C. T., Elliott, D., Gosling, J. T., and Luhmann, J. G.: 2001, *Geophys. Res. Lett.* **28**, 891.
Munro, R. H., Gosling, J. T., Hildner, E., MacQueen, R. M., Poland, A. I., and Ross, C. L.: 1979, *Sol. Phys.* **61**, 201.
Neugebauer, M., and Goldstein, R.: 1997, in N. Crooker, J. A. Joselyn, and J. Feynman (eds.), *Coronal Mass Ejections*, AGU, Washington D.C., 245.
Parker, E. N.: 1957, *Astrophys. J. Supp.* **3**, 51.
Pevtsov, A. A., Canfield, R. C., and Latushko, S. M.: 2001, *Astrophys. J.* **549**, L261.
Pinter, S.: 1973, *Nature Phys. Sci.* **243**, 96.
Rao, U. R., McCracken, K. G., and Bukata, R. P.: 1967, *J. Geophys. Res.* **72**, 4325.
Reinard, A.: 2005, *Astrophys. J.* **620**, 501.
Riley, P., Schatzman, C., Cane, H. V., Richardson, I., and Gopalswamy, N.: 2006, in press.
Roussev, I. I., and Sokolov, I. V.: 2006, *Geophys. Monogr. Ser.* **162**, in press.
Russell, C. T., and Shinde, A. A.: 2005, *Sol. Phys.* **229**, 323.
Rust, D. M.: 1999, *Geophys. Monogr. Ser.* **111**, 221.
Schwenn, R., dal Lago, A., Huttunen, E., and Gonzalez, W. D.: 2005, *Ann. Geophys.* **23**, 625.
Sheeley, N. R., Howard, R. A., Koomen, M. J., Michels, D. J., Schwenn, R., Muhlhauser, K.-H., *et al.*: 1985, *J. Geophys. Res.* **90**, 163.
Sonnet, C. P., Colburn, D. S., Davis, L., Smith, E. J., and Coleman, P. J.: 1964, *Phys. Rev. Lett.* **13**, 153.
Srivastava, N., and Venkatakrishnan, P.: 2004, *J.Geophys. Res. A* **109**, 10103.
Tousey, R.: 1973, *Space Res.* **13**, 713.

Tsurutani, B. T., and Gonzalez, W. D.: 1997, *Geophys. Monogr. Ser.* **98**, 77.
von Steiger, R., Geiss, J., Gloeckler, G., and Galvin, A. B.: 1995, *Space Sci. Rev.* **72**, 71.
von Steiger, R., and Zurbuchen, T.: 2003, in *Proc. International Solar Cycle Studies (ISCS) Symposium*, Vol. 535 of *ESA SP*, p. 835.
Wang, Y. M., Ye, P. Z., Wang, S., Zhou, G. P., and Wang, J. X.: 2002, *J. Geophys. Res.* **107**, 1340, doi:10.1029/2002JA009244.
Webb, D. F., Cliver, E. W., Crooker, N. U., St. Cry, O. C., and Thompson, B. J.: 2000, *J. Geophys. Res.* **105**, 7491.
Wilson, R. M., and Hildner, E.: 1986, *J. Geophys. Res.* **91**, 5867.
Wu, Chin-Chun, Lepping, R. P., and Gopalswamy, N.: 2003, in *Proc. International Solar Cycle Studies (ISCS) Symposium*, Vol. 535 of *ESA SP*, p. 429.
Wu, Chin-Chun, Lepping, R. P., and Gopalswamy, N.: 2006, *Sol. Phys.*, in press.
Xie, H., Gopalswamy, N., Manoharan, P. K., Lara, A., Yashiro, S., and Lepri, S.: 2006, *J. Geophys. Res.* **111**, A01103.
Yurchyshyn, V., Wang, H., Goode, P. R., and Deng, Y.: 2001, *Astrophys. J.* **563**, 381.
Zhang, G., and Burlaga, L. F.: 1988, *J. Geophys. Res.* **93**, 2511.
Zhang, J., Dere, K. P., Howard, R. A., and Bothmer, V.: 2003, *Astrophys. J.* **6582**, 520.
Zwickl, R. D., Asbridge, J. R., Bame, S. J., Feldman, W. C., and Gosling, J. T.: 1982, *J. Geophys. Res.* **87**, 7379.

GEOEFFECTIVITY OF CORONAL MASS EJECTIONS

H. E. J. KOSKINEN[1,*] and K. E. J. HUTTUNEN[†]
University of Helsinki, Department of Physical Sciences, P.O. Box 64, FIN-00014 Helsinki, Finland
*(*Author for correspondence: E-mail: Hannu.E.Koskinen@helsinki.fi)*

(Received 18 July 2005; Accepted in final form 9 January 2006)

Abstract. Coronal mass ejections and post-shock streams driven by them are the most efficient drivers of strong magnetospheric activity, magnetic storms. For this reason there is considerable interest in trying to make reliable forecasts for the effects of CMEs as much in advance as possible. To succeed this requires understanding of all aspects related to CMEs, starting from their emergence on the Sun to their propagation to the vicinity of the Earth and to effects within the magnetosphere. In this article we discuss some recent results on the geoeffectivity of different types of CME/shock structures. A particularly intriguing observation is that smoothly rotating magnetic fields within CMEs are most efficient in driving storm activity seen in the inner magnetosphere due to enhanced ring current, whereas the sheath regions between the shock and the ejecta tend to favour high-latitude activity.

Keywords: coronal mass ejections, magnetic clouds, interplanetary shocks, magnetic storms, geomagnetic indices

1. Introduction

The first coronal mass ejection (CME) was detected using the white-light coronagraph on the OSO 7 satellite on December 14, 1971 (Tousey, 1973). The pioneering OSO 7 observations of 20 CMEs were followed by observations on Skylab, P78-1 (Solwind), and SMM in the 1970s and 1980s. The term "coronal mass ejection" describes the observation of a large amount of mass, some $10^{12} - 10^{13}$ kg, being detected in the corona to leave the Sun without implying any physical interpretation, e.g., of its origin.

The CMEs are often but not always found to be closely related to solar flares at a nearby region in the solar atmosphere. The first flare was observed on September 1, 1859 by Carrington (1859) and Hodgson (1859). For solar-terrestrial connections it is interesting that Carrington noted that a magnetic storm commenced about 17 hours and 40 min after the appearance of the flare. According to large number of different records, this event has remained in the top tier of solar-terrestrial events to the present times (Cliver and Svalgaard, 2004). Based on magnetic recording in Bombay, this storm has been claimed to be, by far, the strongest *Dst* storm during the last 150 years (Tsurutani *et al.*, 2003). Gradually, this and subsequent coincidences

[1]also at Finnish Meteorological Institute, Space Research Unit
[†]presently at University of California in Berkeley, Space Sciences Laboratory

Space Science Reviews (2006) 124: 169–181
DOI: 10.1007/s11214-006-9103-0

of solar flares and magnetic storms led to the concept that the solar flares indeed were the sources of strong magnetic activity on Earth (Hale, 1931).

The discovery of CMEs started slowly to change this picture. An increasing fraction of solar-terrestrial scientists began to appreciate the fundamental role of the CMEs and the interplanetary shocks as the drivers of major non-recurrent magnetospheric storms, but a large fraction of the magnetospheric and ionospheric community continued to refer to flares as the primary cause of magnetic storms. This continued until the 1990s, as noted by Gosling (1993) in his landmark article "The Solar Flare Myth".

A critical turning point was reached in January 1997. A group of scientists were following real-time data from SOHO instruments when they observed a bright halo around the occulting disk of the LASCO coronagraph on January 6, 1997, which they interpreted as an earthward-directed CME. They alerted the space weather service at NOAA of a possibly approaching storm, but as there were no other indications of an approaching storm, the alert was ignored. The interplanetary shock at 1 AU was observed by WIND on January 10 at 0010 UT. Thus the CME was rather slow and caused a medium-size storm in the magnetosphere ($Dst_{\mathrm{min}} = -78$ nT and $Kp_{\mathrm{max}} = 6$). What made this event so important was that it led to an accumulation of a large flux of relativistic electrons in the outer radiation belt, which was suggested to have been the cause of the loss of the geostationary telecommunication satellite Telstar 401 on January 11. Next day, this was in the news world wide, and the importance of CMEs finally penetrated throughout the magnetospheric research community. The event was discussed in a series of papers in a special issue of *Geophysical Research Letters* (vol 25(14), 1998).

After leaving the corona CMEs are called interplanetary CMEs (ICMEs) or ejecta. Signatures of ejecta have been discussed and summarized extensively, e.g., by Zwickl *et al.* (1983), Gosling (1990) and Neugebauer and Goldstein (1997). Magnetic field and solar wind plasma signatures of ICMEs include enhanced magnetic field strength and low variation of magnetic field, unusually low proton temperatures and bidirectional flux of suprathermal electrons. Plasma compositional anomalies, such as increased helium to proton ratio and enhancements of minor ions, also indicate the presence of an ICME. The above mentioned studies, however, point out that there is no unambiguous way to identify an ICME. Most ICMEs exhibit only some of the features and different signatures do not necessarily coincide spatially.

Magnetic clouds (Burlaga *et al.*, 1981) are an important subset of ejecta, in particular for our study, as they contain the strongest magnetic fields in the solar wind. Their signatures are the smooth rotation of the magnetic field direction over a large angle, large magnetic field intensity and low proton beta. The fraction of magnetic clouds of all ejecta varies with the phase of the solar cycle. During the ascending phase of solar cycle 23 almost all ejecta around solar minimum were magnetic clouds, whereas near maximum the fraction decreased below 20% (Richardson and Cane, 2003).

An ICME moving at supersonic (super-Alfvénic) speed with respect to the background solar wind produces a shock front ahead of it. The sheath regions between the shock and the ejecta form another class of highly geoeffective solar wind structures. The role of the shocks in Sun-Earth connections is twofold. First, shocks are efficient accelerators of protons and helium nuclei. The enhanced levels of energetic ions caused by both solar flares and CME-driven shocks provide hazardous environment not only near Earth but also elsewhere in the solar system. Second, from the magnetospheric viewpoint the most important feature of the ICMEs and shocks is their capability of shaking the entire magnetosphere through variable plasma pressure and by the dynamics related primarily to the direction of the interplanetary magnetic field (IMF). A fast plasma cloud can increase the dynamic plasma pressure by a large factor pushing the dayside magnetopause from its nominal distance of about $10 R_E$ inside the geostationary orbit at $6.6 R_E$. This can already be serious for spacecraft that are designed to remain inside the magnetosphere and, e.g., use the geomagnetic field for attitude control. However, as we will discuss below, the direction of the IMF is the most decisive factor.

2. Propagation from the Sun to the Earth

Even a relatively complete knowledge of CME properties in the corona is only the first step toward a reliable prediction of its geoeffectivity. We have very little observations on details of any specific ICME after it has left a coronagraph's field-of-view. An example of this is the disappearance of the bright kernel often seen in the coronagraph images. The kernel is likely cool matter from the erupted prominence. However, this structure is difficult to identify in plasma observations at 1 AU. A useful signature of prominence matter is high He^+/He^{++} ratio (Schwenn *et al.*, 1980; Zwickl *et al.*, 1983; Skoug *et al.*, 1999). However, such events are rare. According to Lynch *et al.* (2003) from the entire time period (before mid-2002) during which reliable composition data were available from WIND and ACE less than 10 events were observed with unusually low charge state ratios. A specific feature of the January 1997 event was very dense cool plasma near the trailing edge of the ICME, which may have been a remnant of the filament matter (Burlaga *et al.*, 1998; Webb *et al.*, 2000).

Observation of the evolution of an ICME is a formidable problem (Gopalswamy, 2006). With *in-situ* satellite observations we get data along a single line through a moving ICME. With more than one spacecraft the situation is improved, but owing to the very large size of the ICMEs the measurement points need to be well separated from each other. Radio wave observations in the megahertz range (e.g., WAVES onboard WIND observing at frequencies below 14 MHz) provide a global means of following the propagation of some CME-driven shocks into the solar wind (e.g., Reiner and Kaiser, 1999; Gopalswamy *et al.*, 2001). While this method gives valuable information of the early motion of some shocks, it has not yet matured to the level where it could be used for predicting the geoeffectivity of the shock and the CME.

3. Storms Seen in Geomagnetic Indices

The strength of magnetospheric perturbations is usually described using some of the several magnetic activity indices (Mayaud, 1980). Different indices emphasize different features of magnetic activity and one has to be careful when considering which aspects one wants to highlight when selecting an index to organize the data. Here we use two of the most popular indices in this context: *Dst* and *Kp*.

The *Dst*-index is a weighted average of the deviation from the quiet level of the horizontal (H) magnetic field component measured at four low-latitude stations around the world. The westward ring current flowing around the Earth at the distance of about 3–4 R_E is the main source of the *Dst*-index. During a magnetospheric storm the ring current is enhanced, which causes a negative deviation in H. Consequently, the more negative the *Dst*-index is, the stronger the storm is said to be. In the present discussion we set the threshold between weak and moderate storms to -50 nT in *Dst*, moderate storms range from -50 to -100 nT. Storms stronger than -100 nT are called intense and stronger than -200 nT big. The *Dst*-index is calculated once an hour. A similar 1-minute index derived from a partly different set of six low-latitude stations ($SYM{-}H$) is also in use.

However, the magnetometers used to calculate *Dst* are also influenced by current systems other than the ring current. High solar wind pressure pushes the magnetopause closer to the Earth both increasing the magnetopause currents and moving them closer to the observation point. Thus a pressure pulse impinging on the day-side magnetopause causes a positive deviation in the H-component. If the solar wind parameters are known, the effect can be cleaned away from the *Dst*-index (so-called pressure corrected *Dst*-index). More difficult is to estimate the effect of the dawn-to-dusk directed magnetotail current. During strong activity the tail current enhances significantly and the current sheet moves closer to the Earth, enhancing the *Dst*-index. It is a controversial issue in magnetospheric physics how to handle this effect

Another widely used index is the planetary K-index *Kp*. Each magnetic observatory has its own perturbation index K, which indicates the range of the maximum perturbation during a 3-hour interval. *Kp* is an average of the K-indices from 13 mid-latitude stations. *Kp* is expressed in a scale of one-thirds: 0, 0+, 1−, 1, ..., 8+, 9−, 9. As *Kp* is based on mid-latitude observations, it is more more sensitive to high-latitude auroral currents than is the *Dst*-index.

Figure 1 illustrates the evolution of the storm strength in a case where two CMEs were released successively from the same region on the Sun. The CMEs were observed as halos by LASCO on March 28, 2001, at 1300 UT and on March 29 at 1030 UT. The times refer to the first observation of the CME edge in the LASCO

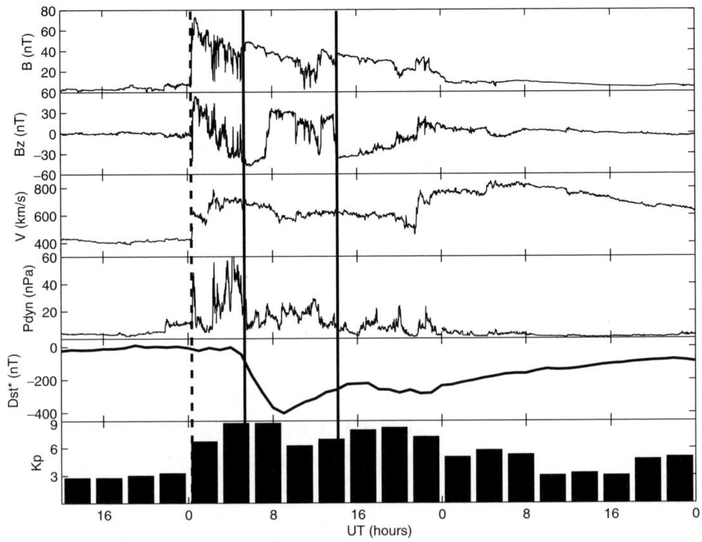

Figure 1. A double-peaked storm driven by two consecutive ICMEs. Upstream solar wind data from ACE are given in the top four panels: IMF intensity, IMF north-south component, plasma velocity, and dynamical pressure. The fifth panel gives the pressure corrected *Dst*-index and the lowermost panel the *Kp*-index. The shock is indicated by the dashed line and the arrivals of the magnetic clouds by the solid lines. The data interval is from 30 March, 2001, 12 UT to 1 April, 2001, 24 UT.

C2 coronagraph. A strong (X1.7) flare was associated with the latter (March 29 at 1015 UT). In such cases the latter ICME often has a higher velocity than the former and catches it on the way out to 1 AU. In this case the initial speed difference was estimated to be 350 km/s and the time interval between the arrival times at 1 AU had reduced to about 8.5 h.

In Figure 1 the shock (dashed line) preceding the first CME as well as the beginning of the smooth magnetic cloud structures (solid lines) are well visible. The end of the first CMEs is more unclear. Note that the total magnetic field remains at a relatively high level. Thus the Alfvén speed is large and no clear shock structure before the second magnetic cloud can be identified.

These ICMEs led to the third biggest storm of cycle 23: $Kp_{\max} = 9-$ and $Dst_{\min} = -387$ nT. This was an example of a double-peaked storm. More typically the first *Dst*-minimum of a double-peaked storm is due to the sheath region and the second to the ejecta (Kamide *et al.*, 1998).

This event illustrates that there is no one-to-one correlation between the storm strengths given by *Kp* and *Dst*. The *Kp*-activity reached storm level soon after the interplanetary shock had hit the magnetopause. Note that the *Dst*-index in Figure 1 is given pressure-corrected, which has removed the shock effect on its trace. A strong *Kp*-storm was in progress while the flow and magnetic field were fluctuating in the sheath region for several hours before the main phase of the

Dst-storm commenced. This happened slightly before the arrival of the magnetic cloud, which is characterized by the smooth evolution of the IMF Z-component. After the first cloud the *Dst*-index started to recover slowly as did the *Kp*-index. While the *Dst*-storm was recovering, the solar wind parameters became again more variable in the region between the two magnetic clouds and the *Kp*-index started to enhance again, whereas the second minimum in *Dst* was reached only after the arrival of the second cloud.

In the following we define a magnetic storm as sheath-associated if 85% of the *Dst* minimum occurred while the dayside magnetosphere was embedded in the CME sheath region. For a magnetic cloud-associated storm we require that during a magnetic cloud passing the magnetosphere *Dst* reached the intense storm level of -100 nT and that for two-step storm developments the time difference between two consecutive *Dst* minima was at least 3 hours. Note that the impact of a pre-existing ring current population is not particularly important in creating a stronger intensification later in a storm (Kozyra *et al.*, 2002).

Huttunen *et al.* (2002) investigated the difference of the *Kp*- and *Dst*-responses to different solar wind drivers during 1996–1999. They found that the fast post-shock streams and sheath regions had a relatively stronger effect on the *Kp*-index, whereas the effects of ejecta favoured the *Dst*-index. This tendency was emphasized further in by Huttunen and Koskinen (2004) who compared the evolution of several magnetic indices (*Dst*, $SYM{-}H$, $ASY{-}H$, AE, and *Kp*) during magnetic cloud and sheath region storms. The difference in the response of magnetic indices is most clear when sheath regions and magnetic clouds, i.e., not all ejecta, are compared because solar wind dynamic pressure and the magnetic field direction behave most differently within these structures.

In Figure 2 we have collected the maximum *Kp* and minimum *Dst* indices of all intense storms ($Kp_{\max} \geq 7-$ or $Dst_{\min} < -100$ nT) during 1997–2003 that we could associate clearly with a sheath region or with a magnetic cloud. From these data it is evident that most of the large *Kp* storms were sheath storms as well as all large *Kp* – smaller *Dst* events, whereas large *Dst* – smaller *Kp* events were mostly associated with magnetic clouds.

We do not yet have a complete explanation what causes this difference in the index response. Our hypothesis is that it is related to the fact that *Kp* is more sensitive to auroral zone current systems than *Dst*. This is supported by our investigation of the storm response in the $SYM{-}H$ and auroral electrojet indices (Huttunen and Koskinen, 2004), where we illustrated using four sample events that the high *Kp*-activity really was due to strongly enhanced auroral activity and not just an artifact produced by the procedure to derive the *Kp* index.

This leads us to propose the following scenario: The irregularities in the sheath region cause perturbations in the magnetopause low-latitude boundary layer, which enhance the Region 1 current system that couples to the auroral current systems in the ionosphere. Consequently the auroral activity is enhanced, which shows up more strongly in *Kp* than in *Dst*. On the other hand, the smooth rotation of the magnetic

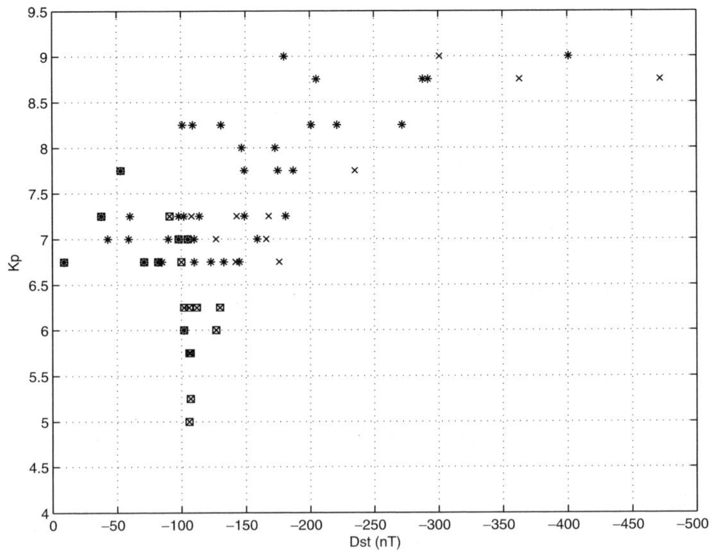

Figure 2. *Kp*- and *Dst*-indices for all intense storms ($Kp_{max} \geq 7-$ or $Dst_{min} < -100$ nT) during 1997–2003 that we could associate clearly with a sheath region (asterisks) or with a magnetic cloud (crosses). The rectangles identify storms that did not fulfil the intense storm condition: $Kp \geq 6-$ over three 3-h intervals, by Gosling *et al.* (1991).

cloud field does not cause the same effect on the high-latitude current systems but strengthens the large-scale magnetospheric convection which leads to ring current build-up and enhanced *Dst*-effect. Note that this is not an either-or question, as we deal with relatively large storms where both ring current and auroral current systems are activated. The large-scale convection also enhances the auroral currents. In fact, we have not encountered *Dst*-storms without significant high-latitude activity but there are examples of *Kp*-storms with very weak *Dst*-response. Several examples of this are seen in Figure 2.

Gosling and McComas (1987) suggested that the draping of the magnetic field lines around the ejecta could cause prolonged periods of southward IMF, which might be an important factor in simulating geomagnetic activity. The importance of the sheath regions as efficient storm drivers was demonstrated by Tsurutani *et al.* (1988), but their significance not been widely appreciated before the more extensive analyses of *in situ* observations from solar cycle 23 (Wu and Lepping, 2002; Huttunen *et al.*, 2002).

Huttunen and Koskinen (2004) showed that 45% of 53 intense ($Dst < -100$ nT) storms were caused by a sheath region (i.e., the sheath caused at least 85% of the *Dst* depression). When the threshold was changed to $Dst < -150$ nT, already 60% of the remaining storms were sheath-driven (Figure 3). The number of events is too small to make statistical conclusions of this feature, but the study clearly shows the importance of sheath regions as storm drivers.

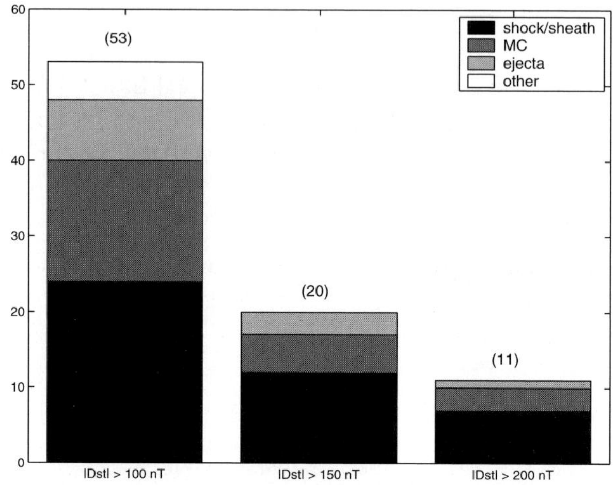

Figure 3. Drivers of intense *Dst*-storms (from Huttunen and Koskinen, 2004).

4. What Determines the Geoeffectivity of ICMEs

The concept of reconnection was introduced to magnetospheric physics by Dungey (1961) who proposed that the magnetospheric convection was driven by magnetic reconnection on the dayside magnetopause. A few years later (Fairfield and Cahill, 1966) demonstrated a statistical correlation between the direction of the IMF and the intermittent magnetospheric activations observed best in the auroral zone, the substorms.

In reconnection theory the reconnection rate is expressed in terms of the reconnection electric field, which in this case is $E = vB_s$, where v is the solar wind speed and B_s the southward component of the IMF. A fast ICME can enhance both factors. The velocities of the shock, the ICME and the post-shock streams can easily be more than twice the background solar wind speed. The IMF in the sheath region between the shock and the ejecta is strongly compressed. If the IMF ahead of a fast ICME already has a southward component, the shock increases it typically by a factor of 3–4. This way the sheath region can drive a storm even if the ICME itself does not hit the magnetosphere. The southward IMF component may be further amplified by draping of the magnetic field around the ICME (Gosling and McComas, 1987), which can lead to a southward IMF component even in cases where the pre-existing IMF is slightly northward. This mechanism works also when the ICME is too slow for shock formation.

Finally, the strongest long-lasting southward magnetic fields in the solar wind are found within magnetic clouds that according to their definition exhibit flux-rope structure determined by the eruptive magnetic structure on the Sun. For example, on November 20, 2003, the north-south magnetic field component of an ICME reached -53 nT, which is an order of magnitude more than typical IMF at 1 AU.

The south component of the IMF is the most critical parameter to determine the geoeffectivity of an ICME. High speed and pressure also perturb the magnetosphere, but they do not cause large-scale storm development if the IMF does not turn to the south. Depending on the background solar wind conditions and on the magnetic structure of the ICME a large number of alternative storm evolutions can take place (Tsurutani *et al.*, 1988; Wu and Lepping, 2002; Zhang *et al.*, 2004; Huttunen *et al.*, 2005).

If the shocked IMF has northward orientation, the sheath region is not yet geoeffective and the storm main phase will not begin until a southward IMF arrives with the magnetic cloud. On the other hand, if the background IMF has even a small southward component, it is enhanced in the sheath and in case of strong enough southward IMF the sheath can drive a strong storm alone.

In cases of well-defined flux-ropes the orientation of the flux-rope axis and the direction the magnetic field is wound varies (Bothmer and Schwenn, 1994). If the inclination of the flux-rope from the ecliptic plane is small, the magnetic structure is bipolar and can arrive with northward (NS) or southward (SN) magnetic field ahead, which obviously give different evolution for the storm. For example, if a southward sheath field is followed by an NS-type cloud with sufficiently strong and long-lasting southward IMF, a double-peaked *Dst*-storm or even two separate storms may follow. Double- or multiple-peaked storms may also take place if several CMEs have been released toward the Earth from the same active region on the Sun.

A flux-rope can also have a large inclination with respect to the ecliptic. In such cases the IMF can have a unipolar orientation either northward (N) or southward (S) throughout the passage of the flux-rope. In the northward case the ICME will most likely pass the Earth with only minor perturbations, whereas the continuously southward case may lead to a really strong storm.

Figure 4 shows the results of an analysis of 73 magnetic cloud events identified in Wind and ACE observations during solar cycle 23 (Huttunen *et al.*, 2005). Unipolar southward clouds always caused at least a medium-size storm ($Dst < -50$ nT), whereas in northward cases only sheath regions caused storms. Note that about one third of the bipolar, either NS or SN, clouds did not lead to a medium-size or larger storm, which is a problem for forecasters. These results are consistent with those by Wu and Lepping (2002) based on 34 magnetic clouds in WIND data during 1995–1998 and by Zhang *et al.* (2004) based on 104 magnetic clouds in ACE data from January 1998 to April 2002. The slight differences in relative percentages likely are due to different selection criteria for a magnetic cloud. Note that Zhang *et al.* (2004) included weak storms (-30 nT $> Dst > -50$ nT) in their study.

5. The Strongest Storms of Cycle 23

Statistically the Earth's magnetic environment is most active somewhat before or after the sunspot maximum. In October 2003, when the overall solar activity was

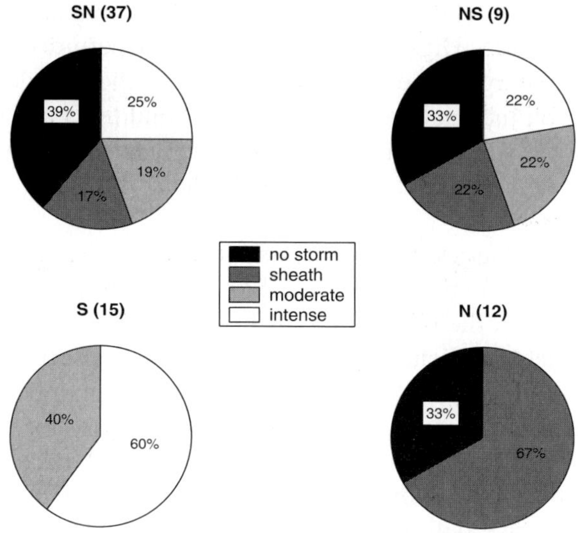

Figure 4. The effect of the flux rope type to the geoeffectivity. Numbers in the parenthesis indicate the total number of magnetic clouds in each category. Color codes are: black – no medium-size or larger storms ($Dst > -50$ nT), dark gray – sheath region storm, light gray – moderate magnetic cloud storm, white – intense magnetic cloud storm (From Huttunen *et al.*, 2005).

descending, a period of very strong activity took place (the so-called Halloween events). Several big flares, the largest one reaching X17 on October 28 with an associated halo CME, led to very disturbed conditions. The maximum *Kp* was 9 and the minimum of the (preliminary) *Dst*-index was –401 nT.

The energetic particle fluxes were so intense that several spacecraft, including SOHO, were switched to the safe mode in order to protect sensitive electronics. Large number of satellite anomalies were reported and the effects reached all the way to the ground. For example, in the Gothenburg region in Sweden the electric distribution system went down and anomalously strong geomagnetically induced current was observed in the Finnish natural gas pipeline system.

The source region on the Sun remained active and erupted again one week later on November 4, 2003, now on the western limb. This was the strongest recorded X-ray flare (X28) so far. The consequences of this event did not reach the Earth and we will never know how severe the following storm would have been.

After these magnificent events the strongest *Dst* storm of cycle 23 came as a little surprise. A large filament disappeared from the solar disk between 0740 and 0800 UT on November 18, 2003. At the same time two medium size (M-class) flares were detected at 0752 and 0831 UT and two fast CMEs were identified in LASCO images (0806 UT, 1223 km/s; 0850 UT, 1660 km/s).

However, there was no indication that a particularly strong event would be expected until the shock arrived at 1 AU on November 20, 2003, at 0727 UT (Figure 5). The shock was followed by a particularly strong magnetic cloud. The

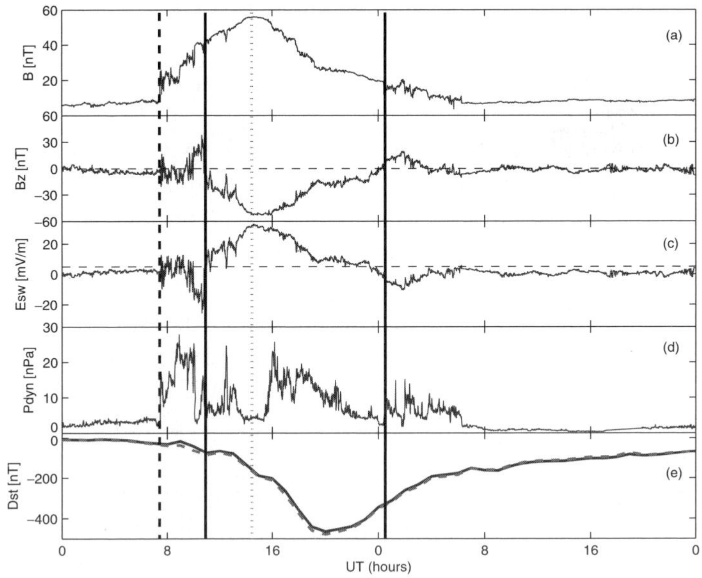

Figure 5. Solar wind parameters from ACE and the *Dst*-response during the November 20 storm. The panels from top to bottom are: IMF intensity, IMF north-south component, east-west component of the solar wind electric field ($E = -v_x B_z$), dynamic pressure and the *Dst* index. In the *Dst* panel also the pressure corrected *Dst* is given by the dashed line. The vertical solid lines indicate the time ACE was within the magnetic cloud (From Huttunen *et al.*, 2005).

maximum magnetic field intensity was about 60 nT and the southward component reached 53 nT. In addition the plasma density of the cloud was large. The resulting storm intensity was 9– in the *Kp* index (interval 15–21 UT) and $Dst_{\text{min}} = -472$ nT (at 20 UT; preliminary value).

The maximum *Kp* or minimum *Dst* are not the only ways of categorizing storms. For practical space weather purposes also the duration of the storm is of great interest. In this sense the Halloween storm was certainly more severe than the November 20 event. During the Halloween storm the *Kp* index was 9 during 9 hours, 9– during another 9 hours and remained above 5– for 60 hours (2.5 days), whereas the November 20 storm was at its highest *Kp* level only 6 hours.

6. Discussion and Conclusions

The different response of different magnetospheric activity indices to sheath regions and ejecta as storm drivers (Huttunen *et al.*, 2002; Huttunen and Koskinen, 2004) is quite interesting. We interpret it to be due to different response of the auroral and ring current systems to the different drivers, as the drivers are structurally quite different. Most notably the magnetic field of a magnetic cloud has a smooth structure of a flux-rope whereas the large-scale field of the sheath is quite unsteady, but it is unclear how these differences are transferred to the magnetosphere. There have been some attempts to associate upstream solar wind fluctuations with the magnetospheric activity but with rather modest success (e.g., Borovsky and Funsten, 1998).

For practical forecasting purposes we need to learn more what determines the CME parameters on the Sun and whether or not it will be possible to recognize them in advance. To reliably predict the magnetic structure of earthward directed ICMEs from solar observations has shown to be a quite challenging task (e.g., McAllister *et al.*, 2001). Also the evolution and propagation of the ICMEs must be better understood. The present propagation time estimates based on halo CME data yield large errors in estimated shock arrival times. Further timing uncertainty is due to the fact that in some cases the storm commences already soon after the shock arrival but in other it does not happen until the rear end of the magnetic cloud reaches the magnetosphere (northward B in the sheath and a NS-type magnetic cloud).

The upcoming STEREO mission of NASA will be very important. The CMEs will be possible to follow visually longer distances from the Sun than earlier and the stereoscopic view will give much better data of the three-dimensional structure of individual CMEs. It is also advantageous to be able to see the Earthward directed CMEs from an angle, as the intensity of halo observations is weaker causing large uncertainties to the initial velocity determination. The *in situ* observations on both sides of the Earth will lead to much better information of the global structure of ICMEs hitting the Earth. On the other hand, when the two STEREO spacecraft will have moved too far from the Earth, they will no longer encounter the solar wind affecting the Earth's magnetosphere. Thus it will be of utmost importance to maintain also the *in situ* upstream observations.

Acknowledgements

We are grateful to R. Schwenn and V. Bothmer for collaboration in our CME studies on LASCO data. The support from the solar-terrestrial physics and space weather groups at the Finnish Meteorological Institute is warmly acknowledged. Much of the results presented here were obtained through the SWAP project supported by the Academy of Finland within the space research programme Antares. The data plots in this review are based on data available in the CDAWeb (Wind and ACE) and at the WDC C2 in Kyoto. The organizers and participants of the very inspiring ISSI workshop are also gratefully acknowledged.

References

Borovsky, J. E., and Funsten, H. O.: 2003, *J. Geophys. Res.* **108**(A6), 1246, doi: 10.1029/2002JA009601.

Bothmer, V., and Schwenn, R.: 1994, *Space Sci. Rev.* **70**, 215.

Burlaga, L., Fitzenreiter, R., Lepping, R., Ogilvie, K., Szabo, A., Lazarus, A., *et al.*: 1998, *J. Geophys. Res.* **103**(A1), 277.

Burlaga, L., Sittler, E., Mariani, F., and Schwenn, R.: 1981, *J. Geophys. Res.* **86**, 6673.
Carrington, R. C.: 1859, *Mon. Not. R. Astron. Soc.* **XX**, 13.
Cliver, E. W., and Svalgaard, L.: 2004, *Solar Phys.* **224**, 407.
Dungey, J. W.: 1961, *Phys. Rev. Lett.* **6**, 47.
Fairfield, D. H., and Cahill Jr., L. J.: 1966, *J. Geophys. Res.* **71**, 155.
Gopalswamy, N., Yashiro, S., Kaiser, M. L., Howard, R. A., and Bougeret, J.-L.: 2001, *J. Geophys. Res.* **106**(A12), 29219.
Gopalswamy, N.: 2006, *Space Sci. Rev.*, this volume, doi: 10.1007/s11214-006-9102-1.
Gosling, J. T.: 1990, in C. T. Russell, E. R. Priest, and L. C. Lee, (eds.), *Physics of Magnetic Flux Ropes*. Geophys. Monogr. **58**, 3518.
Gosling, J. T.: 1993, *J. Geophys. Res.* **98**, 18937.
Gosling, J. T., and McComas, D. J.: 1987, *Geophys. Res. Lett.* **14**, 335.
Gosling, J. T., McComas, D. J., Phillips, J. L., and Bame, S. J.: 1991, *J. Geophys. Res.* **96**, 7831.
Hale, G. E.: 1931, *Astrophys. J.* **73**, 379.
Hodgson, R.: 1859, *Mon. Not. R. Astron. Soc.* **XX**, 15.
Huttunen, K. E. J., and Koskinen, H. E. J.: 2004, *Annales Geophysicae* **22**, 1729.
Huttunen, K. E. J., Koskinen, H. E. J., and Schwenn, R.: 2002, *J. Geophys. Res.* **107**(A7), doi: 10.1029/2001JA900171.
Huttunen, K. E. J., Schwenn, R., Bothmer, V., and Koskinen, H. E. J.: 2005, *Annales Geophysicae* **23**, 625.
Kamide, Y., Baumjohann, W., Gonzalez, W., Tsurutani, B. T., Daglis, I. A., Brekke, A., *et al.*: 1998, *J. Geophys. Res.* **103**, 6917.
Kozyra, J. U., Liemohn, M. W., Clauer, C. R., Ridley, A. J., Thomsen, M. F., Borovsky, J. E., *et al.*: 2002, *J. Geophys. Res.* **107**(A8), doi: 10.1029/2001JA000023.
Lynch, B. J., Zurbuchen, T. H., Fisk, L. A., and Antiochos, S. K.: 2003, *J. Geophys. Res.* **108**(A6), doi: 10.1029/2002JA0009591.
Mayaud, P. N.: 1980, *Geophys. Monogr, 22*, AGU, Washington, D.C.
McAllister, A. H., Martin, S. F., Crooker, N. U., Lepping, R. P., and Fitzenreiter, R. J.: 2001, *J. Geophys. Res.* **106**, 29185.
Neugebauer, M., and Goldstein, R.: 1997, in N. Crooker, J. A. Joselyn, and J. Feynman, (eds.), *Coronal Mass Ejections*, Geophys. Monogr. **99**, 245.
Reiner, M. J., and Kaiser, M. L.: 1999, *J. Geophys. Res.* **104**(A8), 16979.
Richardson, I. G., and Cane, H. V.: 2003, *J. Geophys. Res.* **108**(A4), 1356, doi: 10.1029/2002JA009817.
Skoug, R. M., Bame, S. J., Feldman, W. C., Gosling, J. T., McComas, D. J., Steindber, J. T., *et al.*: 1999, *Geophys. Res. Lett.* **26**(2), 161.
Schwenn, R., Rosenbauer, H., and Mühlhäuser, K.-H.: 1980, *Geophys. Res. Lett.* **7**, 201.
Tousey, R.: 1973, *Space Res.* **XIII**, 713.
Tsurutani, B. T., Gonzalez, W. D., Tang, F., Akasofu, S. I., and Smith, E. J.: 1988, *J. Geophys. Res.* **93**, 8519.
Tsurutani, B. T., Gonzalez, W. D., Lakhina, G. S., and Alex, S.: 2003, *J. Geophys. Res.* **108**(A7), 1268, doi: 10.1029/2002JA009504.
Webb, D. F., Cliver, E. W., Crooker, N. U., St. Cyr, O. C., and Thompson, B. J.: 2000, *J. Geophys. Res.* **105**, 7491.
Wu, C. C., and Lepping, R. P.: 2002, *J. Geophys. Res.* **107**(A10), 1314, doi: 10.1029/2001JA000161.
Zhang, J., Liemohn, M. W., Kozyra, J. U., Lynch, B. J., and Zurbuchen, T. H.: 2004, *J. Geophys. Res.* **109**, A09101, doi: 10.1029/2004JA010410.
Zwickl, R. D., Asbridge, J. R., Bame, S. J., Feldman, W. C., Gosling, J. T., and Smith, E. J.: 1983, in M. Neugebauer, (ed.), *Solar Wind Five*, NASA Conf. Publ., 2280, 711.

RING CURRENT DYNAMICS

IOANNIS A. DAGLIS
Institute for Space Applications and Remote Sensing, National Observatory of Athens, Penteli, GR-15236 Athens, Greece
(E-mail: daglis@space.noa.gr)

(Received 13 December 2005; Accepted in final form 6 June 2006)

Abstract. This chapter reviews the current understanding of ring current dynamics. The terrestrial ring current is an electric current flowing toroidally around the Earth, centered at the equatorial plane and at altitudes of ~10,000 to 60,000 km. Enhancements in this current are responsible for global decreases in the Earth's surface magnetic field, which have been used to define geomagnetic storms. Intense geospace magnetic storms have severe effects on technological systems, such as disturbances or even permanent damage of telecommunication and navigation satellites, telecommunication cables, and power grids. The main carriers of the ring current are positive ions, with energies from ~1 keV to a few hundred keV, which are trapped by the geomagnetic field and undergo an azimuthal drift. The ring current is formed by the injection of ions originating in the solar wind and the terrestrial ionosphere into the inner magnetosphere. The injection process involves electric fields, associated with enhanced magnetospheric convection and/or magnetospheric substorms. The quiescent ring current is carried mainly by protons of predominantly solar wind origin, while active processes in geospace tend to increase the abundance (both absolute and relative) of O^+ ions, which are of ionospheric origin. During intense geospace magnetic storms, the O^+ abundance increases dramatically. This increase has been observed to occur concurrently with the rapid intensification of the ring current in the storm main phase and to result in O^+ dominance around storm maximum. This compositional change can affect several dynamic processes, such as species-and energy-dependent charge-exchange and wave-particle scattering loss.

Keywords: Magnetic storms, ring current substorms, magnetosphere-ionosphere coupling, particle acceleration, inner magnetosphere, plasma sheet

1. Introduction

This chapter summarises the current understanding of ring current dynamics, with an emphasis on recent advances through the analysis of spacecraft observations and their combination with modelling and simulation studies.

The structure of the chapter is as follows:

– Brief historical outline
– Basic properties and structure of the ring current
– Origin of ring current particles
– Ring current formation
– Ring current evolution and decay
– Concluding remarks

Space Science Reviews (2006) 124: 183–202
DOI: 10.1007/s11214-006-9104-z

2. Brief Historical Outline

The trapped energetic particles in the inner magnetosphere were the first component of magnetospheric plasma, and space plasma in general, to be discovered at the very beginning of the space era. The milestone discovery was made through the measurements of Geiger-Mueller tubes of the group of James Van Allen on board the Explorer I spacecraft (Van Allen *et al.*, 1958). Van Allen correctly interpreted those measurements as the result of intense corpuscular radiation (Van Allen, 1959).

The foundations, however, of modern magnetospheric research had already been laid in the 1930s by Chapman and Ferraro (1930, 1931), in their effort to explain the great disturbances in the Earth's surface magnetic field that begin nearly simultaneously all over the Earth and reduce the daily mean value of the horizontal intensity. Chapman and Ferraro (1930, 1931) proposed that a transient stream of outflowing solar ions and electrons reached the Earth and leaked into the magnetosphere. The charged particles would then drift around the Earth, creating a current whose field would oppose the main geomagnetic field. This is amazingly close to what we know today. The only key element of the theory that has changed is the existence of a continuous rather than transient stream of ionized gas from the Sun.

This stream was later named "solar wind" by Eugene Parker (1958) and its existence was confirmed from observations made by the Venus-heading Mariner 2 spacecraft (Neugebauer and Snyder, 1962; Snyder and Neugebauer, 1964).

Hence, the principal idea of the Chapman and Ferraro theory was that the physical reason for the characteristic storm-time magnetic perturbations on the Earth's surface is a huge "ring current" in space circling the Earth. This idea was further elaborated by Singer (1956, 1957) and was eventually confirmed by *in situ* spacecraft measurements, starting with Explorer I.

The ability of the geomagnetic field to trap charged particles was experimentally verified by the Argus experiment, which was proposed by Nicholas C. Christofilos in 1957 and carried out in 1959 (Christofilos, 1959). Christofilos, an unconventional Greek scientist who had been working as an engineer designing elevator systems in Athens before migrating to the US in 1953, had actually communicated to the US Army in the early 1950s that many charged particles, due to the dipole magnetic field, could be trapped around the Earth. He further proposed that an artificial radiation belt, due to beta decay, could be created by exploding a nuclear bomb at high altitude. This proposal evolved into Argus – the very first active experiment in space.

Measuring the intensity of the ring current, or quantifying its effects through an appropriate index, also has its own history. Based on the work of Moos (1910), Chapman (1919) demonstrated the global effects of geospace magnetic storms and named the storm-time variation of the horizontal component of the geomagnetic field "*Dst* variation", meaning "disturbance storm-time". The characteristic average variation of *Dst* led Chapman to regard the storm geomagnetic variations as a unity with a beginning, middle, and end, leading to the classic graphical representation of the measurable ring current effects on the surface of the Earth.

In accordance with Singer's and Chapman's ideas on the reason for the magnetic perturbations on the Earth's surface, the *Dst* index was conceived as measure of the ring current. The concept was based on the assumption that the global decrease of the geomagnetic H-component is solely due to an external westward electric current system (the ring current), which encircles the Earth symmetrically (Akasofu and Chapman, 1961). As we will discuss in Section 6, this paradigm has been questioned both by spacecraft observations and by simulations: the storm-time ring current is often highly asymmetric in the main phase and becomes symmetric only in the late recovery phase (Daglis *et al.*, 2003).

3. Basic Properties and Structure of the Ring Current

The ring current can be envisioned as a toroidal-shaped electric current that flows westward around the Earth, with variable density at geocentric distances between $\sim 2R_E$ and $\sim 9R_E$. Although all geomagnetically trapped charged particles in the inner magnetosphere contribute to this current, it's the ions in the medium-energy range of $\sim$10 keV to a few hundreds of keV that substantially contribute to the total current density, as we will see later on. Electrons contribute little to the ring current due to their negligible energy density.

Geomagnetically trapped charged particles, which gyrate around and bounce along the ambient field line due to the Lorentz force (see Figure 1), are also subject to drift motions due to the gradient and curvature of the magnetic field (Baumjohann and Treumann, 1996). In the guiding centre approximation, the instantaneous position of a charged particle is broken down into its cyclotron motion (gyration) around the magnetic field line and the displacement of the centre of this motion (called the guiding centre) relatively to the magnetic field – which in turn

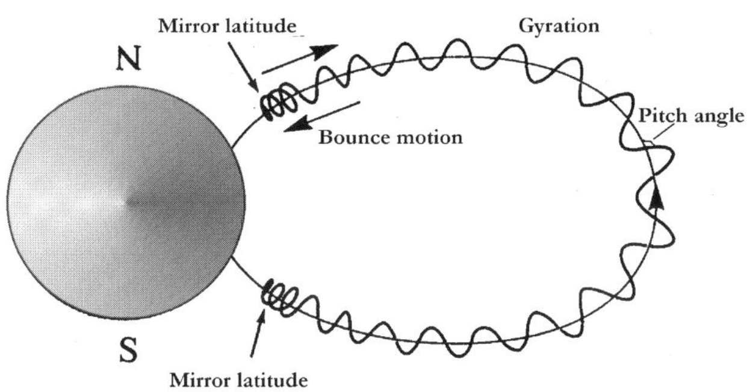

Figure 1. Cyclotron motion (gyration) and bounce motion of a charged particle along a geomagnetic field line.

consists of two components, a bounce motion back and forth along the field line between "mirror points" and a slow drift motion around the Earth. The large difference in the respective time scales of these motions makes a mathematical separation possible and leads to the consideration of the adiabatic invariants.

Adiabatic invariants are quantities that can be considered as constants of the particle motion under the condition that variations of the magnetic field are relatively small when compared to the spatio-temporal scales associated with the particle motion. There are three adiabatic invariants applicable to charged particles that are trapped in the geomagnetic field, and there is a one to one correspondence to the three motions mentioned above (Northrop, 1963).

The first adiabatic invariant or magnetic moment μ of the particle is equal to:

$$\mu = \frac{mv_{\perp}^2}{2B} \tag{1}$$

The first adiabatic invariant is estimated by calculating the magnetic moment produced by a current equivalent to the cyclotron motion of the particle. μ is a constant of the cyclotron motion in spatially and temporally varying magnetic fields B, as long as the rate of change of B is smaller than the gyrofrequency ω_c of the particle, and as long as their gyroradius is comparable to or smaller than the magnetic field line curvature radius.

The second adiabatic invariant is equal to:

$$J = \oint p_{\|} dl = 2 \int_{m1}^{m2} mv_{\|} dl \tag{2}$$

where $p_{\|}$ is the component of the particle momentum along the magnetic field and the integral is along the field line. J is associated with the bounce motion of the particle along field lines and between mirror points m_1 and m_2 in a region of converging magnetic field. The gradient of **B** parallel to the field direction produces the force driving the particle away from regions of increasing field strength. J is invariant as long as the electromagnetic field variations have frequencies ω much smaller than the bounce frequency ω_b.

The third adiabatic invariant is associated with the particle's azimuthal drift around the Earth. This drift is due to magnetic field gradients and field line curvature. In a pure dipole magnetic field it will produce drift surfaces that are figures of revolution of lines of force around the dipole axis.

As the particle bounces, it also drifts around the Earth, moving on a closed three-dimensional drift shell around the magnetic field axis. The third adiabatic invariant is the conserved magnetic flux encircled by the particles' periodic drift shell orbits:

$$\Phi = \oint v_d r d\psi \tag{3}$$

where v_d is the sum of all perpendicular drift velocities, ψ is the azimuthal angle, and the integration is taken over a complete drift path of the particle (Daglis *et al.*,

1999b). Φ is invariant when the frequency ω of electromagnetic field variations is much smaller than the drift frequency ω_d.

The total effect is a collective azimuthal drift, which is oppositely directed for ions and electrons: electrons move eastward and most ions (with energies above a relatively low threshold; see De Michelis *et al.*, 1997) move westward. This drift constitutes a net charge transport; the current associated with the charge transport is the ring current.

The elementary currents $\mathbf{j}_G$, $\mathbf{j}_\nabla$ and $\mathbf{j}_c$, resulting from the gyration, the gradient drift motion, and the curvature drift motion, can be expressed in terms of particle pressure perpendicular ($P_\perp$) and parallel ($P_\parallel$) to the magnetic field, as first established by Parker (1957).

The current due to gyration effects within the particle distribution is

$$\mathbf{j}_G = \frac{\mathbf{B}}{B^2} \times \left(\nabla P_\perp - \frac{P_\perp}{B} \nabla B - \frac{P_\perp}{B^2} (\mathbf{B} \cdot \nabla) \mathbf{B} \right) \tag{4}$$

The current due to particle drift driven by the magnetic field gradient is

$$\mathbf{j}_\nabla = P_\perp \frac{\mathbf{B} \times \nabla B}{B^3} \tag{5}$$

while the current due to particle drift driven by the magnetic field curvature is expressed as

$$\mathbf{j}_C = \frac{P_\parallel}{B^4} \mathbf{B} \times (\mathbf{B} \cdot \nabla) \mathbf{B} \tag{6}$$

The three terms on the right side of (Equation 4) represent currents due to the particle pressure gradient, the magnetic field gradient and the magnetic field line curvature, respectively. Because the drift and gyration terms driven by the magnetic field gradient (in Equations 4 and 5) are equal and opposite, the total current does not depend on gradients of the magnetic field:

$$\mathbf{j} = \mathbf{j}_\nabla + \mathbf{j}_C + \mathbf{j}_G = \frac{\mathbf{B}}{B^2} \times \left(\nabla P_\perp + \frac{P_\parallel - P_\perp}{B^2} (\mathbf{B} \cdot \nabla) \mathbf{B} \right) \tag{7}$$

In the case of an isotropic ($P_\parallel = P_\perp$) distribution or a straight magnetic field line geometry, the magnetic field configuration plays no role in the current build-up. The current system is then established only by particle pressure gradients.

The quiet-time ring current population is distributed over the L-range $\sim$2 to 9, with average current density values of $\sim$1 to 4 nA/m^2 (De Michelis *et al.*, 1997). The storm-time ring current density increases over its entire radial extent, and may exceed current density values of $\sim$7 nA/m^2 (Lui *et al.*, 1987).

Dessler and Parker (1959) and Sckopke (1966) showed that the disturbance ΔB of the equatorial surface geomagnetic field during geospace magnetic storms is proportional to the energy of the ring current particles:

$$\frac{\Delta B}{B_0} = \frac{2E}{3E_m} \tag{8}$$

where B_0 is the average surface geomagnetic field intensity at the magnetic equator (~0.3 Gauss), E is the total energy of the ring current particles and $E_m = B_0^2 R_E^3/3 \simeq 10^{18}$ J, is the energy of the Earth's dipole field above the Earth's surface.

The generalized Dessler-Parker-Sckopke (DPS) relation includes moreover terms from internal sources and boundary sources:

$$\frac{B_D}{B_0} = \frac{2E + M - \oint \mathbf{R} \cdot \mathbf{n} d\sigma}{3E_m} \tag{9}$$

where B_D is the field decrease due to the combined magnetic field from all sources: the ring current, the magnetopause current and the magnetotail current (Daglis *et al.*, 1999b). M is the total magnetic energy inside the magnetosphere, that is the volume integral of $B_D^2/2\mu_0$. The unit vector $\mathbf{n}$ is the outward pointing normal, and $\mathbf{R}$ is

$$\mathbf{R} = \left(p + \frac{B^2}{2\mu_o}\right)\mathbf{r} + \rho(\mathbf{V} \cdot \mathbf{r})\mathbf{V} - \frac{(\mathbf{B} \cdot r)\mathbf{B}}{\mu_0} \tag{10}$$

In (Equation 10) (Daglis *et al.*, 1999b) p is the thermal pressure, $\mathbf{r}$ is the radius vector from the center of the Earth, ρ is the mass density of the solar wind in the plasma mantle, $\mathbf{V}$ is the flow velocity of the mantle plasma, and $\mathbf{B}$ is the total magnetic field vector (dipole field plus magnetospheric field). The pressure is assumed to be isotropic.

The DPS formula considers only the cross-magnetic field currents in the magnetosphere ($j_\perp$), while any field-aligned or ionospheric closure currents from $\nabla j_\perp$ were omitted from its derivation. In other words, the DPS formula is based on the assumption of a nondivergent, symmetric ring current. As we will see later on in Section 6 this does not necessarily hold.

Theoretical and observational work in the past three decades have established the general location and the driving forces of this current system. Successive spacecraft measurements confirmed the existence of the ring current and showed that it is a permanent, though fluctuating in its intensity feature. Early measurements indicated that the ring current is dominated by ions (presumably protons) with energies of about 50 keV (Frank, 1967). The development of novel mass-resolving space instrumentation in the 1970s made it possible to distinguish between ionic species in space (Shelley *et al.*, 1972). However, the detailed composition and energy of the ring current were not clarified until the Active Magnetospheric Particle Tracer Explorer (AMPTE) mission of the late 1980s. A few years later, the Combined Release and Radiation Effects Satellite (CRRES) confirmed and expanded the conclusions of the AMPTE mission.

4. Sources of Ring Current Particles

The plasma sheet population, which is the main source of ring current particles, is supplied by the ionosphere and the solar wind. Hence, the ultimate main sources of ring current particles are the solar wind and the terrestrial ionosphere. Another source, the plasmasphere, may be involved as well, but this will not be discussed here, because this possibility has not been assessed in detail. The plasma sheet population is, in turn, supplied ultimately by the ionosphere and the solar wind. Hence, we will consider that the two ultimate major sources of the ring current are the terrestrial ionosphere and the solar wind. Electrons contribute little to the ring current due to their negligible energy density (Baumjohann, 1993).

As already mentioned, until the late 1980's, the scarceness of ion composition measurements for the important ring current energy range left uncertainties regarding the sources of ring current ions. During the first two decades of space research, the solar wind was the practically undisputable particle source of the hot magnetospheric population, with only two notable exceptions (Dessler and Hanson, 1961; Axford, 1970). The solar origin paradigm persisted despite the discovery of precipitating energetic O^+ ions in the early 1970's (Shelley *et al.*, 1972).

Several studies during the 1970s and the 1980s (Geiss *et al.*, 1978; Peterson *et al.*, 1981; Strangeway and Johnson, 1984) demonstrated that magnetospheric hot plasma contains a significant component of O^+. However, the conclusive composition measurements covering the whole energy range important for the storm-time ring current, were performed by the AMPTE mission. AMPTE demonstrated that magnetospheric O^+ ions originating in the ionosphere are important terrestrial agents in the storm-time geospace (Krimigis *et al.*, 1985; Hamilton *et al.*, 1988; Daglis *et al.*, 1993).

The next breakthrough was achieved through measurements by the Combined Release and Radiation Effects Satellite (CRRES), which operated around the 1990 solar maximum. Daglis (1997a) demonstrated that in all large geospace magnetic storms of the CRRES lifetime, ionospheric O^+ was the dominant ion species in the ring current. Hamilton *et al.* (1988) had found an O^+ dominance in the inner ring current only, which could be partly due to the known earthward gradient of O^+ density, for just one intense storm that was observed by AMPTE near the solar minimum of 1985. Daglis (1997a) demonstrated that O^+ dominated not only in the inner ring current, but also in the outer ring current for all large storms in 1991. Furthermore, Daglis (1997a) showed that the more intense the storm is, the more dominant is the O^+ contribution to the ring current. Moreover, Daglis (1997a) showed that the Dst index magnitude and the O^+ contribution to the ring current increase concurrently. This feature was present in all moderate to large storms during 1991, and it was also noticed in the storm of February 1986 by Hamilton *et al.* (1988).

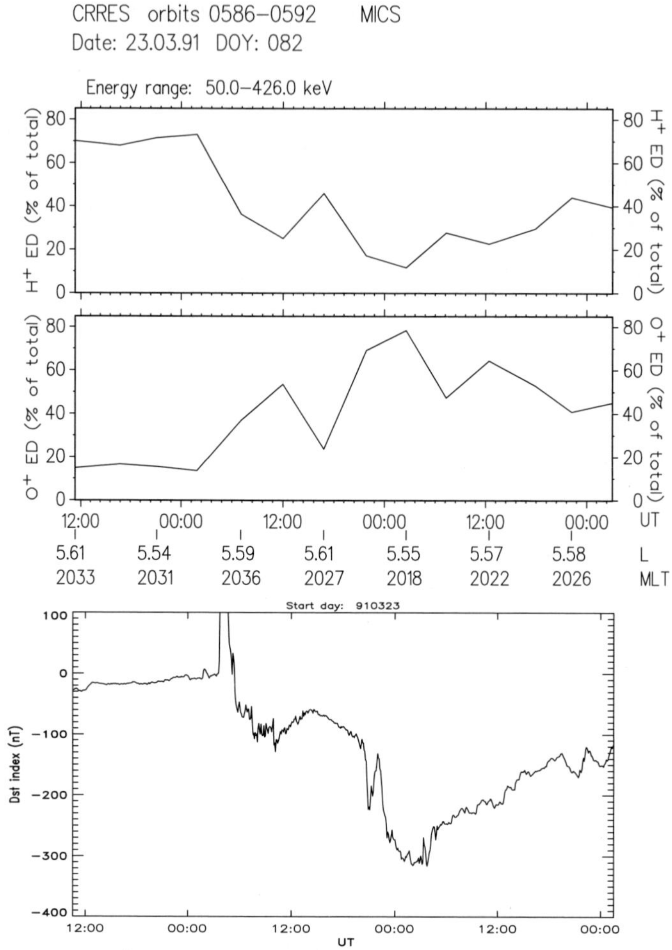

Figure 2. The great magnetic storm of March 24, 1991. From top to bottom: time profile of the H^+ and O^+ contribution to the total ion energy density in the ring current region at $L = 5-6$ (as measured by CRRES-MICS), and time profile of 5-min *Dst* index. The main features to be observed are the dominance of O^+ during storm maximum, as well as the concurrent increase of the $|Dst|$ level and of the O^+ contribution to the total ion energy density (adapted from Daglis *et al.*, 1999a).

During the great storm in March 1991, the increase of the O^+ abundance was overwhelming (Figure 2). O^+ clearly dominated around the maximum of the main phase. Its contribution was more than 65% of the total energy density in the L-range 5 to 7, while it exceeded 80% in the L-range 5 to 6 (Daglis *et al.*, 1999a). Daglis (1997a) emphasized on the especially interesting concurrent increase of the O^+ contribution and the *Dst* magnitude. After the storm sudden commencement at 0341 UT of March 24, the O^+ contribution rose from the ~10% level to the ~40% level, while at the same time *Dst* dropped to about –100 nT. A period of transient

Dst recovery and O^+ decrease followed. During the main phase of the storm both *Dst* and O^+ reached their peaks. A similar pattern had been observed for the great storm of February 1986 (Hamilton *et al.*, 1988). Figure 2 shows that the O^+ contribution remained at an extraordinary high level (above 40%) for a rather long time period (in fact more than 30 hrs).

It is noteworthy that MICS was underestimating the O^+ contribution to the total energy density, because of its high energy threshold of 50 keV. With the manifestation of storm-time dominance of ionospheric O^+ by the CRRES observations, the source issue can be considered to be resolved (cf. Table I, Daglis *et al.*, 1999b). We now know that typical values of O^+ contribution to the ring current energy density are ~6% during quiet time and more than ~50% during large storms.

However, there is still a remaining uncertainty regarding the source of H^+. Magnetospheric H^+ ions originate both in the ionosphere and in the solar wind; this complicates the identification of the dominant source. For the quiet-time ring current, Gloeckler and Hamilton (1987) had estimated that ~35% of H^+ ions in the outer ring current (L = 5–7) and ~75% of H^+ ions in the inner ring current (L = 3–5) are of ionospheric origin. For the storm-time ring current Gloeckler and Hamilton (1987) estimated that ~30% of H+ ions in the outer ring current and ~65% of H^+ ions in the inner ring current are of ionospheric origin. Solar wind H^{++} ions usually contribute less than ~4% of the ring current, except in the case of great storms.

5. Ring Current Formation

The overall picture of ring current formation and intensification is understood quite well. Ions originating in the solar wind and the ionosphere are transported through the magnetotail and the plasma sheet (Möbius *et al.*, 1987; Baker *et al.*, 1996) and directly from the ionosphere (Kaye *et al.*, 1981; Daglis *et al.*, 1991; Grande *et al.*, 1992; Daglis *et al.*, 1994) into the inner magnetosphere during time intervals of enhanced convection and during substorms. However, there is an ongoing dispute as to how important the substorm transport of ions is for the ring current build-up.

The basic transport and acceleration process for ions moving from the magnetotail and the plasma sheet to the inner magnetosphere is the $\mathbf{E} \times \mathbf{B}$ drift imposed by the large-scale electric field in the night-side magnetosphere. In the magnetotail, ions presumably gain energy while they move from regions of weaker to stronger magnetic field, conserving their first adiabatic invariant. While approaching the inner magnetosphere, the ions are transported across magnetic field lines primarily by gradient and curvature drift, as well as by $\mathbf{E} \times \mathbf{B}$ drift in a complicated combination of potential and induction electric fields. The large-scale potential electric fields in the night-side magnetosphere are due to prolonged southward interplanetary magnetic fields, while the impulsive induced electric fields are due to magnetic field reconfigurations during substorms. The substorm-effects dispute mentioned in the beginning of the section is linked to the relative importance of the large-scale convection electric field and the substorm-associated impulsive electric fields in the energisation and transport of ions into the ring current.

There have been numerous studies either proposing or opposing the view that substorms play an active and important role in the ring current build up. Sydney Chapman and Syun-Ichi Akasofu thought of substorms as being the key elements of a magnetic storm and thus named them "sub-storms" to suggest this idea (Akasofu and Chapman, 1961; Chapman, 1962). In their picture, substorms have the role of magnetic pumps, each one inflating the inner magnetosphere with hot plasma: during substorm expansion, induction electric fields accelerate magnetospheric particles and inject them into the inner magnetosphere, where they become trapped and accumulatively form the ring current. According to the classic Chapman-Akasofu substorm paradigm, geospace magnetic storms occur when successively-occurring substorms deliver hot plasma to the inner magnetosphere faster than it can be dissipated.

This classic storm-substorm relation paradigm has come under scrutiny in recent years. Several studies have addressed the issue, without however achieving conclusive evidence (Kamide, 1992; Kamide *et al.*, 1998; Ebihara and Ejiri, 2000). The origins of the storm-substorm dispute can be found in correlation studies between auroral electrojet indices AE and the Dst index. Early studies had indicated a causal relationship between substorms and storms, but investigations that followed have questioned this relationship.

Such investigations have suggested that substorm occurrence is incidental to the main phase of storms, and that ion transport into the ring current is accomplished solely by enhanced large-scale magnetospheric convection.

To mention just a couple of these investigations, Iyemori and Rao (1996) identified a total of 28 geospace magnetic storms "containing" substorm expansion onsets and showed that after substorm onset there was no storm development noticeable in the average $SYM-H$ value ($SYM-H$ is a high-resolution, alternative to Dst, ring current intensity index). The authors concluded that the ring current development was not the result of the frequent occurrence of substorms, but the result of enhanced convection caused by large southward IMF. Along the same line, McPherron (1997) noted that visual inspections of AL and Dst time series during storms generally show that substorms occur at times when there is no ring current development (actually when there is no Dst decrease) or when Dst is recovering.

With regard to the apparent lack of Dst response to substorm expansion, Daglis *et al.* (2000) examined the intense storm of June 5, 1991 and reached different conclusions. Daglis *et al.*, examined the second main phase of this particular storm with respect to substorm occurrence, Dst variations and compositional variations. They showed that the Dst change rate increased after substorm occurrence. The relative abundance of O^+ ions in the ring current also increased after substorm occurrence. Distinct, large compositional changes have been shown to accompany the main phase of intense storms (Daglis, 1997a). This is a crucial feature that will be discussed further on.

Furthermore, with regard to the *Dst* response to substorm expansion, it was theoretically suggested by Siscoe and Petschek (1997) and observationally shown by Ohtani *et al.* (2001) that the effect of the tail current reduction following substorm expansion onset can overcompensate for that of the associated intensification of the ring current through the substorm-injected ions. This is due to the fact that the tail current contributes to *Dst* by 20% or even more (Turner *et al.*, 2000; Ohtani *et al.*, 2001).

Also in response to the anti-substorm school of thought, Sun and Akasofu (2000) suggested that it is more appropriate to examine the relationship between the corrected ring current intensity Dst^* and the upward field-aligned current density, instead of the standard *Dst* and *AE* indices. Sun and Akasofu (2000) proposed the new approach in order to accommodate the dominance of ionospheric ions in the ring current during intense storms (Daglis, 1997a; Daglis, 1997b; Daglis *et al.*, 1999a). Using the Method of Natural Orthogonal Components (Sun *et al.*, 1998), Sun and Akasofu (2000) showed that the directly driven component (DD) of the upward field-aligned currents is poorly correlated to the corrected *Dst* index (correlation coefficient of 0.33), while the unloading component (UL) correlates much stronger (correlation coefficient of 0.81). This indicates that the upward field-aligned currents during substorms play an important role in the formation of the ring current. Sun and Akasofu further concluded that the poor correlation between DD and *Dst* indicates that the formation of the ring current is not the result of enhanced convection. The strong correlation between UL and *Dst*, on the other side, is consistent with the observational evidence that the magnetosphere-ionosphere coupling plays an important role in the ring current growth. As already mentioned compositional changes are a crucial feature and will be discussed a few paragraphs later.

From the above, it is clear that the main issue of dispute is whether or not the occurrence of substorms has significant effects on storm-time ring current enhancement. To quantify the substorm influence on the basis of statistical correlations between indices is only a first-level approach. We can do better than that – on the basis of physical processes and their impacts. There are two major ways for substorms to affect ring current enhancement.

i. Substorms enhance ionospheric outflow (Daglis *et al.*, 1994; Daglis and Axford, 1996) which in turn increases plasma sheet density (Nosé *et al.*, 2005). Simulations indicate that high plasma sheet density can be the crucial parameter turning a moderate storm into an intense one. In other words, ionospheric outflow can significantly modify the geoeffectiveness of southward IMF through the plasma sheet density enhancements that are reached as its result (Daglis *et al.*, 2003).

ii. Substorm-induced impulsive electric fields play a substantial role in the energization of storm-time ring current particles (Daglis and Kamide, 2003; Daglis *et al.*, 1998; Metallinou *et al.*, 2005). This topic has been addressed in several comprehensive computer simulations, which will be discussed briefly here.

Chen *et al.* (1994) used spike-like enhancements of the convection electric field to simulate the effect of individual substorms. The authors concluded that the substorm contribution was subtle, and possibly negative to the development of a ring current.

However, the actual electric fields in the inner magnetosphere are much different than simple transient enhancements of the convection electric field. Wygant *et al.* (1998) showed that during the large March 1991 storm, the large-scale electric field repeatedly penetrated earthward, maximizing between $L = 2$ and $L = 4$ with magnitudes of 6 mV/m. Furthermore, Wygant *et al.* (1998) also noted that strong impulsive electric fields with amplitudes of up to 20 mV/m (i.e., more than three times the largest convection electric field) were observed during magnetic field dipolarizations in the inner magnetosphere, i.e. during substorm expansions or intensifications. Apparently, substorm-induced electric fields are episodic, but on the other side they are much stronger than the convection electric field.

While Fok *et al.* (1996) had suggested that substorm contribution was subtle and possibly negative to the development of a ring current, Fok *et al.* (1999) modified their model to implement a realistic influence on the plasma input by substorm-dipolarization electric fields in the inner plasma sheet. The result was that substorm-associated induced electric fields significantly enhance the ring current by redistributing plasma pressure earthward. Fok *et al.* (1999) concluded that global convection and substorm dipolarizations cooperate to inject plasma energy more deeply into the magnetosphere than either would individually.

In a similar effort to assess the significance of substorm-induced electric fields, a series of other studies addressed their efficiency in ion acceleration and transport into the inner magnetosphere (Delcourt, 2002; Daglis *et al.*, 1998; Metallinou *et al.*, 2005). The results show that the inclusion of substorm-induced electric fields in a test-particle simulation considering H^+ and O^+ seed ions originating from the auroral ionosphere and the plasma sheet, renders ion acceleration much more efficient.

These results advocate the significance of substorm occurrence for the ring current build-up, at least with regard to its O^+ content, which is an important aspect of ring current dynamics. Massive outflow and preferential acceleration of ionospheric O^+ ions is outstanding during intense storms, when the oxygen to proton energy density ratio can reach values of up to 400% (Daglis, 1997b; Daglis *et al.*, 1999a). Evidence for substorm-driven enhancements of the O^+ content of the ring current is also found in energetic neutral atom images of storm by the IMAGE satellite (C:son Brandt *et al.*, 2003; Mitchell *et al.*, 2003; Ohtani *et al.*, 2005).

6. Ring Current Evolution and Decay

As mentioned in Section 2, the general morphology of the global decrease of the geomagnetic H-component during magnetic storms, as reflected in the *Dst*-index profile, led prominent scientists to postulate the development of a westward electric current system (the ring current), which encircles the Earth symmetrically (Akasofu and Chapman, 1961). However, this paradigm has been questioned by spacecraft observations (Hamilton *et al.*, 1988; Fok *et al.*, 2003), detailed analysis of ground observations (Grafe, 1999), and simulations (Liemohn *et al.*, 1999; Ebihara and Ejiri, 2000): the ring current usually exhibits local-time asymmetry in the main phase of storms and becomes symmetric only in the late recovery phase. Recently it was shown that solar wind dynamic pressure enhancements may further enhance the ring current asymmetry during the storm main phase (Shi *et al.*, 2005).

It is noteworthy that although observations in the past had shown a dusk-side asymmetry (Stüdemann *et al.*, 1987; Ebihara and Ejiri, 2000), recent observations by the High-Energy Neutral Atom (HENA) imager on the IMAGE satellite have shown post-midnight enhancements rather than pre-midnight enhancements of storm-time ring current protons (C:son Brandt *et al.*, 2002). Ebihara and Fok (2004) showed that such the post-midnight flux enhancements can be caused by at least two mechanisms: i. a deformation of the convection electric field due to the shielding electric field; ii. local-time dependence of the plasma sheet density.

Recent modelling results indicate that the main phase ring current, especially during intense storms, is highly asymmetric with ions making one pass through the inner magnetosphere on open-drift paths (Liemohn *et al.*, 2001). An open-drift path geometry has important implications for the evolution and decay of the ring current – in particular the two-phase decay often observed during intense magnetic storms.

In a comparative study of a solar-maximum and a solar-minimum intense storm, Daglis *et al.* (2003) showed that the two cases had ring current asymmetry in common. Dial plots of energy density in the equatorial plane clearly illustrate the ring current asymmetry in Figure 3 (middle row). Observed and modeled Dst^* are displayed in the top panels of the figure. The time intervals of the equatorial dial plots are indicated on these panels with respect to the Dst^* development for each storm period. During both events, the ring current turned from being highly asymmetric in the main phase (time intervals 1 and 2) to being progressively less asymmetric in the early recovery phase (time interval 3) to being weaker but symmetric in the late recovery phase (time interval 4).

The ring current asymmetry is a direct consequence of the character of the ion drift paths as they evolve. The bottom row of dial plots in Figure 3 displays the percentage of ion energy on open drift paths as a function of location in the equatorial plane. During the main phase and throughout much of the early recovery phase, the convection electric field is strong and the magnetospheric configuration allows the majority of ring current ions to move along open drift paths to the dayside magnetopause. Not until the late recovery phase are the majority of ions trapped on closed drift paths as the convection electric field weakens. More specifically, during the storm main phase as much as 90% of the simulated Dst^* can be due to ions on open drift paths, decreasing to less than 10% during the late recovery phase.

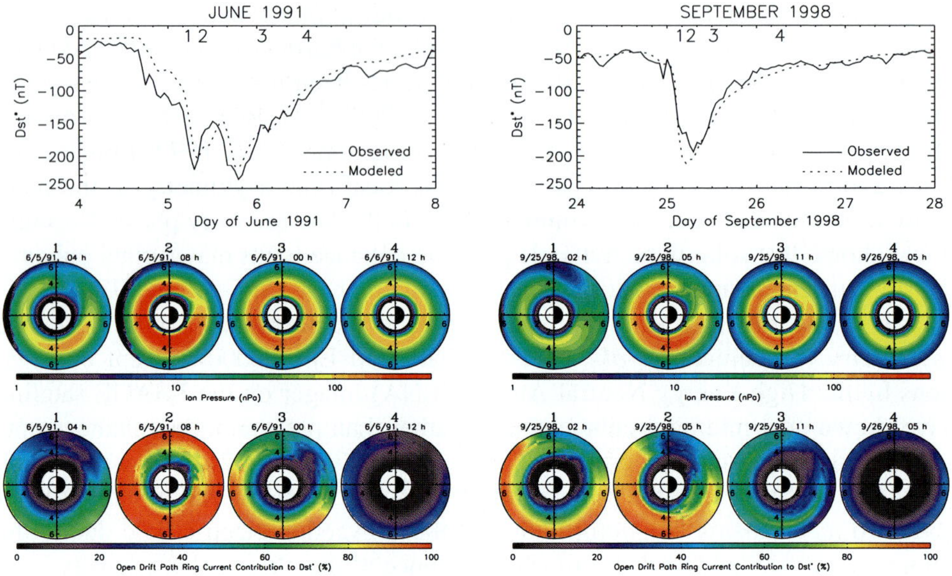

Figure 3. A comparative study of the June 1991 (left) and September 1998 (right) intense magnetic storms showed that the ring current was asymmetric in the main phase of both storms. The upper panels show the observed Dst^* (pressure-corrected Dst) for each storm along with the model Dst^* derived through the Dessler-Parker-Sckopke relation by summing the total ring current energy within the model volume. The ring current energy density (center) and the percentage of the ring current energy on open drift paths (bottom) are plotted in the equatorial plane with noon to the left, midnight to the right, dawn at the top and dusk at the bottom of each dial plot (Daglis *et al.*, 2003).

Drifting along closed paths, the ions become distributed more uniformly in local time and a symmetric ring current develops. However, by this time, the bulk of the ring current energy has already been dissipated: $|Dst^*|$ is at significantly less than half its maximum value by the start of the late recovery phase.

The topological changes described above are important in the growth and decay of the symmetric/partial ring current system. During the main and early recovery phases, when the ring current is topologically connected to the inner plasma sheet, changes in the plasma sheet ion distributions drive the partial ring current characteristics. Observed populations at the outer boundary move into and through the inner magnetosphere following drift paths controlled by the large-scale electric field pattern.

These plasma sheet ions are energized as they move into low L-values and then de-energized as the move outward toward the dayside magnetopause. During the main-phase development of the ring current, ions lost at the dayside boundary are more than compensated by new particles moving in from the magnetotail; therefore Dst grows. However, as magnetic activity subsides, the electric field weakens and the plasma sheet density decreases. These two events both produce a ring current decay, but in different ways.

Consider a weakening electric field with a constant plasma sheet density. When the electric field is strong, the topological separatrix between open and closed drift paths is formed deep within the inner magnetosphere. The weakening convection electric field drives a conversion of open to closed drift paths as the separatrix dividing these regions moves outward. If the electric field changes instantaneously to a lower value, the separatrix moves outward instantaneously. All ions are trapped and no significant change occurs in the ring current energy.

However, if the electric field weakens gradually, a part of the energetic ions are able to drift to the dayside magnetopause before they become trapped on closed drift paths. New plasma moving in on the nightside to replace these ions moves along different drift paths in the weaker electric field, penetrating less deeply with weaker adiabatic energization, and a net energy loss occurs. The amount and time-scale of the energy loss and the strength of the trapped ring current during storm recovery, depend on the time scale of the electric field decrease.

Now consider a decrease in the plasma sheet density with a fixed convection strength. The higher density plasma, moving out of the dayside magnetopause on open drift paths, is gradually replaced by lower density plasma moving through the nightside boundary. To completely replace the higher density with lower density plasma and come to a new quasi-steady-state Dst^* value, takes a time-scale on the order of the average drift time from the nightside plasma sheet to the dayside magnetopause. The actual flow-out loss time-scale is a combination of time-scales associated with the electric field decline and those associated with plasma sheet density changes.

The conversion from the fast "flow out" losses associated with open drift paths to the slower "charge-exchange" losses associated with closed drift paths may be responsible for the two-phase decay seen in some storms.

However, things may differ from storm to storm, as the study of Mitchell *et al.* (2001) on IMAGE imaging of two different storms showed: while the ring current ions during the stronger storm drifted primarily on closed paths, the weaker storm was much more dependent on open drift path dynamics. Hence, there is still a lot of work to be done on the relative importance of the convective drift loss in ring current decay.

Apart from the controversy on the dominant decay mechanism around storm maximum and early recovery phase, charge exchange is still regarded as the major loss process of the storm-time ring current in total. Coulomb collision processes also contribute to ring current loss and have been considered in relevant models (Fok *et al.*, 1995). However, an important ring current loss is due to pitch-angle scattering of particles into the loss cone by wave-particle interactions. The inclusion of pitch-angle scattering by plasma waves in large-scale models has lead to better estimates of the high-energy proton flux and pitch-angle distribution (Fok *et al.*, 1995; Fok *et al.*, 1996).

The most widely studied interactions concern electromagnetic ion cyclotron (EMIC) waves. The time scales for scattering of ions into the loss cone during resonant interactions with EMIC waves can be rapid (Lyons and Thorne, 1972). Therefore, lifetimes of ring current protons resonating with EMIC waves are shorter than charge exchange lifetimes at high (>40 keV) energies (Jordanova *et al.*, 1996). This is appealing since studies of the ring current energy balance (Gonzalez *et al.*, 1989) suggest that energy loss time scales during the main phases of intense-to-great geomagnetic storms may reach values as low as 0.5–1.0 hours (Feldstein *et al.*, 1994), which is too short to be the result of charge exchange or Coulomb collision processes.

The effects of EMIC wave scattering on ring current evolution were examined (Kozyra *et al.*, 1997) for the 2–6 November 1993 storm with the Michigan RAM code (Jordanova *et al.*, 1997), combined with detailed calculations of path integrated wave gain from the HOTRAY code (Horne, 1989; Thorne and Horne, 1992). The extra ion losses due to the interaction with waves caused an additional ∼10 nT recovery of Dst (accounting for ∼10% of the total ring current energy) for the one-hour period in which the effects of waves were added to the simulation of the storm.

More recently, Jordanova *et al.* (2001) calculated wave growth regions, wave amplitudes and associated proton precipitation for the 14–16 May 1997 storm. Wave growth dynamically affected the global patterns of proton precipitation during the storm. Precipitation moved to lower L values during the storm main phase and drifted outward again during storm recovery. In agreement with earlier results, the most intense precipitation due to scattering by EMIC waves was associated with the duskside plasmapause during the main and early recovery phases of the storm. Furthermore, the position of the precipitation zones was in qualitative agreement with the observed local time position of enhanced midlatitude precipitation during the April 1981 magnetic storm studied by Søraas *et al.* (1999).

7. Concluding Remarks

This chapter reviewed the current understanding of ring current dynamics. The terrestrial ring current is an electric current flowing toroidally around the Earth, centered at the equatorial plane and at altitudes of ∼10,000 to 60,000 km. A wealth of observational and computer simulation studies have been instrumental in achieving vital progress in our understanding of the origin of ring current ions, their acceleration and transport into the inner magnetosphere, and their eventual loss.

In spite of historical paradigms, it has been well established that O^+ ions of ionospheric origin are an significant component of the ring current and their contribution increases with solar activity and storm intensity and even becomes dominant around the maximum phase of intense storms. The large-scale magnetospheric convection electric field presumably is the main driver of ion transport into the inner magnetosphere and of the subsequent ring current buildup. The relative contribution and importance of the substorm-induced impulsive electric fields is yet to be quantified. Storm-time ring current evolution and decay is controlled by the concerted synergy of several factors: a variety of internal loss processes combined with gradually reduced inflow of plasma from the plasma sheet and the magnetotail, and convective outflow of plasma through the dayside magnetopause.

The ring current, in its growth, evolution and decay, is the collective result of solar, interplanetary, magnetospheric and ionospheric drivers, and as such, it is a perfect representative of the complex and mind-boggling solar-terrestrial connection. Synergy between sophisticated computer simulations and multi-point observations in space and on the ground will be instrumental in achieving knowledge closure of ring current dynamics. In recent years ring current modelling has been improving swiftly, as our knowledge of the inner magnetospheric electric and magnetic fields and plasma populations has been advancing. A step that will further advance the level of inner magnetosphere and solar-terrestrial modelling, will be the development of magnetosphere-ionosphere coupling models using self-consistent calculations of the fields, waves, and plasmas.

Acknowledgements

I would like to thank the editors for their invitation to join the author team of this book and particularly Rudolf von Steiger for his patience with my late contribution. I very much enjoyed the workshop in Bern, meeting many distinguished colleagues and having the opportunity of fruitful discussions with them. ISSI is to be congratulated for the continuous effort in bringing together international teams and producing excellent books.

References

Akasofu, S.-I., and Chapman, S.: 1961, *J. Geophys. Res.* **66**, 1321.

Axford, W. I.: 1970, in: B. M. McCormac (ed.), *Particles and Field in the Magnetosphere*. Norwell, Mass.: D. Reidel, pp. 46–59.

Baker, D. N., Pulkkinen, T. I., Toivanen, P., Hesse, M., and McPherron, R. L.: 1996, *J. Geomagn. Geoelectr.* **48**, 699.

Baumjohann, W.: 1993, *Space Sci. Rev.* **64**, 141.

Baumjohann, W., and Treumann, R. A.: 1996, *Basic Space Plasma Physics*. London: Imperial College Press.

Chapman, S.: 1919, *Proc. Roy. Soc. London* **A95**, 61.

Chapman, S.: 1962, *J. Phys. Soc. Japan* **17,** Suppl. A-I, 6.

Chapman, S., and Ferraro, V. C. A.: 1930, *Nature* **126**, 129.

Chapman, S., and Ferraro, V. C. A.: 1931, *Terr. Magn. Atmos. Electr.* **36**, 77.

Chen, M. W., Lyons, L., and Shultz, M.: 1994, *J. Geophys. Res.* **99**, 5745.

Christofilos, N. C.: 1959, *J. Geophys. Res.* **64**, 869.

C:son Brandt, P., Mitchell, D. G., Demajistre, R., Roelof, E., Ohtani, S., Jahn, J.-M., *et al.*: 2003, in: A. S. Sharma, Y. Kamide., and G. S. Lakhina (eds.), *Disturbances in Geospace: The Storm-*

Substorm Relationship, Geophys. Monogr. Ser., Vol. 142, doi 10.1029/142GM11. Washington, D. C.: AGU, pp. 103.

C:son Brandt, P., Ohtani, S., Mitchell, D. G., Fok, M.-C., Roelof, E. C., and Demajistre, R.: 2002, *Geophys. Res. Lett.* **29**, 1954, doi:10.1029/2002GL015160.

Daglis, I. A.: 1997a, in: B. T. Tsurutani, W. D. Gonzalez, Y. Kamide, and J. K. Arballo (eds.), *Magnetic Storms, Geophys. Monogr. Ser.,* Vol. 98. Washington, D. C.: AGU, pp. 107.

Daglis, I. A.: 1997b, *Eos Trans. AGU* **24**, 245.

Daglis, I. A., and Axford, W. I.: 1996, *J. Geophys. Res.* **101**, 5047.

Daglis, I. A., Delcourt, D., Metallinou, F.-A., and Kamide, Y.: 1998, *IEEE Trans. Plasma Science* **32** (4), 1449.

Daglis, I. A., and Kamide, Y.: 2003, in: A. S. Sharma, Y. Kamide, and G. S. Lakhina (eds.), *Disturbances in Geospace: The Storm-Substorm Relationship, Geophys. Monogr. Ser.,* Vol. 142, doi 10.1029/142GM11. Washington, D. C.: AGU, pp. 119–129.

Daglis, I. A., Kamide, Y., Mouikis, C., Reeves, G. D., Sarris, E. T., Shiokawa, K., *et al.*: 2000, *Adv. Space Res.* **25** (12), 2369.

Daglis, I. A., Kasotakis, G., Sarris, E. T., Kamide, Y., Livi, S., and Wilken, B.: 1999a, *Phys. Chem. Earth* **24**, 229.

Daglis, I. A., Kozyra, J. U., Kamide, Y., Vassiliadis, D., Sharma, A. S., Liemohn, M. W., *et al.*: 2003, *J. Geophys. Res.* **108**, 1208, doi:10.102/ 2002JA009722.

Daglis, I. A., Livi, S., Sarris, E. T., and Wilken, B.: 1994, *J. Geophys. Res.* **99**, 5691.

Daglis, I. A., Paschalidis, N. P., Sarris, E. T., Axford, W. I., Kremser, G., Wilken, B., *et al.*: 1991, in: J. R. Kan, T. A. Potemra, S. Kokubun, and T. Iijima (eds.): *Magnetospheric Substorms, Geophys. Monogr. Ser.,* Vol. 64. Washington, D. C.: AGU, pp. 323–332.

Daglis, I. A., Sarris, E. T., and Wilken, B.: 1993, *Ann. Geophys.* **11**, 685.

Daglis, I. A., Thorne, R. M., Baumjohann, W., and Orsini, S.: 1999b, *Rev. Geophys.* **37**, 407.

De Michelis, P., Daglis, I. A., and Consolini, G.: 1997, *J. Geophys. Res.* **102**, 14,103.

Delcourt, D. C.: 2002, *J. Atmos. Sol. Terr. Phys.* **64**, 551.

Dessler, A. J., and Hanson, W. B.: 1961, *Astrophys. J.* **134**, 1024.

Dessler, A. J., and Parker, E. N.: 1959, *J. Geophys. Res.* **64**, 2239.

Ebihara, Y., and Ejiri, M.: 2000, *J. Geophys. Res.* **105**, 15,843.

Ebihara, Y., and Fok, M.-C.: 2004, *J. Geophys. Res.* **109**, A12209, doi:10.1029/2004JA010523.

Feldstein, Y. I., Levitin, A. E., Golyshev, S. A., Dremukhina, L. A., Vestchezerova, U. B., Valchuk, T. E., *et al.*: 1994, *Ann. Geophys.* **12**, 602.

Fok, M.-C., Moore, T. E., and Delcourt, D. C.: 1999, *J. Geophys. Res.* **104**, 14,557.

Fok, M.-C., Moore, T. E., and Greenspan, M. E.: 1996, *J. Geophys. Res.* **101**, 15,311.

Fok, M.-C., Moore, T. E., Kozyra, J. U., Ho, G. C., and Hamilton, D. C.: 1995, *J. Geophys. Res.* **100**, 9619.

Fok, M.-C., Moore, T. E., Wilson, G. R., Perez, J. D., Zhang, X. X. C:son Brandt, P., *et al.*: 2003, *Space Sci. Rev.* **109**, 77.

Frank, L. A.: 1967, *J. Geophys. Res.* **72**, 3753.

Geiss, J., Balsiger, H., Eberhardt, P., Walker, H. P., Weber, L., Young,, D. T., *et al.*: 1978, *Space Sci. Rev.* **22**, 537.

Gloeckler, G., and Hamilton, D. C.: 1987, *Phys. Scr.* **T18**, 73.

Gonzalez, W. D., Tsurutani, B. T., Gonzalez, A. L. C., Smith, E. J., Tang, F., and Akasofu, S.-I.: 1989, *J. Geophys. Res.* **94**, 8835.

Grafe, A.: 1999, *Ann. Geophys.* **17**, 1.

Grande, M., Perry, C. H., Hall, D. S., Wilken, B., Livi, S., Søraas, F., *et al.*: 1992, *Substorms 1, Eur. Space Agency Spec. Publ.* **SP-335**, 485.

Hamilton, D. C., Gloeckler, G., Ipavich, F. M., Stüdemann, W., Wilken, B., and Kremser, G.: 1988, *J. Geophys. Res.* **93**, 14,343.

Horne, R. B.: 1989, *J. Geophys. Res.* **94**, 8895.
Iyemori, T., and Rao, D. R. K.: 1996, *Ann. Geophys.* **14**, 608.
Jordanova, V. K., Farrugia, C. J., Thorne, R. M., Khazanov, G. V., Reeves, G. D., and Thomsen, M. F.: 2001, *J. Geophys. Res.* **106**, 7.
Jordanova, V. K., Kozyra, J. U., and Nagy, A. F.: 1996, *J. Geophys. Res.* **101**, 19,771.
Jordanova, V. K., Kozyra, J. U., Nagy, A. F., and Khazanov, G. V.: 1997, *J. Geophys. Res.* **102**, 14,279.
Kamide, Y.: 1992, *J. Geomagn. Geoelectr.* **44**, 109.
Kamide, Y., Baumjohann, W., Daglis, I. A., Gonzalez, W. D., Grande, M., Joselyn, J. A., *et al.*: 1998 *J. Geophys. Res.* **103**, 17,705.
Kaye, S. M., Johnson, R. G., Sharp, R. D., and Shelley, E. G.: 1981, *J. Geophys. Res.* **86**, 1335.
Kozyra, J. U., Jordanova, V. K., Horne, R. B., and Thorne, R. M.: 1997, in: Tsurutani, B. T., Gonzalez, W. D., Kamide, Y., and Arballo, J. K. (eds.), *Magnetic Storms, Geophys. Monogr. Ser.,* vol. 98. Washington, D. C.: AGU, pp. 187.
Krimigis, S. M., Gloeckler, G., McEntire, R. W., Potemra, T. A., Scarf, F. L., and Shelley, E. G.: 1985, *Geophys. Res. Lett.* **12**, 329.
Liemohn, M. W., Kozyra, J. U., Jordanova, V. K., Khazanov, G. V., Thomsen, M. F., and Cayton, T. E.: 1999, *Geophys. Res. Lett.* **26**, 2845.
Liemohn, M. W., Kozyra, J. U., Thomsen, M. F., Roeder, J. L., Lu, G., Borovsky, J. E., *et al.*: 2001 *J. Geophys. Res.* **106**, 10,883.
Lui, A. T. Y., McEntire, R. W., and Krimigis, S. M.: 1987, *J. Geophys. Res.* **92**, 7459.
Lyons, L. R., and Thorne, R. M.: 1972, *J. Geophys. Res.* **77**, 5608.
McPherron, R. L.: 1997, in: Tsurutani, B. T., Gonzalez, W. D., Kamide, Y., and Arballo, J. K. (eds.), *Magnetic Storms, Geophys. Monogr. Ser.,* Vol. 98. Washington, D. C.: AGU, pp. 131.
Metallinou, F.-A., Daglis, I. A., and Delcourt, D. C.: 2005, *Eos Trans. AGU* **86** (52), Fall Meet. Suppl. Abstract SA21A–0293.
Mitchell, D. G., C:son Brandt, P., Roelof, E. C., Hamilton, D. C., Retterer, K. C., and Mende, S.: 2003, *Space Sci. Rev.* **109**, 63–75.
Mitchell, D. G., Hsieh, K. C., Curtis, C. C., Hamilton, D. C., Voss, H. D., Roelof, E. C., *et al.*: 2001, *Geophys. Res. Lett.* **28**, 1151.
Möbius, E., Scholer, M., Klecker, B., Hovestadt, D., Gloeckler, G., and Ipavich, F. M.: 1987, in: Lui, A. T. Y. (ed.), *Magnetotail Physics*. Baltimore, Md.: Johns Hopkins Univ. Press, pp. 231–234.
Moos, N. A. F.: 1910, *Colaba magnetic data 1846 to 1905. Part I: Magnetic data and instruments. Part II: the phenomenon and its description*. Bombay, India.
Neugebauer, M., and Snyder, C.: 1962, *Science* **138**, 1095.
Northrop, T. G.: 1963, *The Adiabatic Motion of Charged Particles*. New York: Wiley Interscience.
Nosé, M., Taguchi, S., Hosokawa, K., Christon, S. P., McEntire, R. W., Moore, T. E., *et al.*: 2005, *J. Geophys. Res.* **110**, A09S24, doi:10.1029/2004JA010930.
Ohtani, S.-I., C:son Brandt, P., Mitchell, D. G., Singer, H., M. Nosé, Reeves, G. D., and Mende, S. B.: 2005, *J. Geophys. Res.* **110**, A07219, doi:10.1029/2004JA010954.
Ohtani, S.-I., Nosé, M., Rostoker, G., Singer, H., Lui, A. T. Y., and Nakamura, M.: 2001, *J. Geophys. Res.* **106**, 21,199.
Parker, E. N.: 1957, *Phys. Rev.* **107**, 924.
Parker, E. N.: 1958, *Phys. Fluids* **1**, 171.
Peterson, W. K., Sharp, R. D., Shelley, E. G., Johnson, R. G., and Balsiger, H.: 1981, *J. Geophys. Res.* **86**, 761.
Sckopke, N.: 1966, *J. Geophys. Res.* **71**, 3125.
Shelley, E. G., Johnson, R. G., and Sharp, R. D.: 1972, *J. Geophys. Res.* **77**, 6104.
Shi, Y., Zesta, E., Lyons, L. R., Boudouridis, A., Yumoto, K., and Kitamura, K.: 2005, *J. Geophys. Res.* **110**, A10205, doi:10.1029/2005JA011019.
Singer, S. F.: 1956, *Bull. Am. Phys. Soc. Series II* **1**, 229 (A).

Singer, S. F.: 1957, *Eos Trans. AGU* **38**, 175.
Siscoe, G. L., and Petschek, H. E.: 1997, *Ann. Geophys.* **15**, 211.
Snyder, C. W., and Neugebauer, M.: 1964, *Space Research* **4**, 89.
Søraas, F., Aarsnes, K., Lundblad, J. A., and Evans, D. S.: 1999, *Phys. Chem. Earth* **24**, 287.
Strangeway, R. J., and Johnson, R. G.: 1984, *J. Geophys. Res.* **89**, 8919.
Stüdemann, W., Wilken, B., Kremser, G., Korth, A., and Fennell, J. F.: 1987, *Geophys. Res. Lett.* **14**, 455.
Sun, W., and Akasofu, S.-I: 2000, *J. Geophys. Res.* **105**, 5411.
Sun, W., Xu, W.-Y., and Akasofu, S.-I: 1998, *J. Geophys. Res.* **103**, 11,695.
Thorne, R. M., and Horne, R. B.: 1992, *Geophys. Res. Lett.* **19**, 419.
Turner, N. E., Baker, D. N., Pulkkinen, T. I., and McPherron, R. L.: 2000, *J. Geophys. Res.* **105**, 5431.
Van Allen, J. A.: 1959, *J. Geophys. Res.* **64**, 1683.
Van Allen, J. A., Ludwig, G. H., Ray, E. C., and C. E. McIlwain: 1958, *Jet Propul.* **28**, 588.
Wygant, J., Rowland, D., Singer, H. J., Temerin, M., Mozer, F., and Hudson, M. K.: 1998, *J. Geophys. Res.* **103**, 29,527.

PLASMASPHERE RESPONSE: TUTORIAL AND REVIEW OF RECENT IMAGING RESULTS

J. GOLDSTEIN
Space Science and Engineering Division, Southwest Research Institute, 6220 Culebra Road, San Antonio TX 78238 USA
(E-mail: jgoldstein@swri.edu)

(Received 22 June 2005; Accepted in final form 23 November 2005)

Abstract. The plasmasphere is the cold, dense innermost region of the magnetosphere that is populated by upflow of ionospheric plasma along geomagnetic field lines. Driven directly by dayside magnetopause reconnection, enhanced sunward convection erodes the outer layers of the plasmasphere. Erosion causes the plasmasphere outer boundary, the plasmapause, to move inward on the nightside and outward on the dayside to form plumes of dense plasma extending sunward into the outer magnetosphere. Coupling between the inner magnetosphere and ionosphere can significantly modify the convection field, either enhancing sunward flows near dusk or shielding them on the night side. The plasmaspheric configuration plays a crucial role in the inner magnetosphere; wave-particle interactions inside the plasmasphere can cause scattering and loss of warmer space plasmas such as the ring current and radiation belts.

Keywords: plasmasphere, inner magnetosphere, imaging, erosion, electric fields, radiation belts

1. Introduction

This tutorial paper reviews some recent space-based imaging observations that have confirmed or improved our understanding of the dynamic global response of the plasmasphere and inner magnetosphere to the effects of the solar wind and interplanetary magnetic field (IMF). The level of discussion is intended to be accessible (with some help from the cited references) to non-specialists and students.

1.1. Plasmasphere Origin

The plasmasphere is a cold (1 eV), dense (10–10,000 cm^{-3}) torus of H^+ (nominally about 80%), He^+ (10–20%), and O^+ (a few to several percent, depending upon geomagnetic activity) (Lemaire and Gringauz, 1998). Figure 1a shows a schematic illustration of the plasmasphere, with a nominal equatorial size of 4 Earth radii (R_E). The plasmasphere is populated by filling from the dayside ionosphere; the sunlit ionosphere leaks up into space along magnetic field lines, slowly filling dayside flux tubes with cold ionospheric plasma (see inset of Figure 1a). Combined with the eastward rotation of the Earth's magnetic field, dayside filling produces a

DOI: 10.1007/s11214-006-9105-y

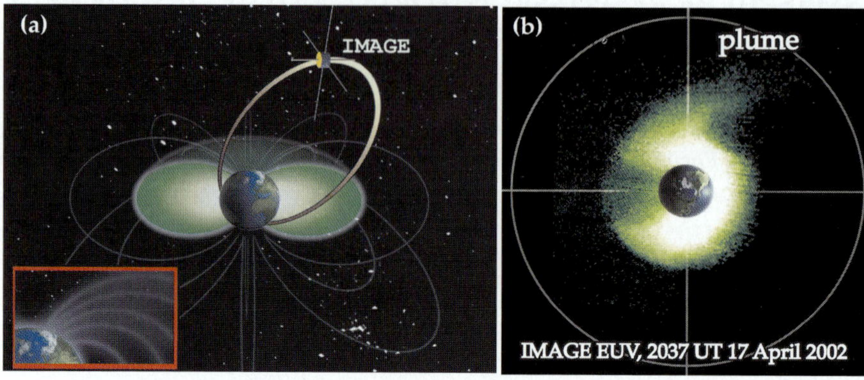

Figure 1. (a) Illustration of idealized plasmasphere torus and schematic IMAGE orbit. The view is from an oblique polar angle. The Earth is drawn in the center, with magnetic field lines drawn in perspective at $L = 4$ and $L = 6$, every 3 hours of magnetic local time (MLT). The plasmasphere is the green torus surrounding the Earth. A cross-section is taken along the noon-midnight MLT meridian. Inset: schematic/conceptual illustration of ionospheric outflow, in which dense ionospheric plasma leaks up into space to populate the magnetic field lines in the plasmasphere. (b) Global He^+ plasmasphere image obtained in 30.4-nm wavelength ultraviolet light by the IMAGE EUV imager, 2037 UT on 17 April 2002. The view is from above the magnetic north pole, looking down on the SM-coordinate magnetic equator. The Sun is to the right; the Earth is illustrated in the center. Geosynchronous orbit (6.6 R_E) and the X- and Y-axes are drawn in. The plasmasphere is shown in false color as the green/white region extending to an average distance of 3.3 Earth radii (R_E). The Earth's shadow extends antisunward, dimming the 30.4-nm emissions. A plasmaspheric plume can be seen extending sunward (to the upper right) from the duskside plasmasphere. EUV images such as this one have provided unambiguous proof of the existence of plumes, confirming predictions that sunward convection erodes the plasmasphere during geomagnetic disturbances (Grebowsky, 1970). (EUV image courtesy of B. R. Sandel).

torus of cold plasma of ionospheric origin. During prolonged periods of very quiet geomagnetic conditions when ionospheric filling is the dominant effect, the plasmasphere can become quite large, reaching beyond geosynchronous orbit ($L = 6.62$, where L is equatorial geocentric distance in units of R_E) and having no distinct outer boundary (Goldstein *et al.*, 2003b).

1.2. Plasmasphere Observations

Observations of the plasmasphere span decades, from first discovery (Carpenter, 1963) to the present (Lemaire and Gringauz, 1998). The earliest measurements of the plasmasphere, obtained by analyzing whistler mode waves from the ground, showed a well-defined outer density gradient (often with a 1–2 order of magnitude density drop) called the plasmapause. Geomagnetic disturbances move the plasmapause inward to smaller L values (Carpenter, 1970; Chappell *et al.*, 1970), and the average plasmapause is larger for duskside magnetic local time (MLT) than for dawnside MLT (Carpenter, 1967). Early models (Grebowsky, 1970) of the plasmasphere offered an explanation for these observations: sunward convection (see Section 2) erodes the outer layers of the plasmasphere, removing plasma and creating a steep plasmapause boundary whose L value is inversely dependent upon geomagnetic activity level and whose MLT shape is influenced by a duskside stagnation region where sunward convection and eastward corotation are oppositely directed. The plasmapause density profile was observed to possess extensive meso-scale (0.1–1 R_E) and fine-scale (<0.1 R_E) structure (LeDocq *et al.*, 1994), including regions of dense plasma that appeared to be completely detached from the main plasmasphere (Chappell, 1974). The convection paradigm explained detached plasma as the single-point observational signature of a two-dimensional plume of sunward-convecting eroded plasma; a spacecraft moving obliquely across L values would see a cross section of this plume that would appear detached from the main plasmasphere. However, the plume interpretation (of the detached plasma observations) was not universally accepted (Chappell, 1974), and it was not until recently that the existence of plumes has been unambiguously confirmed (see Figure 1b, Figures 2a–d, and Section 2).

In the past several years, new techniques have been developed for observing the plasmasphere. From the ground, interpretation of magnetometer data (Dent *et al.*, 2003) and signals from GPS satellites (Foster *et al.*, 2002) provide proto-tomographic capabilities. From space, magnetospheric imaging achieves a global perspective previously only provided by models. The Imager for Magnetopause-to-Aurora Global Exploration (IMAGE) satellite (Burch *et al.*, 2001a) was launched in 2000 with two plasmasphere instruments onboard. The radio plasma instrument (RPI) (Reinisch *et al.*, 2001) uses active radio wave sounding to determine remote electron density, and has yielded some needed information about the density dependence along magnetic field lines (Reinisch *et al.*, 2004). The extreme ultraviolet (EUV) imager (Sandel *et al.*, 2001) routinely obtains full global images of the Earth's plasmasphere by remote-sensing solar 30.4-nm light that has been resonantly scattered by plasmaspheric He^+ ions. Figure 1b shows an EUV plasmasphere image from 2037 UT on 17 April 2002. The brightness of the green-white part of the image is proportional to the line-of-sight integrated He^+ column abundance. The visible portion of the plasmasphere in EUV images corresponds to total (electron) number density above about 40 cm^{-3} (Goldstein *et al.*, 2003c; Moldwin *et al.*, 2003). Plasmasphere images are obtained by EUV every 10 min with a nominal spatial resolution (pixel size) of 0.1 R_E or better. One of the first ideas validated by images such as that in Figure 1b is that plasmaspheric plumes form (as a result of erosion) during geomagnetic disturbance times (Grebowsky, 1970). The plasmasphere depicted in Figure 1b is mostly circular except near dusk (i.e., top of the figure), where a plume is evident. Plumes such as that of Figure 1b have been seen following every erosion event witnessed by EUV, and simultaneous (or near-simultaneous) in situ observations have verified that the plumes in EUV images are identical to detached plasma regions (Spasojević *et al.*, 2003; Garcia *et al.*, 2003; Goldstein *et al.*, 2004b). Plasmaspheric imaging has indeed confirmed the existence of plumes (though it is still probable that much lower density blobs of completely detached plasma do exist in the magnetosphere).

2. Inner Magnetospheric Convection

2.1. Dayside Magnetopause Reconnection (DMR)

During times of geomagnetic disturbance, sunward plasma convection (or advection), plays a crucial role in plasmaspheric dynamics. Perhaps the most fundamental cause of inner magnetospheric convection is dayside magnetopause reconnection (DMR). The magnetopause is the boundary between the geomagnetic field and the interplanetary magnetic field (IMF), nominally found at subsolar distance 10 R_E. When the IMF at the magnetopause is oriented opposite (southward) to the geomagnetic field, these oppositely-directed fields can undergo reconnection, a process that causes dayside geomagnetic field lines to become joined to the IMF lines, which then are dragged antisunward (along with the prevailing solar wind flow) into the stretched out magnetospheric tail (magnetotail). This magnetic flux transfer drives sunward convective flows in the inner magnetosphere (Dungey, 1961). Associated with this sunward convection is a solar-wind electric (E) field that points from dawn to dusk, with magnitude given by the product of the solar wind speed (V_{SW}) and the Z-component of the IMF ($B_{z,IMF}$). The zero-order influence seems to be the polarity of $B_{z,IMF}$, which acts as a switch, turning convection on for southward IMF ($B_{z,IMF} < 0$) and off for northward IMF ($B_{z,IMF} > 0$).

2.2. Plasmasphere Erosion

Plasmasphere images indicate there is an excellent correlation between $B_{z,IMF}$ polarity and the behavior of the plasmasphere (Goldstein *et al.*, 2003a; Spasojević *et al.*, 2003; Goldstein *et al.*, 2002; Goldstein *et al.*, 2003d). The plasmaspheric effect of an enhancement in DMR-driven convection depends upon the magnitude of the convection increase as well as the state of the plasmasphere at the onset of enhanced convection. The most dramatic plasmasphere erosion events are precipitated by exceptionally large convection enhancements that follow prolonged intervals of quiet geomagnetic conditions. If the convection increase is mild, and/or the plasmasphere has very recently been eroded, little or no erosion may occur.

2.2.1. *Erosion: Phases of Plume Evolution*

In plasmasphere images (and consistent with other observations), erosion follows a repeatable 4-phase pattern that was predicted by convection-based models (Grebowsky, 1970; Spiro *et al.*, 1981). Figures 2a through 2d depict a typical erosion event, witnessed by IMAGE EUV on 18 June 2001 (Goldstein and Sandel,

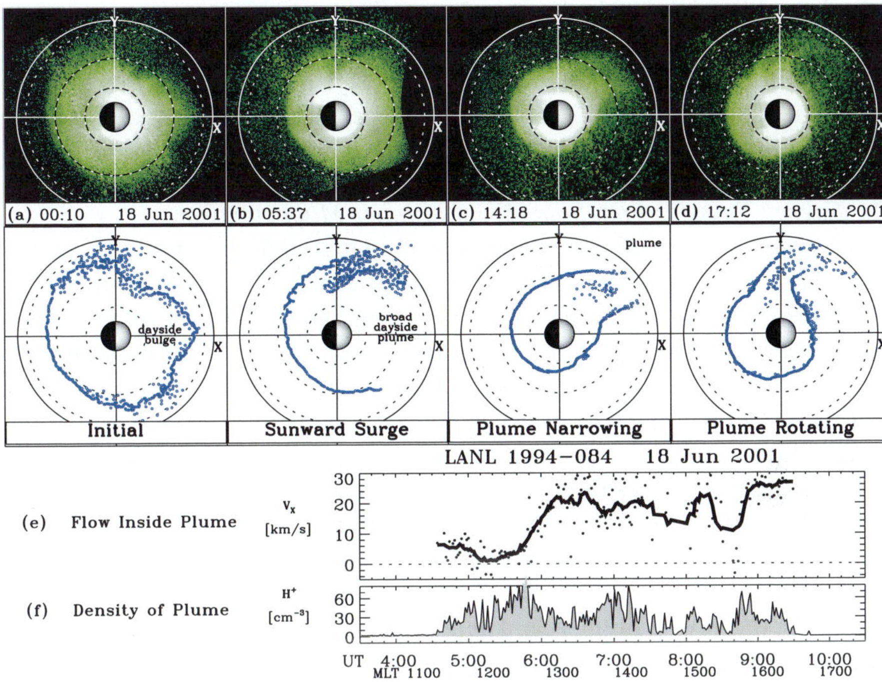

Figure 2. (a–d) Top row: EUV plasmasphere images from a typical erosion event on 18 June 2001 (Goldstein and Sandel, 2005). The format is similar to that of Figure 1b; the Sun is to the right and Earth is the half-shaded circle in the center. Dotted circles are drawn at $L = 2, 4$, and 6; the solid circle indicates geosynchronous orbit. Bottom row: Extracted plasmapause points from the images directly above. During the erosion of 18 June, plasmaspheric plasma moved sunward. The nightside moved inward (Earthward) by about 1 R_E, and the dayside bulge (see panel a) surged sunward to form a broad dayside plume (panel b). Over time, the plume narrowed in MLT (panel c) and then rotated eastward (see panel d) when the convection strength weakened. (e and f) Data from the Magnetospheric Plasma Analyzer (MPA) onboard Los Alamos National Laboratory (LANL) geosynchronouse satellite 1994–084, obtained during the erosion event of 18 June 2001. Consistent with a convective interpretation of the 18 June EUV images, the LANL in situ measurements show the presence of dayside (1120–1630 MLT) plasmaspheric (10–80 cm^{-3}) plasma that is flowing sunward (i.e., positive V_X) due to enhanced convection. The bold line in (e) is a 14-minute running average of the dots. (EUV images courtesy of B. R. Sandel; LANL data courtesy of M. F. Thomsen).

2005). The top row shows EUV plasmasphere images; the bottom row shows plasmapause locations extracted from the images. The general result of the erosion was the sunward displacement of the plasmapause. The initial nightside plasmapause moved inward ($+X$-direction, or sunward) by about 1 R_E, and the dayside bulge of the initial plasmasphere (Figure 2a) surged sunward to form a broad dayside plume which subsequently narrowed in MLT (Figure 2c) and then rotated eastward (Figure 2d) when the convection strength decreased This 4-phase pattern of evolution (initial, sunward surge, plume narrowing, plume rotating) represents the canonical development of plasmaspheric plumes during erosion events

(Goldstein and Sandel, 2005). Figures 2e and 2f show in situ measurements from the Los Alamos National Laboratory (LANL) Magnetospheric Plasma Analyzer (MPA), obtained on 18 June 2001, taken on the dayside. These data clearly support a convective model interpretation of EUV images, for they confirm that cold dense plasma does indeed flow sunward inside plasmaspheric plumes.

2.2.2. *Energy Transfer from Solar Wind to Inner Magnetosphere*

Plasmaspheric imaging allows separation of spatial and temporal effects, which has been important in studies of the timing of erosion events. The first erosion event witnessed by IMAGE EUV occurred during 0450–0830 UT on 10 July 2000 (Goldstein *et al.*, 2003a). During the erosion, the nightside plasmapause moved about 2 R_E inward of its starting position, and the plasmapause motion was driven by southward IMF; during northward IMF the plasmapause speed was zero. For this event there was a time delay of 30 minutes between the arrival of southward IMF at the magnetopause and the subsequent inward motion of the nightside plasmapause. Similar time delays (10–30 minutes) have consistently been observed during EUV-witnessed erosion events (Goldstein *et al.*, 2003b; Spasojević *et al.*, 2003; Goldstein *et al.*, 2004b; Goldstein and Sandel, 2005). Time-delayed convection is also experienced by the ring current (Goldstein *et al.*, 2003a), and by implication, the entire inner magnetosphere. (See Section 3 for more about the ring current and other plasma regions.) What causes this delay? Although it is reasonable to assume some delay is required for the global convection field to reconfigure itself (Coroniti and Kennel, 1973), this explanation has not been verified. This question needs an answer if we are to fully understand the way solar wind energy is imparted to the inner magnetosphere.

Tracking the speed of the plasmapause boundary during erosion can provide an estimate for the electric field associated with the erosion, assuming $E \times B$ motion of the cold plasma (Carpenter *et al.*, 1972). Careful analysis of EUV images during erosion events has yielded 1D and 2D maps of equatorial plasmapause electric fields (Goldstein *et al.*, 2004c; Goldstein *et al.*, 2004a; Goldstein and Sandel, 2005; Goldstein *et al.*, 2005b). From EUV E-field estimates, approximately 10–12% of the solar wind electric (E) field E_{SW} is felt at the plasmapause. This result is consistent with model predictions (Volland, 1973; Maynard and Chen, 1975) that only a fraction of E_{SW} is transmitted inside geosynchronous orbit, possibly owing to less-than-perfect reconnection efficiency (i.e., not all southward IMF lines reconnect). There have been few missions to measure the innermost magnetospheric E-field (Wygant *et al.*, 1998); in this regard plasmasphere imaging has provided a much-needed additional data source.

2.3. SUBSTORMS

The substorm is a critical magnetospheric process that is only partially understood even after decades of research (Akasofu, 1964; Goldstein *et al.*, 2005b). Substorms are believed to occur when excess magnetic flux in the magnetotail is suddenly released (Baker *et al.*, 1996). In this scenario the global nightside magnetic field reconfigures from a tail-like geometry (indicative of stored magnetic flux and high magnetic tension) to a more dipolar geometry (indicative of the release of magnetic tension). This magnetic dipolarization causes rapid sunward motion of geomagnetic field lines, which induces a global electric field (Aggson *et al.*, 1983) that transports plasma earthward. Early studies of the plasmaspheric effects of substorms suggested that the substorm induction E-field reduces the nightside plasmapause L (Carpenter and Stone, 1967; Carpenter and Akasofu, 1972). These predictions were recently confirmed by plasmaspheric imaging (Goldstein *et al.*, 2004a; Goldstein *et al.*, 2005b); a substorm that occurred at 1900 UT on 17 April 2002 caused ripples to propagate along the plasmapause, eastward and westward from pre-midnight MLT. The motion of the ripples was consistent with the interpretation that a sunward-propagating impulse swept past the plasmasphere, distorting the plasmapause shape during its passage. In contrast with DMR-driven convection events which produce a net reduction of the plasmapause L, the substorm-triggered plasmapause motion was only temporary; after the passage of the disturbance, the plasmapause returned to its starting location/shape. The plasmapause distortion was found to be strongly correlated both with auroral signatures of the substorm and with intensification and distortion of the ring current (Goldstein *et al.*, 2005b). This correlation implies that the substorm was the cause of the plasmapause (and ring current) distortion and also indicates strong coupling among different plasma populations. Imaging of plasmasphere and ring current (see Section 3.1), allows determination of causal relationships and global spatial/temporal properties of the propagating impulse.

3. Intra-Magnetospheric Plasma Coupling

The inner magnetosphere is a complex, electrodynamically coupled, self-modifying system; individual plasmas such as the plasmasphere, ring current, ionosphere, and radiation belts evolve interdependently and in many cases physically overlap. Interested readers are directed to the references cited in this section for a more complete survey of how imaging has improved our understanding of global intra-magnetospheric coupling (Burch *et al.*, 2001a,b; Goldstein *et al.*, 2005b).

3.1. Ring Current Imaging

The ring current (Daglis *et al.*, 1999) is a magnetically-confined plasma composed of warm (1–100 keV) ions (H^+, O^+) and electrons in the inner magnetosphere. In this energy range ions and electrons are subject to oppositely-directed magnetic drifts (see Figure 3a); ion pressure gradients produce a westward-flowing diamagnetic current. A major loss term for ring current ions is charge exchange,

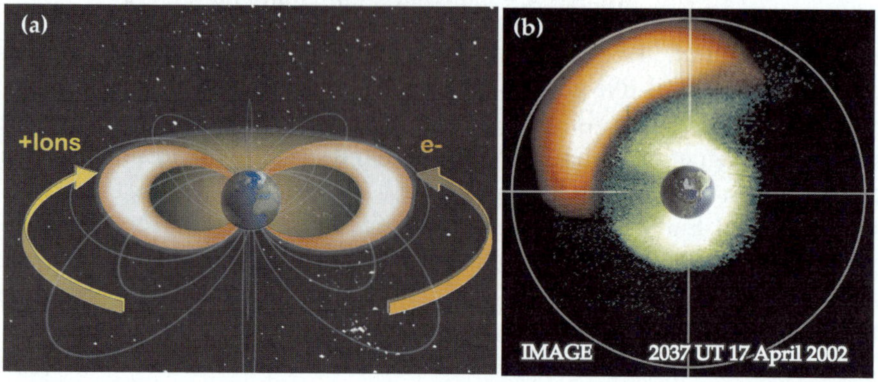

Figure 3. (a) Illustration of idealized ring current, format identical to that of Figure 1a. The ring current is the orange torus surrounding the Earth. Westward (eastward) magnetic drift of ions (electrons) indicated by the yellow (orange) curved arrow. (b) Global composite image of the inner magnetosphere (Pulkkinen *et al.*, 2005). IMAGE HENA proton pressure (10–60 keV, 0.5–0.8 nPa) image has been overlaid onto Figure 1b. The HENA image shows the partial ring current that has been injected by a substorm. The plasmasphere and ring current are roughly spatially complementary, although there is some overlap near dusk, at the eastern edge of the plasmaspheric plume. (HENA image courtesy of P. C. Brandt; EUV image courtesy of B. R. Sandel).

in which a warm ion accepts an electron from a nearby cold neutral particle in the Earth's exosphere, thereby producing an energetic neutral atom (ENA) which is not magnetically confined. The IMAGE high-energy neutral atom (HENA) imager (Brandt *et al.*, 2002) remotely detects escaping ENAs in the energy range 10–60 keV. Mathematical inversion of ENA images yields H^+ ring current pressure distributions (DeMajistre *et al.*, 2004).

During quiet times the ring current is roughly symmetric (as depicted schematically in Figure 3a), but geomagnetic disturbances produce strong partial (asymmetric) ring currents with pressure localized in MLT (Daglis *et al.*, 1999). Figure 3b shows a HENA proton pressure distribution obtained at 2037 UT on 17 April 2002, overlaid onto the EUV plasmasphere image of Figure 1b. This image was obtained near the end of a substorm that affected both ring current and plasmasphere (Section 2.3). As a result of the substorm, a strong partial ring current formed (Figure 3b) in the pre-midnight MLT sector where the plasmapause was similarly distorted by magnetic dipolarization.

3.2. Electrodynamics of Ring Current and Ionosphere Coupling

Currents flow continuously in closed loops. Owing to its finite azimuthal extent, the westward-directed partial ring current cannot close at low latitudes, and so is instead diverted along field lines to close in the ionosphere (Vasyliūnas, 1970). The field-aligned currents (FACs) that couple the dynamics of the ring current and the

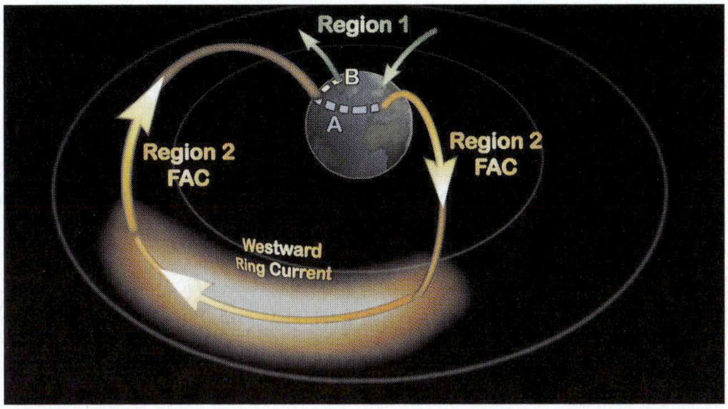

Figure 4. Cartoon of global current system linking the ring current to the ionosphere. Current must flow in closed loops, so the partial westward ring current gets diverted along field lines to form the region 2 field aligned current (FAC) system. Current closure is achieved in the ionosphere, either via an eastward current (path A) linking the two branches of the region 2 FACs, or via a northward current (path B) that connects the region 2 FAC to the more poleward region 1 auroral zone currents.

ionosphere, called region 2 (R2), are depicted in Figure 4. On the duskside, R2 FACs flow from the western edge of the ring current (RC) down into the ionosphere. On the dawnside, R2 FACs flow up from the ionosphere to connect with the eastern edge of the partial RC. Ring-current-ionosphere coupling is of fundamental importance because it modifies the inner magnetospheric electric field. The nature of that modification depends on how the R2 FACs are closed in the ionosphere.

3.2.1. *Shielding, Undershielding, and Overshielding*

If the duskside (into the ionosphere) R2 FAC is connected to the dawnside (out of the ionosphere) R2 FAC by an eastward-flowing ionospheric current (path A in Figure 4), this generates an eastward (dusk-to-dawn) electric field. This eastward *E*-field opposes the dawn-to-dusk convection electric field, thus shielding the inner magnetosphere from convection (Jaggi and Wolf, 1973). Effective shielding requires the establishment (via Alfvén waves) of this system of R2 and ionospheric currents, which happens on a time scale $\leq$ 1 hour (Kelley *et al.*, 1979; Senior and Blanc, 1984). Changes in convection strength which occur slower than this time scale may be effectively shielded, but sudden changes produce a residual "penetration" *E*-field in the inner magnetosphere (Goldstein *et al.*, 2003d). A sudden southward IMF transition causes undershielding, in which shielding is temporarily unable to counter the newly enhanced convection. During undershielding, the plasmasphere can be eroded, but within an hour the erosion tapers off if effective shielding is established (Goldstein and Sandel, 2005). Following a rapid northward IMF turning, overshielding occurs: convection suddenly decreases, leaving a residual eastward (dusk-to-dawn) *E*-field that drives antisunward convection. Because the ionospheric conductivity is lowest in the midnight-to-dawn MLT sector, it is there that antisunward convection (from overshielding) can often be strongest (Senior and Blanc, 1984; Fejer and Scherliess, 1995), which has been demonstrated to create shoulder-like bulges of the plasmapause (Sandel *et al.*, 2003; Goldstein and Sandel, 2005). Because the solar wind and IMF conditions typically fluctuate on much faster time scales than that required to establish the shielding current system, evidence of perfect shielding (exactly canceling out the convection *E*-field) is rare in global images (Spasojević and Goldstein, 2005).

3.2.2. *Sub-Auroral Polarization Stream (SAPS)*

If the duskside region 2 (R2) FAC is connected to the auroral current system (called region 1) via a poleward-flowing ionospheric current (path B in Figure 4), this generates a northward *E*-field and associated westward flow known as the subauroral polarization stream (SAPS) (Foster and Burke, 2002). Because of the low ionospheric conductivity at subauroral latitudes, the northward SAPS *E*-field can be quite large, and when mapped (along magnetic field lines) to the magnetic equator, produces an intense radial *E*-field located at the inner edge of the ring current, i.e., just outside or overlapping the plasmapause (Goldstein *et al.*, 2003b). The SAPS *E*-field produces strong westward flows that move the duskside plasmapause inward and can create narrow duskside plumes (Foster *et al.*, 2002; Goldstein *et al.*, 2003b; Goldstein *et al.*, 2004a; Goldstein and Sandel, 2005; Goldstein *et al.*, 2005a). Westward SAPS flows are a major influence near dusk, where models that ignore SAPS incorrectly predict the location of a flow stagnation region.

3.3. Hot-Cold Plasma Interactions

This section considers the role of cold plasmaspheric plasma in the dynamics of the warmer particle populations, the ring current and radiation belts.

3.3.1. *Ring Current and Plasmasphere*

The ring current was introduced in Sections 3.1 and 3.2. As depicted in Figure 3b, the plasmasphere (green) and ring current (orange) are roughly spatially complementary, but the two plasmas do overlap in the range 1600–1800 MLT. This overlap can lead to the loss of the ring current. The intermingling of warm ring current ions and cold, dense plasmaspheric plasma favors the growth of electromagnetic ion cyclotron (EMIC) waves (Gary *et al.*, 1995), which can scatter the ring current ions into the ionosphere. Thus, where the ring current encounters the plasmasphere, it can suffer EMIC-wave scattering and dump its particles into the ionosphere, producing distinctive auroral signatures (Spasojević *et al.*, 2004). Overlap between the plasmasphere and ring current is an unstable situation, so that on long enough time scales the plasmasphere and ring current should be spatially complementary.

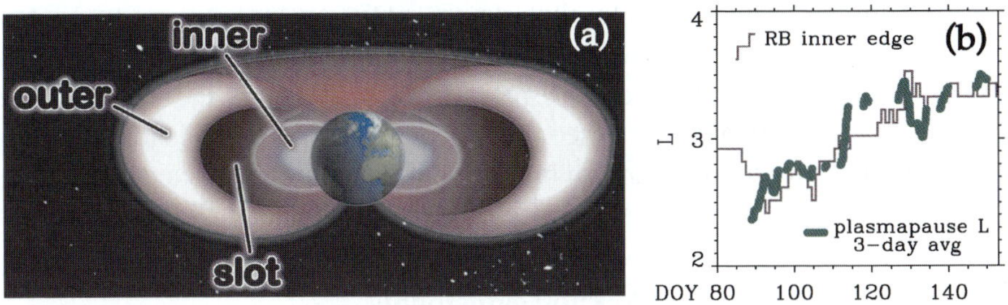

Figure 5. (a) Cartoon of the radiation belts, showing the inner and outer belts and the slot region. (b) Plot illustrating the close relationship between the inner edge of the outer radiation belt (purple), and the 3-day-averaged plasmapause (green), for two months in 2001 (Goldstein *et al.*, 2005c).

3.3.2. *Radiation Belts and Plasmasphere*

The radiation belts (or "Van Allen" belts) are magnetospheric regions of magnetically trapped high-energy ions and relativistic electrons (Van Allen and Frank, 1959). The relativistic electrons are separated into two belts, an inner belt below $L \approx 2$ and an outer belt above $L \approx 3$ (see Figure 5a). Whereas the inner belt is quite stable, unaffected by all but the most severe geomagnetic storms, the outer belt is highly sensitive to geomagnetic conditions. The two belts are normally separated by a "slot" region devoid of relativistic electrons, but during intense storms, the outer electron belt can move inward to penetrate (and rarely, completely fill) the slot region (Baker *et al.*, 1994). A crucial role in the creation of the slot region is played by the plasmasphere, which is typically filled with broad-band whistler mode wave emissions known as plasmaspheric hiss (Thorne *et al.*, 1973). Radiation belt electrons outside of $L \approx 2$ are susceptible to scattering by the hiss wave emissions; thus, for decades it has been believed that pervasive wave-particle interactions (between hiss waves and radiation belt particles) inside the plasmasphere are the cause of the electron losses that maintain the slot region. If this is true, the outer extent of the plasmasphere should on average coincide with the inner extent of the outer electron belts (Russell and Thorne, 1970). Recent studies comparing global plasmasphere images with in situ relativistic electron data have confirmed this prediction (Baker *et al.*, 2004; Goldstein *et al.*, 2005c), as illustrated in Figure 5b, and have shed light on the conditions under which the outer belt can penetrate the slot region. Intense storms produce severe erosions, so that the plasmapause moves inside the nominal slot region; without the usually-present hiss to remove the electrons, the slot region may have the opportunity to be filled in by a persistent population of newly-energized relativistic electrons.

4. Concluding Remarks

The advent of space-based imaging has provided a unique perspective to study the response of the inner magnetosphere to the ever-changing solar wind conditions.

Decades-old hypotheses about how the plasmasphere is eroded by enhanced convection have been confirmed, and important sub-global effects (such as SAPS and shielding) have proven to be a critical part of the behavior of the inner magnetosphere. Imaging has allowed us to see that plasmasphere erosion is just one aspect of the coupled response of the entire inner magnetosphere and ionosphere.

Acknowledgements

I am grateful indeed to have been invited to such an exciting and fruitful workshop. I wish to thank all the 2005 ISSI workshop participants, the members of the organizing committee, and our gracious and generous host institution ISSI. I was delighted to interact with solar and heliospheric physicists, and to make new friends. Scientific discussions with G. Haerendel, J. Linker, D. Baker, I. Daglis, R. Lin, and B. Fleck were particularly rewarding. Special thanks are given to M. Desai and M. Guhathakurta for animated conversation at the lunch table, to S. Gibson for the introduction to Dr. B. T. Klöun, and to T. Zurbuchen for making physicists look good at the Einstein house. Data were most kindly provided by B. R. Sandel (IMAGE EUV), M. F. Thomsen (LANL MPA), P. C. Brandt (IMAGE HENA), and I thank J. Burch and the IMAGE team for the continuing success of the mission. Research support from the NASA Sun-Earth Connections Guest Investigator program under NAG5-12787, and from the NASA IMAGE mission under NAS5-96020, is most gratefully acknowledged.

References

Aggson, T. L., Heppner, J. P., and Maynard, N. C.: 1983, *J. Geophys. Res.* **88**, 3981.

Akasofu, S.-I.: 1964, *Planet. Space Sci.* **12**, 273.

Baker, D. N., Blake, J. B., Callis, L. B., Cummings, J. R., Hovestadt, D., Kanekal, S., *et al.*: 1994, *Geophys. Res. Lett.* **21**, 409.

Baker, D. N., Kanekal, S. G., Li, X., Monk, S. P., Goldstein, J., and Burch, J. L.: 2004, *Nature* **432**, 878.

Baker, D. N., Pulkkinen, T. I., Angelopoulos, V., Baumjohann, W., and McPherron, R. L.: 1996, *J. Geophys. Res.* **101**, 12975.

Brandt, P. C., Ohtani, P. S., Mitchell, D. G., Fok, M.-C., Roelof, E. C., and DeMajistre, R.: 2002, *Geophys. Res. Lett.* **29**(20), 1954.

Burch, J. L., M. S. B., Mitchell, D. G., Moore, T. E., Pollock, C. J., Reinisch, B. W., *et al.*: 2001a, *Science* **291**, 619.

Burch, J. L., Mitchell, D. G., Sandel, B. R., Brandt, P. C., and Wüest, M.: 2001b, *Geophys. Res. Lett.* **28**, 1159.

Carpenter, D. L.: 1963, *J. Geophys. Res.* **68**, 1675.

Carpenter, D. L.: 1967, *J. Geophys. Res.* **72**, 2969.

Carpenter, D. L.: 1970, *J. Geophys. Res.* **75**, 3837.

Carpenter, D. L., and Akasofu, S.-I.: 1972, *J. Geophys. Res.* **77**, 6854.

Carpenter, D. L., and Stone, K.: 1967, *Planet. Space Sci.* **15**, 395.
Carpenter, D. L., Stone, K., Siren, J. C., and Crystal, T. L.: 1972, *J. Geophys. Res.* **77**, 2819.
Chappell, C. R.: 1974, *J. Geophys. Res.* **79**, 1861.
Chappell, C. R., Harris, K. K., and Sharp, G. W.: 1970, *J. Geophys. Res.* **75**, 50.
Coroniti, F. V., and Kennel, C. F.: 1973, *J. Geophys. Res.* **78**, 2837.
Daglis, I. A., Thorne, R. M., Baumjohann, W., and Orsini, S.: 1999, *Rev. Geophys.* **37**, 407.
DeMajistre, R., Roelof, E. C., C:son Brandt, P., and Mitchell, D. G.: 2004, *J. Geophys. Res.* in press.
Dent, Z. C., Mann, I. R., Menk, F. W., Goldstein, J., Wilford, C. R., Clilverd, M. A., *et al.*: 2003, *Geophys. Res. Lett.* **30**(12), 1600.
Dungey, J. W.: 1961, *Phys. Rev. Lett.* **6**, 47.
Fejer, B. G., and Scherliess, L.: 1995, *Geophys. Res. Lett.* **22**, 851.
Foster, J. C., and Burke, W. J.: 2002, *EOS Trans. AGU* **83**, 393.
Foster, J. C., Erickson, P. J., Coster, A. J., and Goldstein, J.: 2002, *Geophys. Res. Lett.* **29**(13), 1623.
Garcia, L. N., Fung, S. F., Green, J. L., Boardsen, S. A., Sandel, B. R., and Reinisch, B. W.: 2003, *J. Geophys. Res.* **108**(A8), 1321.
Gary, S. P., Thomsen, M. F., Yin, L., and Winske, D.: 1995, *J. Geophys. Res.* **100**(A11), 21961.
Goldstein, J., Burch, J. L., and Sandel, B. R.: 2005a, *J. Geophys. Res.* **110**, A09222.
Goldstein, J., Burch, J. L., Sandel, B. R., Mende, S. B., C:son Brandt, P., and Hairston, M. R.: 2005b, *J. Geophys. Res.* **110**, A03205.
Goldstein, J., Kanekal, S. G., Baker, D. N., and Sandel, B. R.: 2005c, *Geophys. Res. Lett.* **32**, L15104.
Goldstein, J., and Sandel, B. R.: 2005, in: J. L. Burch, M. Schulz, and H. Spence (eds.), *Inner Magnetosphere Interactions: New Perspectives from Imaging*. Washington, D. C.: American Geophysical Union, p. 1.
Goldstein, J., Sandel, B. R., Forrester, W. T., and Reiff, P. H.: 2003a, *Geophys. Res. Lett.* **30**(3).
Goldstein, J., Sandel, B. R., Hairston, M. R., and Mende, S. B.: 2004a, *Geophys. Res. Lett.* **31**, L15801.
Goldstein, J., Sandel, B. R., Reiff, P. H., and Hairston, M. R.: 2003b, *Geophys. Res. Lett.* **30**(24), 2243.
Goldstein, J., Sandel, B. R., Thomsen, M. F., Spasojević, M., and Reiff, P. H.: 2004b, *J. Geophys. Res.* **109**, A03202.
Goldstein, J., Spasojević, M., Reiff, P. H., Sandel, B. R., Forrester, W. T., Gallagher, D. L., *et al.*: 2003c, *J. Geophys. Res.* **108**(A4), 1147.
Goldstein, J., Spiro, R. W., Reiff, P. H., Wolf, R. A., Sandel, B. R., Freeman, J. W., *et al.*: 2002, *Geophys. Res. Lett.* **29**(16).
Goldstein, J., Spiro, R. W., Sandel, B. R., Wolf, R. A., Su, S.-Y., and Reiff, P. H.: 2003d, *Geophys. Res. Lett.* **30**(8), 1421.
Goldstein, J., Wolf, R. A., Sandel, B. R., and Reiff, P. H.: 2004c, *Geophys. Res. Lett.* **31**(1), L01801.
Grebowsky, J. M.: 1970, *J. Geophys. Res.* **75**, 4329.
Jaggi, R. K., and Wolf, R. A.: 1973, *J. Geophys. Res.* **78**, 2852.
Kelley, M. C., Fejer, B. G., and Gonzales, C. A.: 1979, *Geophys. Res. Lett.* **6**, 301.
LeDocq, M. J., Gurnett, D. A., and Anderson, R. R.: 1994, *J. Geophys. Res.* **99**, 23661.
Lemaire, J. F., and Gringauz, K. I.: 1998, *The Earth's Plasmasphere*. Cambridge: Cambridge University Press.
Maynard, N. C., and Chen, A. J.: 1975, *J. Geophys. Res.* **80**, 1009.
Moldwin, M. B., Sandel, B. R., Thomsen, M., and Elphic, R.: 2003, *Space Sci. Rev.* **109**, 47.
Pulkkinen, T. I., Tsyganenko, N. A., and Friedel, R. H. W. (eds.).: 2005, *Physics and Modeling of the Inner Magnetosphere (Geophysical Monograph 155).* Washington, DC: American Geophysical Union.
Reinisch, B. W., Huang, X., Haines, D. M., Galkin, I. A., Green, J. L., Benson, R. F., *et al.*: 2001, *Geophys. Res. Lett.* **28**, 1167.

Reinisch, B. W., Huang, X., Song, P., Green, J. L., Fung, S. F., Vasyliunas, V. M., *et al.*: 2004, *J. Geophys. Res.* **109**, A01202.

Russell, C. T., and Thorne, R. M.: 1970, In: *Cosmic Electrodynamics I*. Dordrecht, Holland: D. Reidel Publishing Co., p. 67.

Sandel, B. R., Goldstein, J., Gallagher, D. L., and Spasojević, M.: 2003, *Space Sci. Rev.* **109**, 25.

Sandel, B. R., King, R. A., Forrester, W. T., Gallagher, D. L., Broadfoot, A. L., and Curtis, C. C.: 2001, *Geophys. Res. Lett.* **28**, 1439.

Senior, C., and Blanc, M.: 1984, *J. Geophys. Res.* **89**, 261.

Spasojević, M., Frey, H. U., Thomsen, M. F., Fuselier, S. A., Gary, S. P., Sandel, B. R., *et al.*: 2004, *Geophys. Res. Lett.* **31**, L04803.

Spasojević, M., and Goldstein, J.: 2005, *EOS Trans. AGU* **86**(18)**, Jt. Assem. Suppl.**, Abstract SM42A–03.

Spasojević, M., Goldstein, J., Carpenter, D. L., Inan, U. S., Sandel, B. R., Moldwin, M. B., *et al.*: 2003, *J. Geophys. Res.* **108**(A9), 1340.

Spiro, R. W., Harel, M., Wolf, R. A., and Reiff, P. H.: 1981, *J. Geophys. Res.* **86**, 2261.

Thorne, R. M., Smith, E. J., Burton, R. K., and Holzer, R. E.: 1973, *J. Geophys. Res.* **78**, 1581.

Van Allen, J. A., and Frank, L. A.: 1959, *Nature* **183**, 430.

Vasyliūnas, V. M.: 1970, in: B. M. McCormac (ed.), *Particles and Fields in the Magnetosphere*. Dordrecht, Holland: D. Reidel Publishing Co., p. 60.

Volland, H.: 1973, *J. Geophys. Res.* **78**, 171.

Wygant, J. R., Rowland, D. E., Singer, H. J., Temerin, M., Mozer, F., and Hudson, M. K.: 1998, *J. Geophys. Res.* **103**, 29527.

END-TO-END MODELING OF THE SOLAR TERRESTRIAL SYSTEM

M. WILTBERGER[1,*] and D. BAKER[2]
[1]*National Center for Atmospheric Research 3080 Center Green Place Boulder, CO 80301, USA*
[2]*Laboratory for Atmospheric and Space Physics 1234 Innovation Drive, Boulder, CO 80301, USA*
*(*Author for correspondence: E-mail: wiltbemj@ucar.edu)*

(Received 18 November 2005; Accepted in final form 22 February 2006)

Abstract. Traditionally modeling for space science has concentrated on developing simulations for individual components of the solar terrestrial system. In reality these regions are coupled together. This coupling can be as simple as the driving of the magnetosphere – ionosphere – thermosphere system by the solar wind or as a complicated as the feedback of the ionospheric conductivity and currents on the magnetosphere. As part of the CISM project we are beginning a concentrated effort to compressively model the entire system. This approach includes chains of models. In the first chain physics based numerical models are utilized while in the second chain empirical models are coupled together. The first half of this paper discusses the numerical modeling approach by highlighting the coupling of pairs of regions within the system. In the second section we present results from empirical models which are combined to make long term forecasts of conditions in the geospace environment. It is expected that a validated and reliable forecast model for space weather can be obtained by combining the strongest elements of each chain.

Keywords: space physics, numerical modeling, solar physics, magnetospheric physics, ionospheric physics

1. Introduction

Advances in modeling the Earth's weather and climate have required the development of models which couple our understanding of the interactions between the atmosphere, ocean, and land into a single comprehensive model. The long term success of this approach forms the basis of the modeling approach for solar – terrestrial system laid out in the United States National Space Weather Plan. Instead of treating each region of the system separately they will be linked together into a single model capable of describing the entire system and the interactions between regions.

Development of a comprehensive solar-terrestrial model requires the utilization of models for the solar corona, solar wind, magnetosphere, inner magnetosphere, ionosphere and thermosphere. While other groups have conducted work in coupling all (Groth *et al.*, 2000) or a subset (Raeder *et al.*, 2001) of the components together, this paper highlights the work conducted by the Center for Integrated Space Weather Modeling (CISM). The coupling of physics based models together can be augmented in two ways by working with empirical models. First empirical models can serve as baseline for assessing the accuracy and improvements in the

Space Science Reviews (2006) 124: 217–231
DOI: 10.1007/s11214-006-9106-x

physical models. In addition, an optimal forecast model can be built from linking the strongest components of each model coupling chain into a single comprehensive end-to-end model.

This paper is divided into two main sections. The first section covers the results from the numerical modeling chain. These results are presented in a series of subsections which deal with the details of couplings between regions in a pairwise fashion. In the second section we discuss the empirical modeling chain. The details of making a solar wind prediction from an empirical coronal model are presented in the first subsection. Empirical modeling for the radiation belts is used to describe the techniques required for simulating the response of geospace to solar wind driving. The paper concludes with a few remarks about tools and techniques that increase the utility of end-to-end modeling.

2. Numerical Modeling

To build a numerical model for space weather prediction which extends from the Sun to the Earth, it is necessary to break the region down into manageable pieces. In Figure 1 an initial breakdown of the region into components is shown. A solar corona model is used to simulate the region in which solar transient events, e.g. flares and coronal mass ejections, begin their initial propagation out into the he-

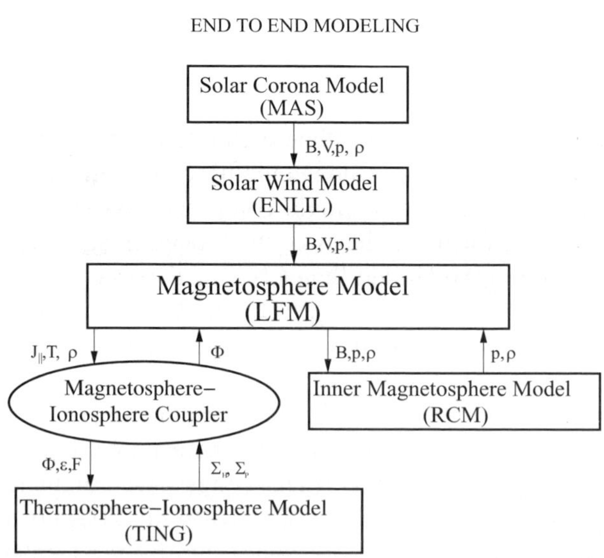

Figure 1. A schematic breakdown of the Sun – Earth system into component modeling regions. The arrows represent information flow with the parameters begin passed between models indicated next to each arrow.

liosphere. It is important to note that this model also plays an important role in specifying the structure of the magnetic topology of the Sun which governs location of the heliospheric current sheet and the location of slow and fast flows in the solar wind. The solar wind model takes information about the state of the plasma near the Sun, typically around 30 solar radii, from the solar corona model and propagates it out into the heliosphere. The coupling between these two models is fairly straight forward and usually involves passing specification of the plasma at a two-dimensional interface from the solar corona model to the solar wind model.

The modeling within the magnetosphere – ionosphere – thermosphere system is considerably more complicated because of the interconnection between these regions. The coupling between the solar wind model and the magnetospheric model requires specification of the plasma parameters along the entire computational domain of the magnetosphere. At resolutions that are practical with modern computers this typically means producing a one-dimensional solar wind stream similar to spacecraft observations, but as computing power and computational techniques advance this will become a three-dimensional problem. The magnetospheric model needs to specify the plasma and magnetic field parameters throughout a region that extends from close to the ionosphere to beyond the magnetopause and bow shock in all three-dimensions. Since taking the physical parameters involves a utilization of a set of models we break this process out into a separate interface coupler as indicated by the ellipse in Figure 1.

The most complicated coupling in this system is the interchange of information between the magnetosphere model and inner magnetosphere model which provides an accurate representation of the ring current and plasmasphere. It involves the specification of three-dimensional magnetic field structure in a subset of the closed field region within the magnetosphere. In this region the magnetosphere model provides an initial guess as to the structure of the plasma pressure and density which is then modified by the inner magnetosphere model and fed back in along the magnetic field lines. Coupling between geospace models is obviously quite tight and must be handled with considerable care in order to ensure the stability of the individual models and the reliability of the results.

Figure 1 is schematic, illustrating only an initial set of models necessary to represent the Solar – Terrestrial system. It can be expanded to include several additional models. Emergence of magnetic flux from beneath the photosphere and its evolution into structures which can erupt as CMEs may require the addition of regional models within the domain covered by the solar corona model. Specification of the radiation belts within the magnetosphere requires tracking highly energetic particles within the magnetosphere. Entry of solar energetic particles into the magnetosphere and thermosphere requires an understanding of the processes controlling their generation, propagation through the heliosphere, and transport through the topology of the magnetic field in geospace. All of these aspects can be thought of as additional rectangles in the modeling schematic which requires the creation of couplers to keep the interfaces clean and to facilitate easily inclusion of other approaches.

As part of the efforts by the CISM, individual models for each of the physical regions have been selected to form the comprehensive physical model. The Magnetohydrodynamics Around a Sphere (MAS) (Linker and Mikic, 1995; Mikic *et al.*, 1999) model is used to model the solar corona. The ENLIL model developed by Odstrcil and Pizzo, 1999, simulates the solar wind propagation through the heliosphere. The Earth's magnetosphere is modeled using the Lyon – Fedder – Mobbary (LFM) code (Fedder *et al.*, 1995; Lyon *et al.*, 2004) with the inner magnetosphere being simulated by the Rice Convection Model (RCM) (Harel *et al.*, 1981; Wolf, 1983; Toffoletto *et al.*, 2003). The Thermosphere Ionosphere Nested Grid (TING) model (Wang *et al.*, 1999) is used to simulate the upper atmosphere and ionosphere. In the remainder of this section we will provide a brief overview of the pairwise couplings that make the chain of models in the solar terrestrial physics system.

2.1. Solar Corona and Solar Wind Coupling

The MAS model is used to simulate the solar corona. It uses magnetohydrodynamic (MHD) equations to model the plasma on a spherical computational grid which extends from the photosphere out to 30 solar radii. The model uses global photospheric magnetic field maps, which can be purely theoretical or obtained from magnetograms, to determine the boundary conditions. More detail about these boundary conditions and their temporal evolution can be found in Mikic *et al.* (1999). As described in Linker *et al.* (1999) the model uses a polytropic equation of state, with $\gamma = 1.05$, which helps produces a density and flow differential on open and closed field lines. Unfortunately this physical simplification does not include effects like Alfven wave heating and results in the model having outflow speeds and gradients between high and low speeds which are considerably smaller than observed. It is important to note that the outer boundary is beyond the critical point so that the flow is supersonic.

The ENLIL solar wind model is designed to study the motion of supercritical plasma and magnetic field evolution throughout the heliosphere. It uses the MHD equations with a polytropic equation of state. In this case it uses a a value of $\gamma = 1.5$. The computational domain extends from a region beyond the critical point near the Sun to as far out into the heliosphere as desired. In order to reduce the computational work load and optimize resolution the latitude extent is limited to a region $\pm 60^\circ$ from the equator.

Since the coupling between these models has been discussed in great detail in Riley *et al.* (2001b), Odstrcil *et al.* (2002) and Odstrcil *et al.* (2004) we will only provide brief overview of the process here. The solar corona model is run for a period of time and the data are transfered to the ENLIL model, where a small spherical shell of a few grid cells is used to determine the MHD state vector at the grid points along

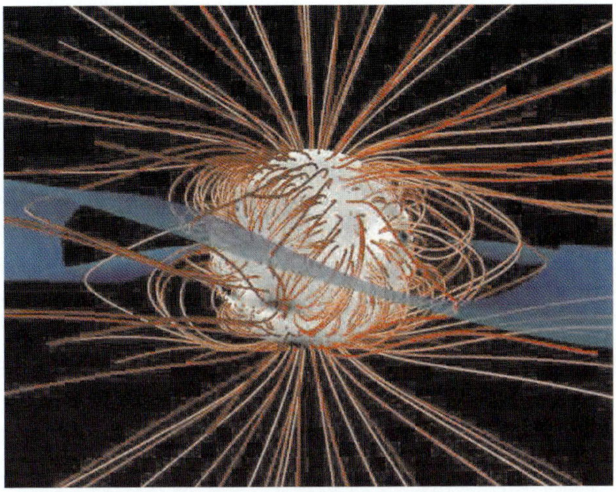

Figure 2. Image from a MAS simulation for Carrington rotation 1923. The event was driven by synoptic magnetic field maps seen on the photospheric surface in the image. Red magnetic field lines show an open region near the poles and streamer belt. The heliospheric current sheet is distorted by the magnetic topology and is colored by the local solar wind velocity. Courtesy Sarah McGregor.

the inner boundary of the computational domain. At this point several additional changes are made to the plasma parameters to improve the physical accuracy of the coupling. First, the coronal model is rotated by a discrete amount between each time step in order to create a Parker spiral distorted heliospheric current sheet as seen in Figure 2. Since the MAS simulations have a solar wind speed that is lower than observed it is also adjusted by a correction factor dependentupon the magnetic field topology. Using this inner boundary condition information allows ENLIL to specify the plasma parameters through the heliosphere, including at the Earth where it is used to drive the geospace models.

2.2. Magnetosphere and Ionosphere - Thermosphere Coupling

To model the magnetospheric plasma configuration another MHD model, the LFM (Lyon *et al.*, 2004), is used. The LFM computational domain covers a roughly cylindrical region embedded in the solar wind that extends from 30 R_E upstream to 300 R_E downstream of the Earth. The radial extent is typically 100 R_E. At the position of the Earth a small sphere is cut out which is typically between 2–3 R_E in radius. The boundary conditions along the outer edges are specified by the solar wind conditions. At the inner boundary MHD parameters are mapped down to ionospheric heights where they are used to determine the cross polar cap potential which is used as the flow boundary condition for the magnetospheric plasma. When operating in a stand-alone mode this calculation is done within a simple two-dimensional electrostatic model for the ionosphere.

The coupled thermosphere-ionosphere system is modeled by the Thermosphere Ionosphere Nest Grid model which solves for the coupled equations for mass, momentum, and energy for neutrals and O^+ in this domain (Wang *et al.*, 1999, 2004). Chemical equilibrium is used to determine the density of other ion species, e.g. NO^+, O_2^+, and the electrons. The model has the capability to include nested grid structures within its regular latitude-longitude grid for improved regional resolution. The vertical domain extends from 97 to 500 km with 25 constant pressure levels. It uses the F10.7 flux to determine the EUV ionization. It needs the energy flux of precipitating electrons and a polar cap convection pattern to be specified. Finally, it needs to have the tides from the lower atmosphere specified at 97 km altitude. In stand-alone operation the magnetospheric parameters are determine from empirical models.

The coupling of these two models forms the Coupled Magnetosphere Ionosphere Thermosphere (CMIT) model which is described in detail in Wiltberger *et al.* (2004) and Wang *et al.* (2004), so once again we will only provide a brief overview. Since the transfer of information between these two models is bidirectional and involves the transformation of different sets of physical parameters, we decided to separate these functions into a discrete coupling module. Currently this module is closely linked with the LFM, but it can removed and placed into other models in a straight forward fashion. This module takes the plasma parameters along the inner boundary of the MHD domain and transforms them into the precipitating electron energy flux using empirical relationships discussed in Slinker *et al.* (1995). Using the previous information from the TING model it makes an initial guess for the ionospheric convection pattern which is passed along to the model with electron information to determine a conductance distribution. The new conductance information is used to compute a new potential pattern which is used as the boundary condition for the MHD domain. There is a large disparity in time steps between these two computational domains, which we utilize to our advantage. In TING a typical time step is two minutes, while in the LFM a typical time step is 0.3 sec. In CMIT, the LFM portion uses the conductance information for a two minute interval while recomputing the potential pattern based upon the current field aligned current configuration. At the end of that interval, another information exchange between the models occurs and the conductivities are updated for the next two minute interval. Originally, we thought that it might be necessary to iterate the exchange between these two models to prevent large changes in the convection pattern from being seen by TING, but that has not turned out to be necessary in practice.

Results from the coupled model are shown in Figure 3. The simulation was driven by solar wind parameters taken from an end-to-end event study described in Luhmann *et al.* (2004). The high density and southward IMF results in a strong compression of the dayside magnetopause which can be seen by the red high density region in the Figure. Reconnection is occurring along the dayside which results in open field lines being created in the polar cap. At the ionospheric altitude a strong

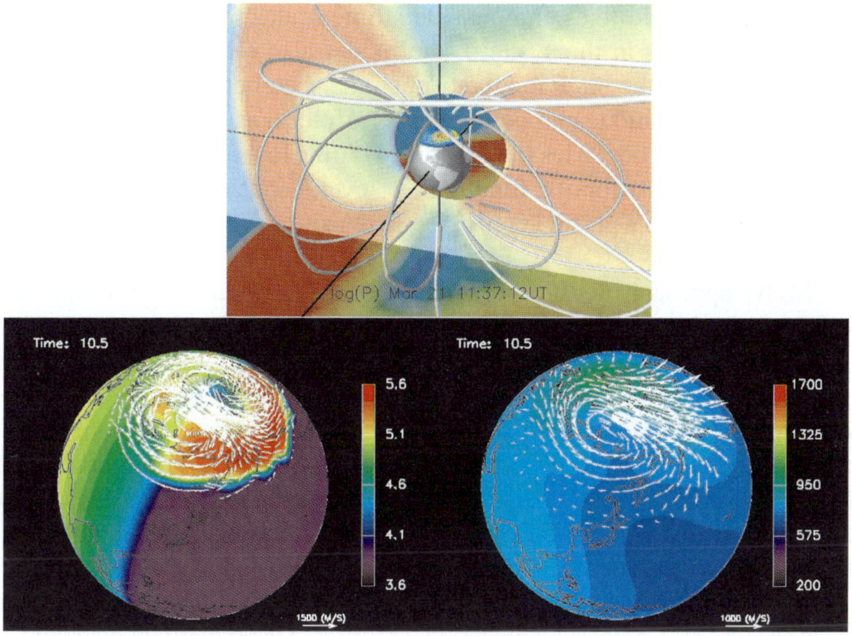

Figure 3. The top panel shows an image from the CMIT simulation which shows the plasma density on equatorial and meridional cut planes through the computational domain. A series of magnetic field lines are drawn in white including several open ones. At the ionospheric altitude the Hall conductance as determined by TING is displayed during the strong driving of magnetic storm. On the bottom left we see the we see the electron density in the E region with the ion drift (convection) pattern superposed on top. On the right hand side we see the F region temperature and the neutral winds.

enhancement to the Hall conductance in the auroral zone is clearly evident. This storm produces strong effects in the thermosphere, e.g. heating of the F region and enhanced neutral winds, which are discussed in Wang *et al.* (2004) and are summarized in the bottom panel of Figure 3. On the left hand side we see ion drifts in the E region of the ionosphere as imposed by the magnetospheric portion of the CMIT model. The color scale shows the electron density with significant enhancements present in the auroral zone. Strong driving results in higher neutral temperatures in the F region as seen the lower right portion of the figure. In addition, the strong ion drifts are driving neutral winds at F region altitudes which show the characteristic rotation enhancement on the dawnside.

2.3. Magnetosphere and Inner Magnetosphere Coupling

The Rice Convection Model uses the Vasyliunas, 1970, approach of using drift and current conservation equations to determine the plasma motion, currents, and electric fields in the inner magnetosphere. It assumes an isotropic distribution of particles on closed field lines and considers time scales for the drifts to be longer than other time scales in the system. The model also requires that the magnetic field configuration in the inner magnetosphere be specified. A detailed description of this numerical technique can be found in Toffoletto *et al.* (2003). The essential point here is that the model uses the flux tube volume to track the drift motion of the slices of the plasma distribution function for several species. This information can be converted back into a plasma pressure and density which is constant throughout the flux tube.

The MHD approach of the LFM and the multi-species kinetic approach of the RCM are in fact quite complementary. The RCM provides an excellent description of the inner magnetosphere including the effects of ionospheric shielding, formation of region 2 currents. However, it does not provide a self consistent description of the magnetic field. The LFM provides a magnetic field configuration which is necessarily consistent with the plasma configuration, but it does not well represent the gradient and curvature drift physics. This results in weak region 2 currents and little enhancement of the ring current during magnetic storms. In the two-way coupled model, the LFM provides RCM with a magnetic field configuration in the inner magnetosphere along with initial values for the plasma pressure and density. The RCM assumes a Maxwellian distribution for the plasma and uses it as time-dependent boundary conditions for a drift calculation. This calculation is not required to preserve the Maxwellian character of the distribution. These plasma parameters are then passed back to the LFM with the assumption that the density and pressure determined by the RCM is constant along each field line. The core of this coupling requires transferring information between the regular spatial grid of the LFM and the flux tube volume representation of the RCM. It is accomplished by tracing field lines on a rectangular grid, with source points that originate from the regular ionospheric grid in the RCM.

A comparison between the stand-alone LFM and the coupled LFM-RCM model is shown in Figure 4. Each model was run for a configuration which has the interplanetary magnetic field turning southward at 4UT and remaining southward for the remainder of the simulation. In the left-hand panel we see the LFM configuration at 6 UT with coupled model results at the same time shown in the right hand panel. In each panel the color contours show the plasma pressure with the same color bar. Magnetic field lines, shown in white, are drawn from the same locations along the X axis. In both models what appears to be a substorm has already occurred. In the stand alone LFM the field in the inner magnetosphere is primarily dipolar, with minimal enhancement of the pressure in the inner magnetosphere. In the coupled model the field shows a more stretched configuration, with significant enhancement of the inner magnetospheric pressure. This pressure spreads out in both local time and radial extent as the interval progresses, indicating the development of a significant ring current in the coupled simulation results.

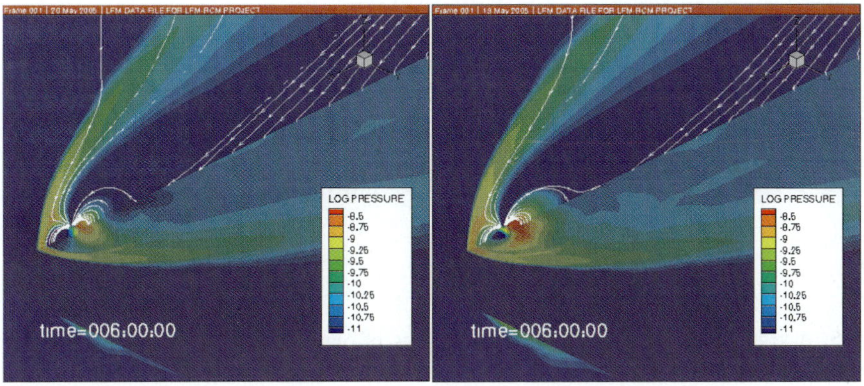

Figure 4. This figure compares results for the pressure (colored contours) and the magnetic field (white lines) configuration between the stand alone LFM (left) and the two-way coupling between the LFM and RCM (right). A clear enhancement of the inner magnetospheric pressure in the inner magnetosphere is present indicative of a improve representation of the ring current. The magnetic field lines show inflation caused by the larger ring current.

3. Empirical Modeling

While building a physics-based numerical model for the Solar-Terrestrial system is a key goal of the CISM project, this effort is augmented by efforts to build a forecast model from empirical, semi-empirical and inverse models. In addition to providing forecasts which may be of utility to the Space Weather community the coupling of these models together will provide a baseline which can be used to assess the accuracy and robustness of the physics-based model.

As with the physics-based models, the key to making predictions about the state of the geospace environment is to understand the conditions in the solar wind. The top path in Figure 5 shows the typical prediction route where measurements of the solar wind density (ρ), velocity (**V**), and magnetic field (**B**) are propagated to the Earth. These measurements can be used to drive models to predict the MeV electron flux (Vassiliadis *et al.*, 2002; Li *et al.*, 2003), magnetic field configuration (Tsyganenko, 1995), ground magnetic field perturbations (Weigel *et al.*, 2003), and the vast array of magnetic indicies (Kp, Ap, Dst, AE) (Boberg *et al.*, 2000; McPherron, 1999; Klimas *et al.*, 1998; Takalo and Timonen, 1997). In the bottom path data input is pushed back to the solar surface where observations of the photospheric magnetic field are used within the Wang – Sheeley – Arge (WSA) method (Arge and Pizzo, 2000) to determine the solar wind speed and limited information about the interplanetary magnetic field at 1 AU. Using this path allows for earlier lead times in the predictions; however as we will see, it greatly decreases the accuracy of the predictions at the Earth. In the remainder of this section we will briefly explore some of these empirical models and their ability to predict the configuration of the Sun-Earth system.

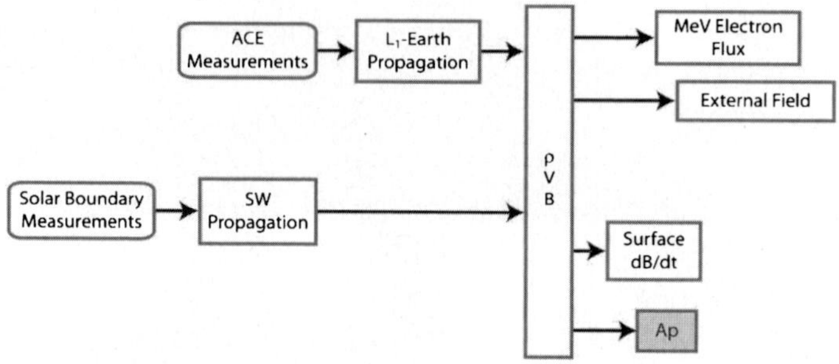

Figure 5. A schematic representation of the coupling of observations to empirical models for predicting the state of the magnetosphere.

3.1. Solar Wind Modeling

The empirical modeling of the solar wind provides information about the ambient solar wind speed and interplanetary magnetic field polarity throughout the heliosphere. It is important to note that this kinematic model provides information about the quasi-steady solar wind flow and will therefore have greater accuracy for recurrent phenomena than for transients. The original Wang and Sheeley (WS) model (Wang and Sheeley, 1992) used ground-based line-of-sight observations of the solar surface magnetic field as input to a potential magnetic field model (PFSS (Schatten *et al.*, 1969) of the coronal magnetic field configuration. The expansion of magnetic field from the photosphere to a spherical surface at 2.5 solar radii is used to drive an empirical model for determining the solar wind speed at this surface (Wang and Sheeley, 1990). The model then assumes that solar wind flow propagates radially outward from this surface with all subsequent interactions ignored.

Several improvements have been made to the original WS model over the years and are incorporated into the WSA version of the model. These improvements are discussed in detail by Arge *et al.* (2004) so we will only provide a brief overview of them here. The first category of modifications include improvements to the way the input photospheric magnetic field is processed. These include accounting for line of sight projection effects, polar field values, line-saturation effects in the creation of global synoptic maps from the raw magnetograms. They also developed techniques for making frequent updates to the synoptic map and the handling of data gaps. The WSA version contains several improvements to the physical model itself. This includes utilization of a slightly more sophisticated propagation model which allows for interactions between the streams. A higher-accuracy empirical model for determining the solar wind speed was created by utilizing a fit to L1 observations. In addition, the model was further improved by including information about the distance between the open field line and the nearest coronal hole boundary (Riley *et al.*, 2001a). These modifications result in significant increases in the prediction efficiency of the the model.

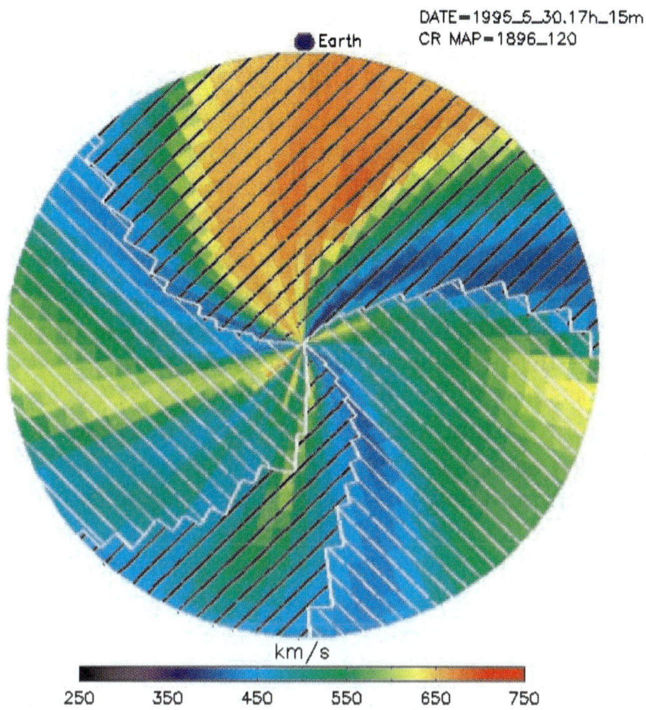

Figure 6. Results from the WSA model for Carrington Rotation 1896. Color indicates speed, while magnetic field polarity toward(away) is indicated by the white(black) lines. Regions of magnetic field toward/away the sun are covered with black/while lines. The location of the Earth is shown with a blue sphere.

Figure 6 shows the results for the WSA model during Carrington rotation 1896. It is clear that a series of high speed streams are reaching the Earth. In other portions of the heliosphere the solar wind speed is quite low. The figure also indicates regions of magnetic field pointed toward (black lines) the Sun and regions of field pointed away (white lines) from the Sun. A time history of the solar wind speed and polarity is obtained by extracting data from the location of the Earth as the model evolves forward in time. It is important to note this time history has a low cadence of one data point every few hours.

3.2. Radiation Belt Modeling

Empirical modeling of the response of geospace to solar wind conditions is applied to many parameters that describe the system. In general, these methods can be broken down into two broad categories. A black box system utilizes techniques like neural networks to transform the input parameters into output conditions without requiring any physical understanding of the system. These models can produce reliable results quickly for complex systems. The accuracy of the data is the major limiting factor, and as such they have difficulty with predictions for events at the limit or outside the range of data used to develop the model. Grey box systems combine qualitative knowledge of the system with data to develop a model for making predictions. These techniques include linear filters and state-space models capable of representing ordinary differential equations (ODE) with coefficients determined from a training data set. The essential characteristics of these techniques are the same whether they attempt to predict Ap or radiation belt flux at geosynchronous orbit. In this section we will limit our discussion to radiation belt flux.

Baker *et al.* (1990) used the linear filter method to predict the average daily flux of electrons, J_e, at geostationary orbit. In a classic application of the grey box philosophy of design they assumed that the dynamics of J_e can be described in part by an ODE with the solar wind velocity, V_{SW}, as a driver. The coefficients of this ODE are derived from a historical data set of V_{SW} and J_e. The result of applying this technique is the generation of an impulse response function (IRF) which describes how the flux responds to a unit increase in solar wind velocity. In this case, the IRF shows that the flux initially experiences a sharp dropout followed by a rise in flux over the two day interval following the impulse. Once the peak is reached the flux levels decay over the next four days until they return to the pre-impulse levels. Vassiliadis *et al.* (2002) applied this technique to a range of L-shells and found a relatively localized peak near L $= R_E$ at one day, and a subsequent peak at 2–3 days that extended from 4–7 R_E consistent with Baker *et al.* (1990) earlier findings.

Baker *et al.* (2004) linked this empirical model for predicting radiation belt fluxes to the WSA model which predicts the value of solar wind speed. They utilized the solar observations for the entire year of 1995 which is near solar minimum and has several recurring high speed streams. These conditions are favorable for the application of the WSA technique. The interval also contained several data gaps and a flat heliospheric current sheet which make predictions more challenging. Comparison of the predictions with L1 measurements have a correlation coefficient of 0.36 and a prediction efficiency (PE) of 0.13. The model clearly captures the trends in the solar wind flow, but has significant errors (± 2 days) in predicting the arrival time of impulses at the Earth. Feeding these solar wind velocity measurements to the empirical model for radiation belt fluxes leads to results with a PE of 0.05 for 3 and 4 days out. This compares to a PE of 0.17 at 3 days and a PE of 0.13 at 4 days for a model based upon old L1 observations. Combining the WSA predictions with observations of solar speed and recent flux levels improves the results to slightly better than the model run on L1 data. This implies that significant limiting factor in the 3 to 4 predictions is the poor PE of the WSA model. It anticipated that as the WSA improves the PE will exceed that of the L1 model.

4. Summary and Conclusions

In this paper we have presented results from numerical and empirical models which have been joined together to make predictions about the state of the coupled

Sun – Earth system. The numerical chain utilizes state-of-the-art regional models and couples them together by passing physical parameters along regions of interface. In some cases these couplings are simple one-way transfers of information along boundaries and in other cases they involve two-way transfers of information in three-dimensional regions of overlap. On the empirical side the coupling of models involves the passing of information from one interface to another. It highlights the fact that resulting model will have an accuracy which is limited by the least accurate model in the coupling chain. As part of the CISM program we envision a close coupling between these two chains. At a minimum, the empirical models can provide a baseline model for assessing the skill of the numerical models as they become operational. In addition, a forecast model will be created which utilizes the strongest components of each chain to make the best possible prediction.

The development of a linked chain of models for the Sun-Earth system highlights the need for a consistent framework for coupling the models together and a set of tools for visualizing and analyzing the results. In Goodrich *et al.* (2004) the details of the CISM framework are discussed. A common method for transferring information between running codes, including the difficult problem of passing information between codes with data distributed amongst several processors, form one pillar of the framework. The second pillar utilizes a package to provide interpolation between various grids by simply defining the location of the grid points at any given point in time. CISM-DX is a package which has been developed for visualizing and analyzing space science model output and observations within a single tool (Wiltberger *et al.*, 2005). It combines the three-dimensional visualization strengths of OpenDX (IBM, 1998) with the Octave (Eaton, 2002) analysis tool. In addition, it includes numerous extensions of each package for space physics applications. The package is released under an open-source licensing agreement and contains examples of analysis which extend from very simple to very complex.

Modeling of the Solar – Terrestrial system does not happen in a vacuum. Our efforts are highly dependent upon the quality and quantity of observations, especially at regions which for the boundaries of the system. Improvements in the specification of the photospheric magnetic field include results from STEREO which will expand our field of view and the Solar Dynamics Observatory which will improve our understanding of the connection between small scale and large scale magnetic structures at the photosphere. Improvements in our physical understanding of the system are also required. Understanding the mechanisms that lead to eruption of CMEs will be essential for making accurate predictions at the Earth. In the terrestrial environment will need to further our understanding of the interaction between waves and particles that lead to acceleration and loss of radiation belt particles. There are many challenges in the future for modeling, but the coupling of these models provides an unprecedented opportunity for understanding the system as a single entity.

Acknowledgements

We wish to thank the members of the CISM team who developed the models and underlying coupling technology which ties them together. In particular we thank, J. A. Linker, P. Riley, J. Lyon, F. R. Toffoletto, W. Wang, A. Burns, S. C. Solomon, R. S. Weigel, T. Guild, and S. McGregor for their specific contributions to this paper. This material is based upon work supported in part by CISM, which is funded by the STC program of the National Science Foundation under Agreement Number ATM-0120950. The National Center for Atmospheric Research is sponsored by the National Science Foundation. Generous support from ISSI for attending the conference in Bern was greatly appreciated.

References

Arge, C. N., Luhmann, J. G., Odstrcil, D., Schrijver, C. J., and Li, Y.: 2004, *J. Atm Solar Terr. Phys.* **66**, 1295.

Arge, C. N., and Pizzo, V. J.: 2000, *J. Geophys. Res.* **105**, 10465.

Baker, D. N., McPherron, R. L., Cayton, T. E., and Kebesadel, R. W.: 1990, *J. Geophys. Res.* **95**, 15133.

Baker, D. N., Weigel, R. S., Rigler, E. J., Vassiliadis, R. L. M. D., Arge, C. N., Siscoe, G. L., *et al.*: 2004, *J. Atm. Solar Tres. Res.* **66**, 1491.

Boberg, F., Wintoft, P., and Lundstedt, H.: 2000, *Physics and Chemistry of the Earth C* **25**, 275.

Eaton, J. W.: 2002, *GNU Octave Manual*. Network Theory Ltd., 1st edition.

Fedder, J. A., Lyon, J. G., Slinker, S. P., and Mobarry, C. M.: 1995, *J. Geophys. Res.* **100**, 3613.

Goodrich, C. C., Sussman, A. L., Lyon, J. G., Shay, M. A., and Cassak, P. A.: 2004, *J. Atm. Solar Tres. Phys.* **66**, 1469.

Groth, C. P. T., DeZeeuw, D. L., Gombosi, T. I., and Powell, K. G.: 2000, *J. Geophys. Res.* **105**(A11), 25053.

Harel, M., Wolf, R. A., Reiff, P. H., Spiro, R. W., Burke, W. J., Rich, F. J., *et al.*: 1981, *J. Geophys. Res.* **86**, 2217.

IBM: 1998, *IBM Visualization Data Explorer User's Guide*. Yorktown Heights, NY USA: IBM, 3rd edition.

Klimas, A. J., Vassiliadis, D., and Baker, D. N.: 1998, *J. Geophys. Res.* **103**(12), 20435.

Li, X., Temerin, M., Baker, D. N., Reeves, G. D., Larson, D., and Kanekal, S. G.: 2003, *Eos Transactions of the AGU* **84**, 369.

Linker, J. A., and Mikic, Z.: 1995, *J. Geophys. Res.* **100**, 25165.

Linker, J. A., Mikic, Z., Biesecker, D. A., Forsyth, R. J., Gibson, W. E., Lazarus, A. J., *et al.*: 1999, *J. Geophys. Res.* **104**, 9809.

Luhmann, J., Solomon, S., Wang, W., and Wiltberger, M.: 2004, *J. Atm. Solar Terr. Phys.* p. doi:10.1016/j.jastp.2004.04.005.

Lyon, J. G., Fedder, J. G., and Mobarry, C. M.: 2004, *J. Atm. Space Phys.* pp. 1333–1350.

McPherron, R. L.: 1999, *Phys. Chem. Earth (C)* **24**((1–3)), 45.

Mikic, Z., Linker, J. A., Schnack, D. D., Lionello, R., and Tarditi, A.: 1999, *Phys. Plasmas* **6**, 2217.

Odstrcil, D. J., Linker, J. A., Lionello, R., Mikic, Z., Riley, P., Pizzo, V. J., *et al.*: 2002, *J. Geophys. Res.* **107**, doi:10.1029/2002JA009334.

Odstrcil, D. J., and Pizzo, V. J.: 1999, *J. Geophys. Res.* **104**, 483.

Odstrcil, D. J., Pizzo, V. J., Riley, J. A. L. P., Lionello, R., and Mikic, Z.: 2004, *J. Atm. Solar Terr. Phys.* **66**, doi:10.1016/j.jastp.2004.04.007.

Raeder, J., Wang, Y., and Fuller-Rowell, T. J.: 2001, In: H. S. and (ed.), *Space Weather*, Vol. 125 of *Geophysical Monograph Series*. AGU, pp. 377–384.

Riley, P., Linker, J., Mikic, Z., and Lionello, R.: 2001a, In: P. Song, H. Singer, and G. Siscoe (eds.), *Space Weather*, Vol. 125 of *Geophys. Mono.* Washington, D.C.: AGU, p. 159.

Riley, P., Linker, J. A., and Mikic, Z.: 2001b, *J. Geophys. Res.* **106**, 15889.

Schatten, K. H., Wilcox, J. M., and Ness, N. F.: 1969, *Solar Physics 9* pp. 442–455.

Slinker, S. P., Fedder, J. A., and Lyon, J. G.: 1995, *Geophys. Res. Lett.* **22**, 859.

Takalo, J., and Timonen, J.: 1997, *Geophys. Res. Lett.* **24**, 2403.

Toffoletto, F. R., Sazykin, S., Spiro, R. W., and Wolf, R. A.: 2003, *Space Sci. Rev.* **107**, 175.

Tsyganenko, N. A.: 1995, *J. Geophys. Res.* **100**, 5599.

Vassiliadis, D., Klimas, A. J., Kanekal, S., Baker, D., and Weigel, R. S.: 2002, *J. Geophys. Res.* **107**, 10.1029/2001JA000506.

Vasyliunas, V. M.: 1970, In: B. M. McCormac (ed.), *Particles and Fields in the Magnetosphere*. Hingham, MA: Reidel, pp. 60–71.

Wang, W., Burns, A. G., Solomon, S., and Killeen, T. I.: 2004, *Adv. Space Res.* p. in press.

Wang, W., Killeen, T. L., Burns, A. G., and Roble, R. G.: 1999, *J. Atm. Solar Terr. Phys.* **61**, 385.

Wang, Y. M., and Sheeley, N. R.: 1990, *Ap. J.* **355**, 726.

Wang, Y. M., and Sheeley, N. R.: 1992, *Ap. J.* **392**, 310.

Weigel, R. S., Klimas, A. J., and Vassiliadis, D.: 2003, *J. Geophys. Res.* **108**, doi:10.1029JA009627.

Wiltberger, M., Wang, W., Burns, A., Solomon, S., Lyon, J. G., and Goodrich, C. C.: 2004, *J. Atm. Solar Terr. Phys.* p. submitted.

Wiltberger, M., Weigel, R. S., Gehmeyr, M., and Guild, T.: 2005, *J. Geophys. Res.* **110**, doi:10.1029/2004JA010956.

Wolf, R. A.: 1983, In: R. L. Carovillano and J. M. Forbes (eds.), *Solar Terrestial Physics*, D. Hingham, MA: Hingham, pp. 303–368.

PARTICLE ACCELERATION BY THE SUN: ELECTRONS, HARD X-RAYS/GAMMA-RAYS

R. P. LIN
Physics Department & Space Sciences Laboratory, University of California, Berkeley, CA 94720-7450, USA
(E-mail: rlin@ssl.berkeley.edu)

(Received 15 June 2006; Accepted in final form 26 July 2006)

Abstract. Observations of hard X-ray (HXR)/γ-ray continuum and γ-ray lines produced by energetic electrons and ions, respectively, colliding with the solar atmosphere, have shown that large solar flares can accelerate ions up to many GeV and electrons up to hundreds of MeV. Solar energetic particles (SEPs) are observed by spacecraft near 1 AU and by ground-based instrumentation to extend up to similar energies, but it appears that a different acceleration process, one associated with fast Coronal Mass Ejections (CMEs) is responsible. Much weaker SEP events are observed that are generally rich in electrons, ^{3}He, and heavy elements. The energetic particles in these events appear to be similar to those accelerated in flares. The Ramaty High Energy Solar Spectroscopic Imager (RHESSI) mission provides high-resolution spectroscopy and imaging of flare HXRs and γ-rays. The observations of the location, energy spectra, and composition of the flare accelerated energetic particles at the Sun strongly imply that the acceleration is closely related to the magnetic reconnection that releases the energy in solar flares. Here preliminary comparisons of the RHESSI observations with observations of both energetic electrons and ions near 1 AU are reviewed, and the implications for the particle acceleration and escape processes are discussed.

Keywords: electrons, solar energetic particles, flares, SEPs, radio, hard X-rays, RHESSI, WIND

1. Introduction

In large solar flares, nuclear γ-ray lines and pion decay emission have been detected that are produced by energetic ions with $\sim$10–100 MeV and GeV energies, respectively, in nuclear collisions with the solar atmosphere (Lin *et al.*, 2003). HXR/γ-ray continuum emissions, produced by bremsstrahlung collisions of energetic electrons with the atmosphere, have been observed up to $\gtrsim$100 MeV. Ions and electrons up to about the same energies have been directly detected by *in situ* space observations (ground-based observations for the most energetic ions) in the interplanetary medium (IPM) near 1 AU in solar energetic particle (SEP) events. Somewhat surprisingly, the energetic ions and electrons that produce the HXR and γ-ray emissions in solar flares appear to be accelerated by a different process than the SEPs observed near 1 AU – those appear associated with fast coronal mass ejections (CMEs) and the shocks they drive, or possibly by high coronal acceleration processes. These extremely energetic particle acceleration phenomena are associated with enormous transient releases of energy, $\gtrsim 10^{32}$ ergs for large solar

Space Science Reviews (2006) 124: 233–248
DOI: 10.1007/s11214-006-9107-9

flares, with accelerated particles containing a significant fraction of the total energy released (Lin and Hudson, 1976).

The Sun exhibits a wide range of acceleration phenomena. Flares of all sizes, ranging down to microflares ($\sim 10^{-6}$ the energy release of the largest flares) that occur every few minutes near solar maximum, appear to accelerate electrons (Lin *et al.*, 2001). For flares, the frequency of occurrence increases rapidly as the size decreases, suggesting the possibility that the total averaged energy release might be significant for the heating of the corona.

Solar type III radio bursts are generated by electrons accelerated in the corona that escape to the IPM. A large fraction, perhaps most, of type III bursts are unaccompanied by flares. In the IPM near 1 AU, impulsive electron events are often detected at $\sim$1–100 keV energies, sometimes down to energies of $\sim$0.1 keV. Such low energy electrons must originate high in the corona since energy losses due to Coulomb collisions limit the amount of coronal material they can traverse (Lin *et al.*, 1996). These impulsive events are closely associated with type III bursts that extend into the IPM. The more energetic ($\gtrsim$30 keV) electrons in these events, however, often show delayed onsets that suggest injections $\sim$10 minutes after the type III burst at the Sun.

The relationship between the energetic particles at the Sun and the energetic particles observed near 1 AU is not understood. Here we compare the energetic particle measurements from the Wind, ACE, and other spacecraft near 1 AU with RHESSI observations that provide detailed information on the energetic particle populations in solar flares. In particular, the 28 October and 2 November 2003 large flares were the first for which RHESSI had γ-ray line observations and SEP ions were observed by Wind and near 1 AU. In addition, we summarize the comparisons between RHESSI hard X-rays observations and energetic electron observations from Wind.

2. RHESSI Hard X-Ray/γ-Ray Observations of Solar Flares

The NASA Reuven Ramaty High Energy Solar Spectroscopic Imager (RHESSI) mission (Lin *et al.*, 2002), launched on February 5, 2002, provides high resolution imaging and spectroscopy from soft X-rays (3 keV) to γ-rays (17 MeV). The RHESSI imager (Hurford *et al.*, 2002) is made up of nine bi-grid rotating modulation collimators (RMCs), each consisting of a pair of widely separated grids mounted on a rotating spacecraft, to provide spatial resolution of 2.3 arcsec to 3 arcmin over the full Sun ($\sim 1^\circ$) field of view. Behind each RMC is a coaxial germanium detector (GeD), cooled to $\sim$80 K to achieve spectral resolution of $\sim$1 keV FWHM for HXRs, increasing to $\sim$4 keV FWHM at $\sim$2 MeV, the best ever achieved for solar measurements.

The first γ-ray line flare observed by RHESSI was the intense GOES class X4.8 flare of 23 July 2002 (Lin *et al.*, 2003). The flare HXR and γ-ray observations

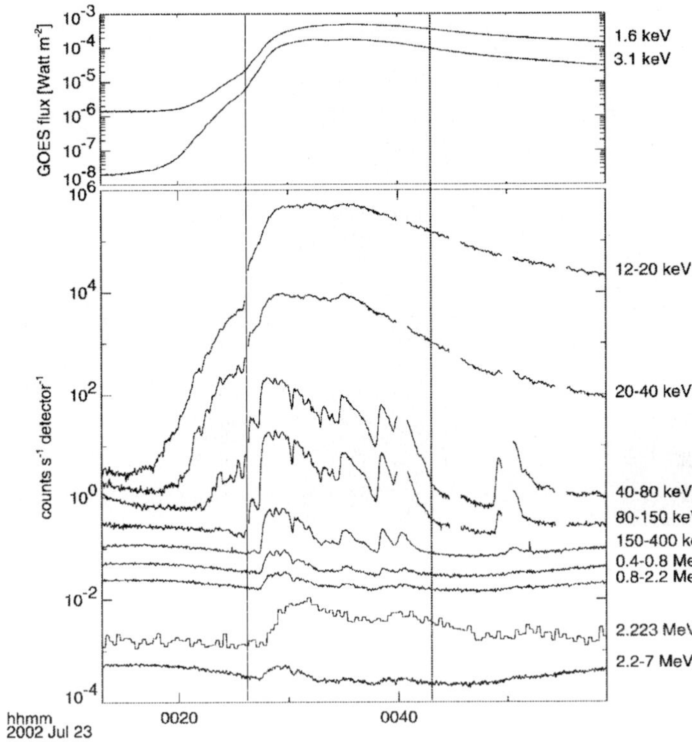

Figure 1. RHESSI HXR and γ-ray count rates (in units of counts s^{-1} per detector) for the 2002 July 23 flare, scaled to fit: 20–40 keV $\times$ 0.3; 40–80 keV $\times$ 0.07; 80–150 keV $\times$ 0.02; 150–400 keV; 400–800 keV $\times$ 0.001; 800–2218 keV $\times$ 0.0005; 2218–2228 keV $\times$ 0.01; and 2228–7000 $\times$ 2×10^{-5}. The thick shutter is inserted at ~0026, 0041, 0044, and 0050 UT and removed at ~0040, 0043, 0049 UT. The vertical lines separate the impulsive phase from the rise and decay. The slow variation of the γ-ray rates through the interval is due to the background from cosmic-ray interactions with the atmosphere and spacecraft (Lin *et al.*, 2003).

(Figure 1) divide naturally into a rise phase (~00:18 to ~00:27 UT) dominated by a coronal HXR source that appears to be non-thermal, an impulsive phase (~00:27 to ~00:43 UT) with continuum and γ-ray line emission extending up to $\gtrsim$7 MeV, and a decay phase ($\gtrsim$00:43 UT) dominated by a superhot (~40 MK) thermal source. The spatial distribution of HXR sources (Krucker *et al.*, 2003) is shown superimposed on TRACE 195 Å images, together with the simultaneous spatially integrated X-ray spectra, in a movie at: `http://hesperia.gsfc.nasa.gov/hessi/presentations/video/`.

A remarkable coronal hard X-ray source with a steep double power-law spectrum above 10 keV (exponents $\gamma_L \sim 5$ and $\gamma_H \sim 6.5$, with break at 20–35 keV) dominates the flare X-ray emission during the rise before the impulsive phase. Little or no footpoint emission (or TRACE emission in the lower corona) is detected through most of this period. The energy in accelerated electrons during

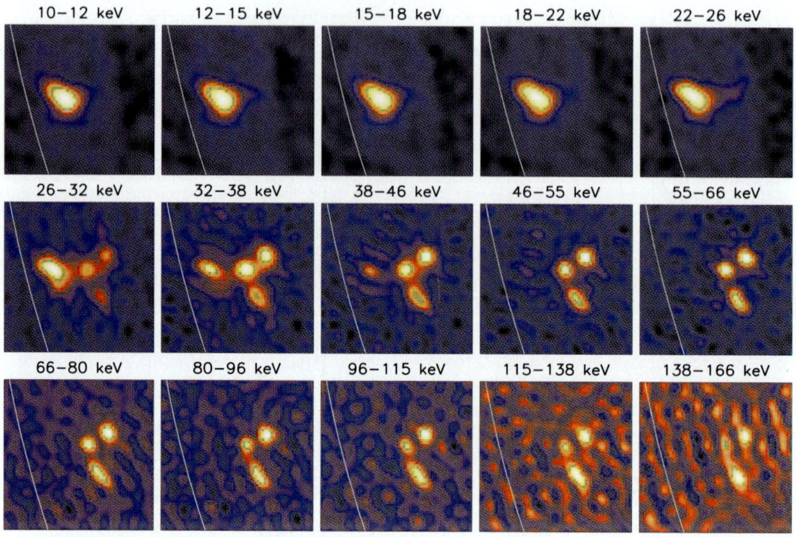

Figure 2. RHESSI X-ray images in different energy bands for a 20 s interval during the impulsive phase of the 23 July 2002 flare. At energies up to 26 keV, only the hot coronal thermal source is visible. Above that energy, the three (possibly more) non-thermal footpoint sources are observed (Emslie *et al.*, 2003).

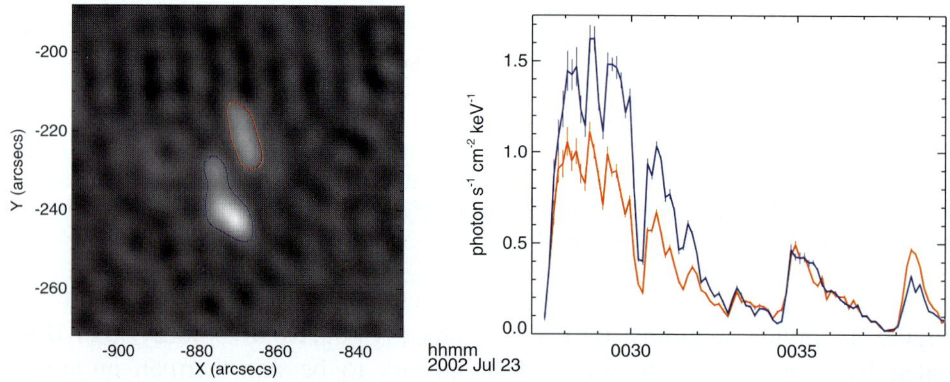

Figure 3. Left panel: RHESSI image at 100–200 keV, showing two main footpoints for the 23 July 2002 flare. The circles define the regions for which the count rates shown in the bottom panel are accumulated. Right panel: the count rates for the two regions are plotted as a function of time. The close correlation in time of the count rates for the two regions strongly suggest that they are opposite ends of a loop that is being fed by a single source of energetic electrons (Krucker *et al.*, 2003).

this time, however, is significant, comparable to that released in the impulsive phase.

During this flare's impulsive phase, RHESSI's detailed hard X-ray imaging spectroscopy show three footpoint sources (Figure 2), together with a coronal superhot ($T \sim 40$ MK) thermal source that dominates below $\sim$30 keV. The temporal

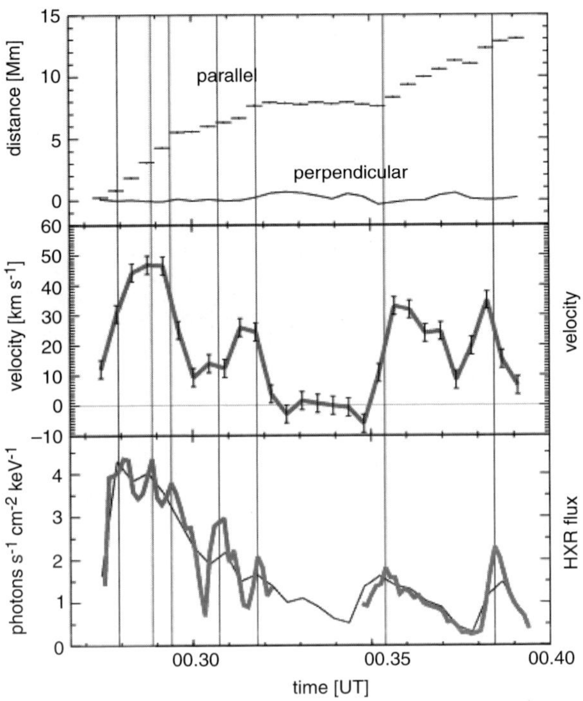

Figure 4. The motion of the two main footpoints for the 23 July 2002 flare observed by RHESSI. Top panel the position of the moving footpoint versus time. Center panel: The speed computed from the top panel versus time. Bottom panel: the hard X-ray flux versus time. A rough correlation is evident between the speed and hard X-ray flux (Krucker *et al.*, 2003).

variations of the hard X-ray fluxes of the different foot points are closely correlated (Figure 3), and the footpoints and coronal source show correlated and systematic motions, strongly suggesting magnetic reconnection in the corona, possibly at or near the thermal coronal source. Furthermore, the footpoint X-ray fluxes are found to be approximately proportional to the footpoint separation speed (Figure 4), as expected if the rate of electron acceleration is proportional to the reconnection rate.

Gamma-ray line and continuum emission show that ions are accelerated to tens of MeV and electrons to $\gtrsim$7 MeV. The prompt de-excitation γ-ray lines of Fe, Mg, Si, Ne, C, and O show mass-dependent red shifts of 0.1–0.8%, implying downward motion of accelerated protons and alphas along magnetic field lines that are not radial, but tilted toward the Earth by $\sim$40° (Smith *et al.*, 2003). RHESSI made the first γ-ray line imaging for a flare (Hurford *et al.*, 2003), showing that the accelerated ions interact in a region near the optical flare (Figure 5). The centroid of the 2.223 MeV neutron capture line emission, however, is located $\sim 20 \pm 6$ arcseconds from the centroids for the 0.3–0.5 and 0.7–1.4 MeV bands that are dominated by electron bremsstrahlung, implying that the acceleration and/or propagation of the accelerated ions must differ from that of the electrons.

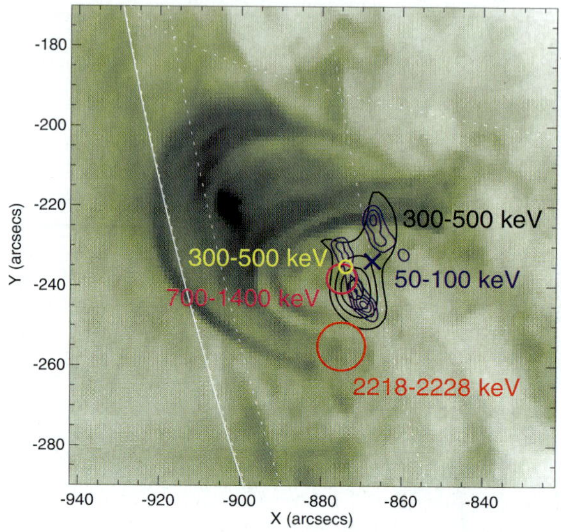

Figure 5. The first image of a flare in a γ-ray line, obtained by RHESSI for the 23 July 2002 flare in the 2.223 MeV neutron-capture line (Hurford *et al.*, 2003). The red circle shows the centroid of the flare-averaged 2.223 MeV line image, superimposed on a TRACE image taken well after the flare. The blue contours show the flare-averaged hard X-ray image at 50–100 keV. The black contour shows the flare-averaged continuum image at 300–500 keV. The yellow and red circles show the centroid of the 300–500 keV and 700–1400 keV continuum, respectively, obtained through the same grids as the 2.223 MeV image.

Assuming that Coulomb collisions dominate the energetic electron and ion energy losses (thick-target) we estimate a minimum of $\sim 2 \times 10^{31}$ ergs is released in accelerated >18 keV electrons during the rise phase, with $\sim 10^{31}$ ergs in ions above 2.5 MeV and about the same in electrons above 30 keV released in the impulsive phase. There could be much more energy in accelerated particles if their spectra extends to lower energies.

Recently, RHESSI imaging of three more flares (28, 29 October and 2 November 2003) in the 2.223 MeV neutron-capture γ-ray line was reported (Hurford *et al.*, 2006). Comparison of imaged and spatially-integrated fluences show that in all these flares (including 23 July 2002) most, if not all, of the emission was confined to compact sources with size scales of tens of arcsec or smaller, that are located in the flare active region. Thus, the γ-ray producing ions appear to be accelerated by the flare process and not by a widespread shock driven by a fast coronal mass ejection. The 28 October 2003 event yielded the first of such images to show double-footpoint γ-ray line sources (Figure 6), similar to what is seen in the hard X-ray image, and strongly supporting a similar flare origin for the ion acceleration. These foot-point sources straddled the flaring loop arcade but were displaced from the corresponding 0.2–0.3 MeV electron-bremsstrahlung emission footpoints by 14 and 17 ± 5 arcsec. Taken together with the previously studied 23 July 2002

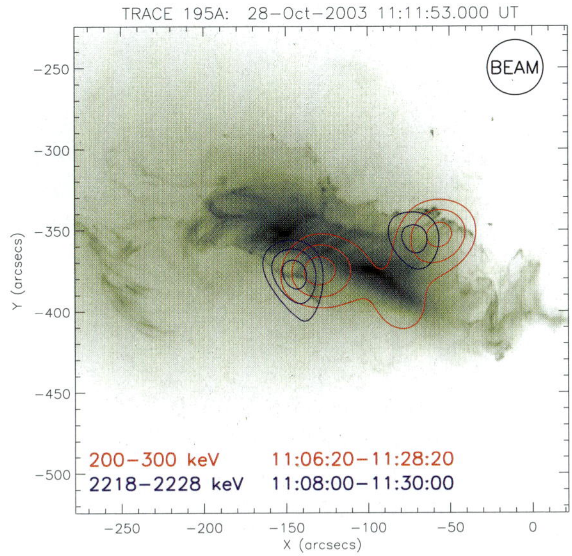

Figure 6. RHESSI γ-ray line image of the 28 October 2003 solar flare (Hurford *et al.*, 2006). The blue contours show the flare-averaged (RHESSI was in shadow for the first few minutes of the flare) 2.223 MeV neutron-capture γ-ray line (produced by 10–100 MeV ions) image, together with the 200–300 keV hard X-ray continuum (produced by energetic electrons) image made through the same grids and using the same analysis procedure. The angular resolution is indicated by beam circle (upper right). The accelerated electrons and ions appear to be separated by $\sim$10,000 km.

event where the centroid of the γ-ray line source is displaced by $\sim 25 \pm 5$ arcsec from the centroid of the electron bremsstrahlung source, this implies spatial differences in acceleration and/or propagation between ions and electrons in solar flares.

In many flares the coronal thermal sources observed by RHESSI above the lower temperature loops seen by TRACE move upward with time, suggesting the energy release site moves to high altitudes with time, in agreement with classical flare models (Sturrock, 1966; Shibata *et al.*, 1995). In the 21 April 2002 flare the onset of the flare and thermal source motions is well correlated with the beginning of the associated CME (Gallagher *et al.*, 2002).

In some flares a second much weaker coronal source is detected above the normal thermal source (Sui and Holman, 2003). The centroids of the two sources, obtained at energies from $\sim$8 to $>$20 keV, show a systematic behavior with the most energetic centroids located closest to the region in between the two sources (Figure 7), with less and less energetic centroids the further away from that point. This is consistent with a temperature gradient away from that point, suggesting a large scale current sheet with energy release from magnetic reconnection in the region in between, again consistent with the Shibata *et al.* (1995) classical flare model.

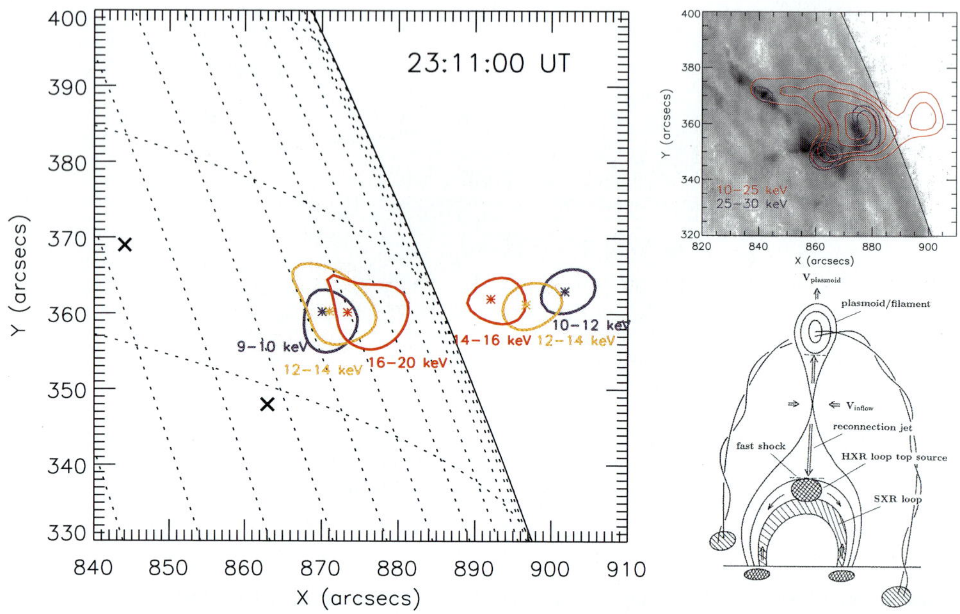

Figure 7. RHESSI imaging of the hot flare plasma above the flare loop, showing evidence for a large scale current sheet. The centroids of the sources at different energies indicate a temperature gradient both upward and downward from the center (Sui and Holman, 2003), as expected in models of flares where magnetic reconnection occurs in the current sheet. The ×'s indicate where the hard X-ray footpoints are located.

3. Energetic Ions at the Sun and SEPs at 1 AU

RHESSI has detected γ-ray line emission from about a dozen solar flares to date, with SEP events detected near 1 AU for three of them. A major SEP event occurred following the 21 April 2002 flare that was well observed by RHESSI (Gallagher *et al.*, 2002), but no γ-ray line emission was detected, indicating that no significant acceleration of energetic ($\gtrsim$10 MeV) ions occurred in the flare. The flare and CME initiation were closely related in time, with the impulsive hard X-ray emission starting the whole process.

The flare on 23 July 2002 was accompanied by a very fast (2180 km/s) and wide CME. Since it was located at S13 E72 near the east limb of the Sun, it was not surprising that no SEPs were detected by the ACE or Wind spacecraft near the Earth. No SEPs were detected by the Mars Global Surveyor (MGS) spacecraft located on the opposite side of the solar system very close to the nominal Parker spiral field from the flare region, however, even though two previous SEP events (July 16 and 19) were detected from the same active region. This suggests that even a very fast and wide CME may not always accelerate SEPs.

SEP events were detected from the 28 October 2003, 29 October 2003, and 2 November 2003 γ-ray line flares observed by RHESSI. For 28 October 2003 and 2

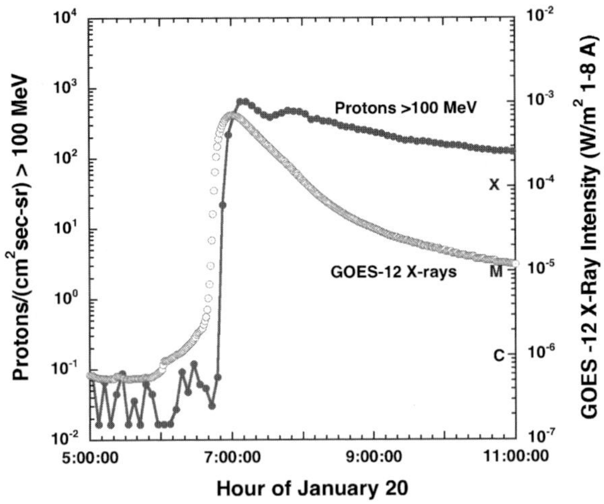

Figure 8. In the solar energetic particle event of 20 January 2005 the 100 MeV protons arrive and reach peak intensity within minutes of the GOES soft X-ray peak for the flare.

November 2003, the spectra of energetic protons were obtained from the observed narrow line fluences of the 2.223 MeV neutron-capture line and the 4.443 MeV carbon line (Ramaty and Murphy, 1987). (During the 29 October 2003 flare, RHESSI passed close to the South Atlantic Anomaly so the background is high and the line fluences uncertain.) These can be compared to the number and power law exponent for the energetic protons observed near 1 AU (using ACE, GOES 10, and SAMPEX spacecraft to provide full energy coverage), integrating over the entire event to obtain the fluences. The observed proton spectra are double power-law with a downward break at ∼20–30 MeV. The integrated fluence will probably overestimate the number of particles accelerated/injected near the Sun, since scattering in the interplanetary medium will allow some particles to cross 1 AU distance more than once (Li *et al.*, 2003).

The spectral exponents for the γ-ray producing protons and the protons at 1 AU are essentially the same for the magnetically well-connected 2 November 2003 event, within measurement uncertainties, but different for the 28 October 2003. Furthermore, very preliminary estimates of the Ne line fluences, which are sensitive to protons down to ∼2.5 MeV, indicate that they are unusually low, suggesting that the γ-ray producing proton spectrum flattens at low energies, similar to what is observed for the protons detected at 1 AU.

More recently RHESSI observed γ-ray line emission from the flare associated with the 20 January 2005 event, the most intense SEP event detected in nearly five decades, at energies above a few hundred MeV. The SEPs arrive within minutes after the flare X-ray peak (Figure 8), raising questions about the role of CME shock acceleration. The very limited CME observations (the SOHO coronagraph was quickly saturated by the penetrating SEPs!) indicate that the CME velocity in this event ranged from ~2500 to ~3500 km/s (Mewaldt *et al.*, 2005; Gopalswamy *et al.*, 2005), implying that the CME was only $\lesssim 2R_s$ above the solar surface when the first GeV protons were released. It is uncertain whether a shock could form and accelerate particles to GeV energies in the short time and distance available (Kahler, 2005). Finally, it is interesting that in the 20 January 2005 event (also magnetically well-connected) both the flare-accelerated proton spectrum from ~10 to ~100 MeV (as derived from RHESSI γ-ray observations) and the proton spectrum at 1 AU (derived from SAMPEX, ACE, and GOES data) are similar, both as hard as or harder than any spectra observed using these techniques (Share, personal communication). The implications of these observations for the relative roles of flare acceleration and shock acceleration in this event are not clear. Given that our current understanding is that the γ-ray producing protons are accelerated by a different process (flares) from the SEP protons (fast CMEs), this is very surprising.

4. Energetic Electrons at the Sun and at 1 AU

Electrons accelerated at the Sun and interacting with the solar atmosphere will produce hard X-rays through bremsstrahlung collisions, and radio emission through wave-particle interactions and through synchrotron emission. As the faster electrons run ahead of the slower ones when the impulsively accelerated electrons escape from the Sun, bump-on-tail distributions will be generated that are unstable to the growth of Langmuir waves. These waves then interact with the ambient plasma or with other waves to produce radio emission at the plasma frequency or its harmonic. As the electrons travel to lower and lower density the radio emission goes to lower frequencies, leading to the characteristic fast drift solar type III radio burst.

If the HXR-producing and escaping electrons come from a single acceleration, a hard X-ray burst should be detected with a near simultaneous type III burst starting at high frequencies. When the type III burst drifts down to near the local plasma frequency at 1 AU, the escaping electrons and Langmuir waves can be directly detected *in situ*. Figure 9a shows an example of this, while Figure 9b shows the flare X-ray spectrum (both thermal and HXR) observed by RHESSI, and the electron spectrum measured by the Wind 3D Plasma & Energetic Particle (3-DP) experiment (Lin *et al.*, 1995). Both spectra fit a double power-law with a downward break at a few tens of keV.

Figure 10 shows a comparison of power-law exponents above the break for the electron spectra observed by Wind at 1 AU, with exponents for the HXR photon spectra observed by RHESSI, for ~15 events that have the timing consistent with a single acceleration (Krucker *et al.*, 2004). The points should fall on the "Thick" target line if the escaping electrons directly sample the accelerated population (without any energy changes), and the accelerated electrons produce the HXRs as they lose all their energy to Coulomb collisions, i.e., if the acceleration occurs high in the corona and some of the electrons escape to the IPM while the rest are trapped

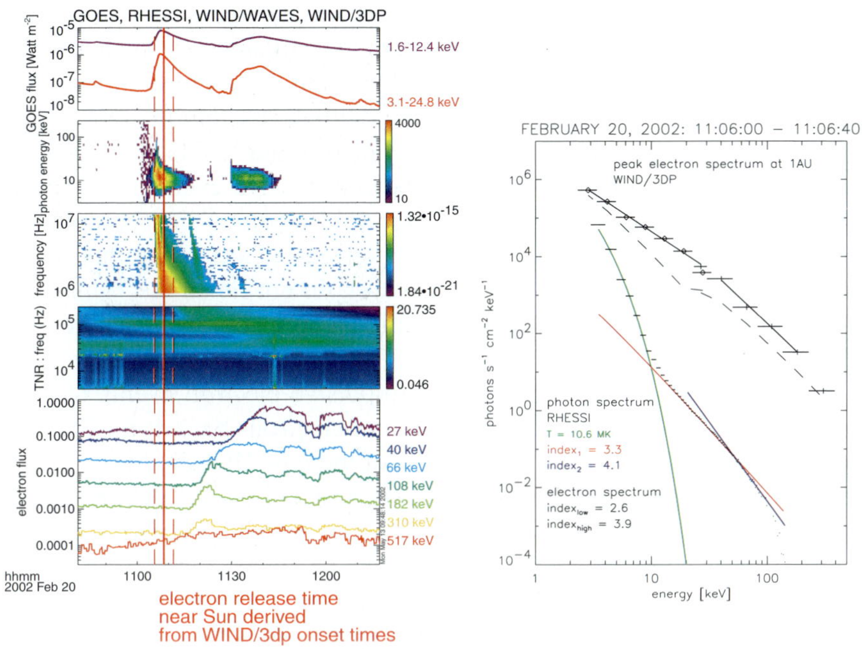

Figure 9. (a) Example of a flare hard X-ray burst observed by RHESSI with corresponding solar type III radio burst and energetic electrons (and Langmuir waves) observed *in situ* by the Wind spacecraft (Krucker and Lin, 2002). Top panel: GOES soft X-rays; second panel: Spectrogram of RHESSI X-rays from 3 to 250 keV; third and fourth panels: radio emission observed by the Wind WAVES instrument; fifth panel: Electrons from ~20 to ~400 keV observed by Wind 3-DP instrument. (b) Top trace: energy spectrum of the electrons observed by Wind 3-D P instrument; bottom trace: X-ray spectrum observed by RHESSI, fitted to a thermal spectral shape at low energies, and to a double power-law at high energies (Krucker and Lin, 2002).

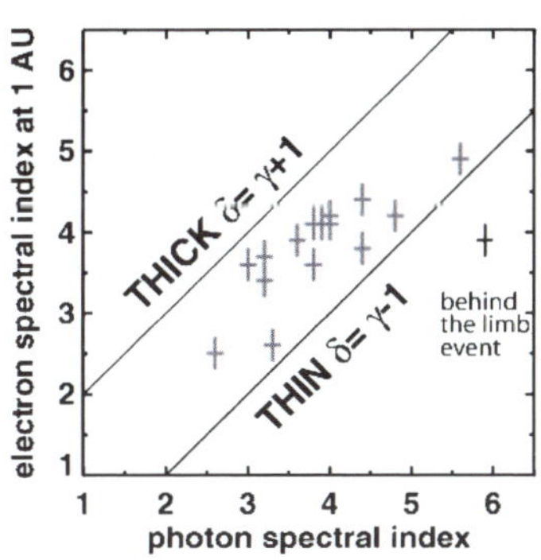

Figure 10. Comparison of power-law exponent for the hard X-ray spectrum at the peak of the burst measured by RHESSI with the power-law exponent for the electron spectrum measured at the time of maximum at each energy. δ and γ are the power-law exponents for the electron and hard X-ray spectra, respectively (Krucker *et al.*, 2004).

in the solar atmosphere. The "Thin" target line would be for the case where the electrons produce the HXRs as they escape, but the collisions are too few to modify the spectrum. The data points appear to show a rough linear correlation with larger electron exponents for larger HXR exponents (with the exception of the behind the limb event), suggesting that the electrons producing the HXRs at the Sun indeed are somehow related to the electrons in these impulsive events observed in the IPM. The RHESSI images typically show the HXRs come from footpoints where the ambient density is high – presumably the electrons are losing their energy to collisions, i.e., thick target. The points, however, do not lie on the "Thick" line or the "Thin" line, indicating that the relationship is more complex than these simple models.

As mentioned above, for many of the impulsive electron events observed at energies of tens of keV, the inferred injection of electrons back at the Sun appears to be delayed by ~10 minutes from the start of the type III radio burst, suggesting acceleration by a coronal or CME shock wave (Krucker *et al.*, 1999; Haggerty and Roelof, 2002). Many impulsive electron events often extend down to below ~1 keV (Lin *et al.*, 1996) and many are detected even in the energy range ~0.1 to ~1 keV (Gosling *et al.*, 2003).

Recently, Wind 3DP observations (which covers from ~1 eV up to $\gtrsim$300 keV electrons) of three scatter-free impulsive electron events with delayed onset at $\gtrsim$30 keV (Haggerty and Roelof, 2002) were carefully analyzed to accurately determine the injection near the Sun (Wang *et al.*, 2006). The event shown in Figure 11 extends down to 0.4 keV, and it shows a rapid, nearly symmetric rise and decay. This indicates essentially scatter-free propagation from the Sun to 1 AU, since scattering would result in a slowly decaying tail in the time profile (Lin, 1974).

A model is applied where the injection time profile is assumed to be triangular, with equal time for rise to the peak and decay back to zero (Figure 12). The injected electrons were assumed to travel ~1.2 AU, the Parker spiral field line length appropriate for the measured ~400 km/s solar wind. Model time profiles were calculated using the measured spectrum of the event and integrating over the width of each energy channel. The injection time and width were then adjusted to fit the observed profile in each energy channel. As can be seen (Figure 12), the injection profiles at energies above ~20 keV are similar, with comparable widths of ~5 minutes. The best-fit injection times are the same at all energies above ~20 keV, confirming that ~1.2 AU is appropriate for the path length. The onset of the injection for >20 keV electrons is clearly delayed by ~8 minutes relative to the type III burst injection (dashed vertical line).

A data gap and poor statistics in the ~4 to 20 keV measurements precluded accurate timing for those energies. The inferred injection profiles for ~0.6 to ~3 keV electrons show onsets starting prior to or at the type III burst injection, early enough that they could be the source of the radio emission. Previous *in situ* observations at 1 AU of the Langmuir waves responsible for type III radio emission show that they occur primarily when ~2–12 keV electrons arrive at the spacecraft (Lin, 1985).

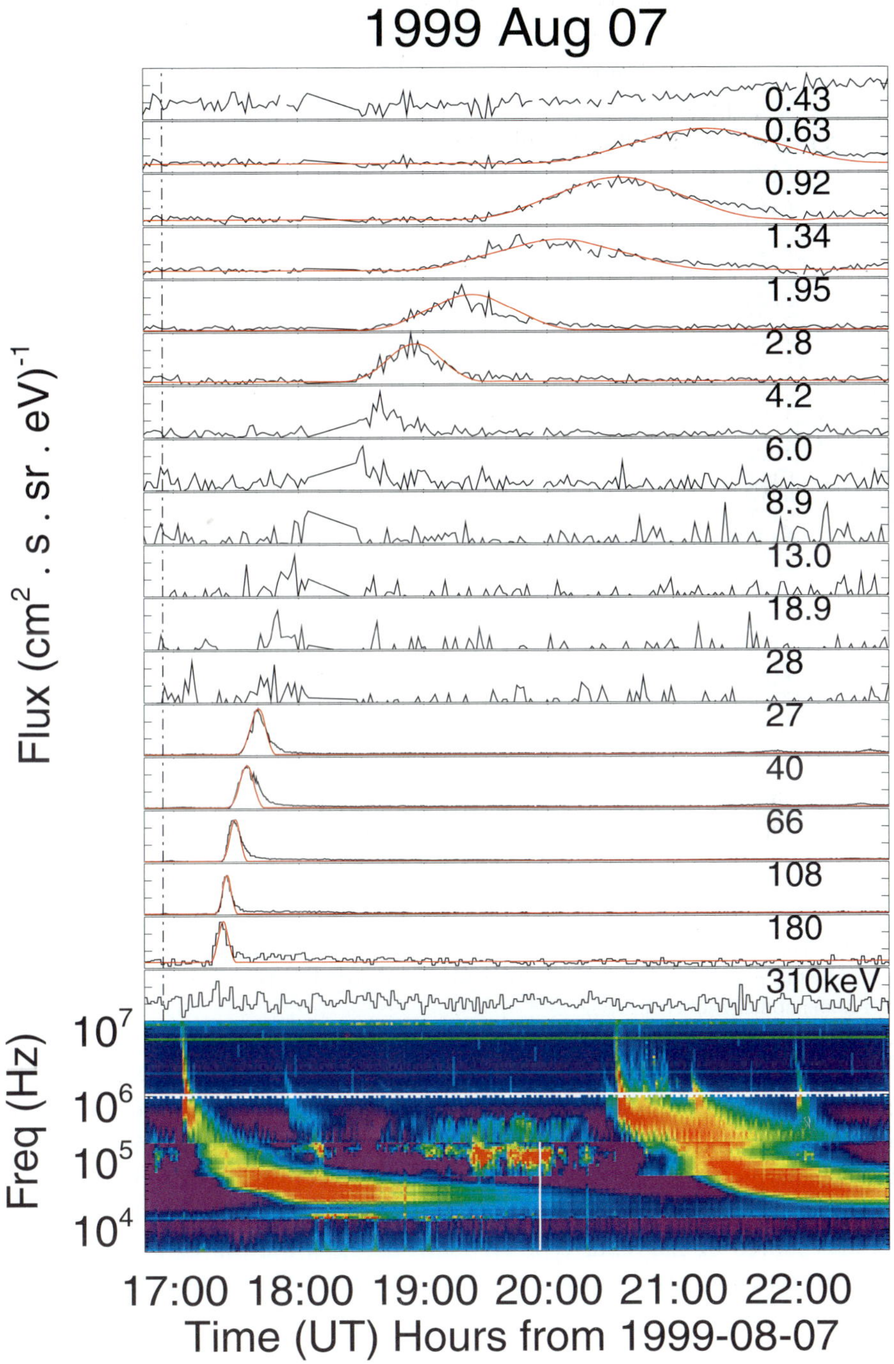

Figure 11. Scatter-free impulsive electron event observed by the Wind 3-D P instrument at energies from 0.1 to 300 keV (top two panels), and radio dynamic spectrum from the Wind WAVES instrument (Wang *et al.*, 2006). The smooth lines show the model fit (see Figure 12).

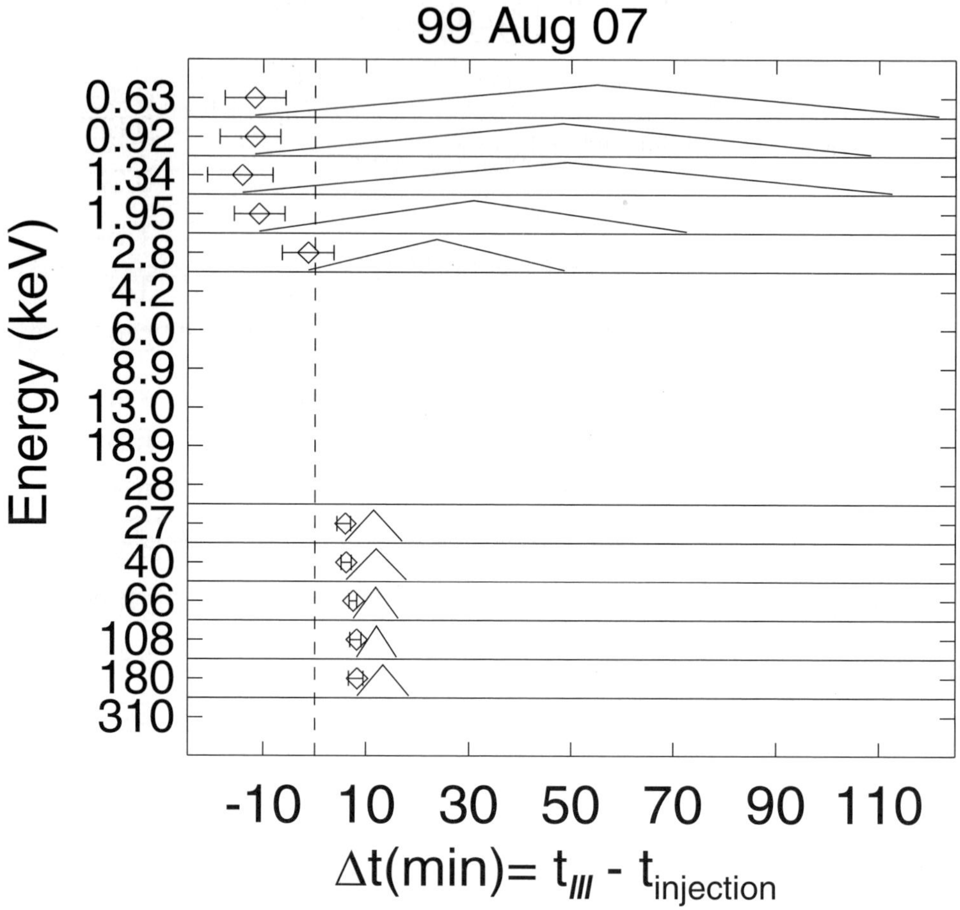

Figure 12. The inferred injection time profiles at the Sun (triangles) from fitting to the time profiles observed by the Wind 3-D P instrument near 1 AU (see Figure 11). The uncertainties in the onset are shown as diamonds with error bars.

The peaks of the injection for ~0.6 to 3 keV electrons, however, are delayed relative to the injection peaks for $\gtrsim$20 keV electrons, and the durations are much longer, ~30–70 minutes. In the same study, the other two scatter-free events with delays at energies $\gtrsim$20 keV also show the same injection characteristics.

Thus, the injection at the Sun of electrons at energies below ~10 keV appears to be the source of the solar type III radio burst, while the delayed injection of $\gtrsim$20 keV electrons points to a second injection, possibly related to a coronal or CME shock wave as suggested by Krucker *et al.* (1999) and Haggerty and Roelof (2002).

It should be noted that many hard X-ray bursts do not have associated type III radio bursts – presumably the electrons are trapped and unable to escape. On the other hand, many type III bursts are not accompanied by hard X-rays – either the electrons are accelerated high in the corona where the ambient density is too low or the number of accelerated electrons is too low for detectable hard X-ray emission. Further more detailed comparisons between RHESSI HXR/γ-ray emission and SEPs observed by ACE, Wind, and other spacecraft will help resolve the relationship between particles at the Sun and in the IPM.

Acknowledgements

I'm pleased to acknowledge useful discussions with D. Smith, S. Krucker, L. Wang at Berkeley; R. Mewaldt at Caltech; and G. Share and R. Murphy at NRL. This research was supported in part by NASA contract NAS5-98033 and grant NAG FDNAG5-11804.

References

Emslie, A. G., Kontar, E. P., Krucker, S., and Lin, R. P.: 2003, *Astrophys. J.* **595**, L107.

Gallagher, P., Dennis, B., Krucker, S., Schwartz, R., and Tolbert, K.: 2002, *Solar Phys.* **210**, 341.

Gopalswamy, N., Xie, H., Yashiro, S., and Usoskin, I.: 2005, *American Geophys. Union Meeting, San Francisco, CA*, Abstract SH23A-0318.

Gosling, J. T., Skoug, R. M., and McComas, D. J.: 2003, *Geophys. Res. Lett.* **30**(13), 1697: 30-1 to 4.

Haggerty, D. K., and Roelof, E. C.: 2002, *Astrophys. J.* **579**, 841.

Hurford, G., Schmahl, E. J., Schwartz, R. A., Conway, A. J., Aschwanden, M. J., Csillaghy, A., *et al.*: 2002, *Solar Phys.* **210**, 61.

Hurford, G., Schwartz, R. A., Krucker, S., Lin, R. P., Smith, D. M., and Vilmer, N.: 2003, *Astrophys. J.* **595**(2), L81.

Hurford, G., Krucker, S., Lin, R. P., Schwartz, R. A., Share, G. H., and Smith, D. M.: 2006, *Astrophys. J.* **644**(1), L93.

Kahler, S.: 2005, *American Geophys. Union Fall Meeting*, San Francisco, CA, Abstract SH21A-07.

Krucker, S., Larson, D. E., Johnson, B. P., and Lin, R. P.: 1999, *Astrophys. J.* **519**, 864.

Krucker, S. and Lin, R. P.: 2002, *Solar Phys.* **210**, 229.

Krucker, S., Hurford, G. J., and Lin, R. P.: 2003, *Astrophys. J.* **595**, L103.

Krucker, S., Lin, R. P., and Kontar, E. P.: 2004, *2004 Joint Assembly*, 17–21 May, Montreal, Canada, paper SH22A-02.

Li, G., Zank, G. P., and Rice, W. K. M.: 2003, *J. Geophys. Res.* **108**(A2), 1082: SSH 10-1 to 10-22.

Lin, R. P.: 1974, *Solar Phys.* **16**, 189.

Lin, R. P. and Hudson, H. S.: 1976, *Solar Phys.* **50**, 153.

Lin, R. P.: 1985, *Solar Phys.* **100**, 537.

Lin, R. P., Anderson, K. A., Ashford, S., Carlson, C., Curtis, D., Ergun, R., *et al.*: 1995, *Space Sci. Rev.* **71**, 125.

Lin, R. P., Larson, D., McFadden, J., Carlson, C. W., Ergun, R. E., Anderson, K. A., *et al.*: 1996, *Geophys. Res. Lett.* **23**, 1211.

Lin, R. P., Feffer, P. T., and Schwartz, R. A.: 2001, *Astrophys. J. Lett.* **557**, L125.

Lin, R. P., and 65 co-authors: 2002, *Solar. Phys.* **210**, 3.

Lin, R. P., Krucker, S., Hurford, G. J., Smith, D. M., Hudson, H. S., Holman, G. D., *et al.*: 2003, *Astrophys. J. Lett.* **595**, L69.

Mewaldt, R. A., Looper, M. D., Cohen, C. M., Mason, G. M., Desai, M. I., Haggerty, D. K., *et al.*: 2005, *2005 Joint Asssembly*, New Orleans, LA, Abstract SH32A-05.

Ramaty, R. and Murphy, R. J.: 1987, *Space Sci. Rev.* **45**, 213.

Shibata, K., Masuda, S., Shimojo, M., Hara, H., Yokoyama, T., Tsuneta, S., *et al.*: 1995, *Astrophys. J. Lett.* **451**, L83.

Smith, D. M., Share, G. H., Murphy, R. J., Schwartz, R. A., Shih, A. Y., and Lin, R. P.: 2003, *Astrophys. J. Lett.* **595**, L81.

Sturrock, P. A.: 1966, *Nature* **211**, 697.

Sui, L. and Holman, G. D.: 2003, *Astrophys. J.* **596**(2), L251.

Wang, L., Lin, R. P., Krucker, S., and Gosling, J. T.: 2006, *Geophys. Res. Lett.* **33**, L03106, doi: 10.1029/2005GL024434.

PARTICLE ACCELERATION IN A THREE-DIMENSIONAL MODEL OF RECONNECTING CORONAL MAGNETIC FIELDS

PETER J. CARGILL[1,*], LOUKAS VLAHOS[2], RIM TURKMANI[1], KLAUS GALSGAARD[3] and HEINZ ISLIKER[2]

[1] *Space and Atmospheric Physics, The Blackett Laboratory, Imperial College, London SW7 2BW, UK*
[2] *Department of Physics, University of Thessaloniki, 54124 Thessaloniki, Greece*
[3] *Niels Bohr Institute, Juliane Maries Vej 30, DK-2100 Copenhagen, Denmark*
*(*Author for correspondence: E-mail: p.cargill@imperial.ac.uk)*

(Received 11 July 2005; Accepted in final form 18 October 2005)

Abstract. Particle acceleration in large-scale turbulent coronal magnetic fields is considered. Using test particle calculations, it is shown that both cellular automata and three dimensional MHD models lead to the production of relativistic particles on sub-second timescales with power law distribution functions. In distinction with the monolithic current sheet models for solar flares, particles gain energy by multiple interactions with many current sheets. Difficulties that need to be addressed, such as feedback between particle acceleration and MHD, are discussed.

Keywords: particle acceleration, solar flares

1. Introduction

The acceleration of charged particles to relativistic energies is a ubiquitous process in plasma physics and astrophysics. It is now known that particle acceleration can occur in energetic events in the galaxy, such as supernovae, at the solar wind termination shock, in the magnetosphere and importantly from the viewpoint of this paper, in solar flares. Particle acceleration mechanisms can be divided into three broadly different classes. Shock acceleration occurs in two forms: drift acceleration where particles interact with electric fields at the shock front, and diffusive acceleration where particles are continually scattered by hydromagnetic waves that are present both upstream and downstream of the shock front (a 1st order process). Turbulent acceleration occurs when particles interact with a spectrum of hydromagnetic waves, gaining energy because there are slightly more head-on than overtaking interactions with the waves (a 2nd order process). Finally, both weak and strong direct electric fields can accelerate particles. For a weak electric field, the magnitude must exceed the Dreicer field, but for any realistic case, the strong field case will apply, since the Dreicer field is usually rather small. Further discussion of the theory can be found in Miller *et al.* (1997).

Space Science Reviews (2006) 124: 249–259
DOI: 10.1007/s11214-006-9108-8

Solar flares have long been known to involve the release of large amounts of magnetic energy: in excess of 10^{32} ergs in some cases, and to be efficient particle accelerators. Following analysis of the initial data from the Ramaty High Energy Solar Spectroscopic Imager (RHESSI) mission (e.g. Lin *et al.*, 2003; Saint-Hilaire and Benz, 2005; Sui *et al.*, 2005), the following results are now widely accepted: (a) up to 10^{37} electrons per sec are accelerated in a very short time (sub-second) to energies between 20 keV and in excess of a few MeV; (b) up to 10^{35} ions per sec are accelerated to energies above 1 MeV; (c) a fraction of these electrons and ions attain relativistic energies; (d) the typical spectra consist of a thermal part at low energies, and a one- or two-part power law at higher energy; (e) acceleration occurs both before the main impulsive phase and well into the decay phase; (f) electrons are accelerated faster than ions. These results have affirmed the longstanding semi-conjecture that energetic particles may account for the largest share of the flare's energy budget. Indeed in some cases up to 50% of the energy released by solar flares seems to go to electrons alone (e.g. Miller *et al.*, 1997). When one adds energetic ions, especially the hard to determine component with energies under a few MeV, the energy conversion requirements are significant.

These results pose formidable challenges for theoretical models. The main one is the need to convert a large amount of magnetic energy into energetic particle kinetic energy. Another is that, despite the high spatial and temporal resolution of current observations, the majority of flares, especially medium and small ones, are still probably not yet adequately spatially resolved. It is plausible that the energy release processes may occur on subscales well beyond our observational means, and indeed theoretical models have shown that small scale energy release can answer many questions, in particular the observed fast timescales (e.g. Miller *et al.*, 1997; Dmitruk *et al.*, 2003; Arzner and Vlahos, 2004; Turkmani *et al.*, 2005). Finally, one needs to address how the small-scale acceleration physics can be incorporated into the large-scale coronal magnetic field configuration associated with a flare.

This paper addresses the last of these points. In Section 2 we discuss the difficulty of developing models that combine the physics of particle acceleration with large-scale magnetic field properties. Sections 3 and 4 present two examples of how one might address these problems using, respectively, cellular automata and MHD models. Section 5 outlines some potentially profitable future lines of investigation.

2. Multiple Scales and Magnetic Complexity in Solar Flares

Observations suggest that the hot plasmas and energetic particles produced in solar flares occupy large coronal structures: length scales between 10^9 and 10^{10} cm are reasonable. These structures are believed to outline the large-scale topology of the coronal magnetic field, although of course direct measurements of the coronal field strength and direction itself are very difficult. Despite this, all our understanding of particle acceleration suggests that it occurs on very small length scales, certainly sub-km, and perhaps much less. Considering the three processes discussed in the introduction, for direct electric field acceleration at a coronal current sheet, the scale is $L \sim \eta/V$. For a velocity of 50 km/s, temperature of 2×10^6 K, density of 10^{10} cm^{-3} and a classical conductivity ($\eta = 10^9/T^{3/2}$ m^{2}s^{-1}), $L \ll 1$ m. In turbulent acceleration a 1 MeV proton will resonate with an Alfvén wave of wavelength ($2\pi V/\Omega_i$) of 100 m, and for the same energy particle, shock acceleration involves regions several times this size, of order 1 km. In addition, particle acceleration is intrinsically a kinetic process: the details of the distribution function and how particles at different parts of the distribution interact with, for example waves, are of importance.

Another issue concerns the spatial distribution of acceleration sites. While the present-day preference for flare particle acceleration at a monolithic current sheet (e.g. Shibata, 1999) may be of relevance for the later stages of the largest (eruptive) flares, it is by no means clear that it is relevant for the impulsive phase, and for smaller (compact) flares. Instead, complex coronal geometries with multiple acceleration sites need to be considered. In the following two sections we discuss why this can be expected from theoretical considerations, but note here that such a scenario implies that a given particle can be accelerated at more than one location, perhaps enhancing the energisation process.

How can theoretical models address these problems? Most models for particle acceleration in flares consider a single accelerating site (current sheet, shock etc.) with simplified background parameters (constant density, magnetic field), permitting analytic or semi-analytic calculations. This often enables very detailed calculations of the kinetic physics in the acceleration processes, but of course ignores the global environment.

In order to model the kinetic physics in a large-scale environment, some workers have carried out full particle simulations of a coronal magnetic field (e.g. Winglee *et al.*, 1991). Provided enough particles can be included, kinetic modelling in regions of varying magnetic field and plasma can be carried out. The difficulty with this approach lies in the compression of scales. Particle codes are required to resolve scales of order the Debye length (for coronal plasma of order 1 mm), whereas computer memory limitation implies that the overall system scale can be only 10^3 Debye lengths. The problems are obvious, yet these simulations are unique in showing how the global coronal electric circuits that are important for maintaining quasi-neutrality can be set up. But it needs to be said that even with the most optimistic projections, such modelling with the correct scales will be beyond the capabilities of computers for decades.

Thus it is clear that modelling acceleration in a global corona involves sacrificing something. Full particle codes retain much kinetic physics, but the length scales are irrelevant to the solar corona. MHD models have the right global scales, but have artificially low magnetic Reynolds numbers and no kinetic physics. The question is whether there is an intermediate approach.

3. Acceleration in Cellular Automata Models

The idea of multiple dissipation sites in flares is longstanding, but the first serious effort at modelling this was by Lu and Hamilton (1991) who took ideas associated with self-organised criticality (SOC) and applied them to the solar corona. They considered a coronal magnetic field that was gradually stressed by the injection of magnetic energy at random points inside the 3D structure of the active region. A series of simple rules governed the behaviour of the field, and dissipation was deemed to occur when the local current reached a threshold. Lu and Hamilton showed that (a) if the right conditions were met, the triggering of dissipation at a single point could lead to a "spreading" of dissipation across a large coronal volume and (b) the distribution of event size as a function of energy followed a power law similar to that observed in flares. Extensive debate followed, much of which centred on how to satisfy Maxwell's equations (e.g. Vlahos *et al.*, 1995; Vassiliadis *et al.*, 1998; Isliker *et al.*, 2000). However, a fundamental question remains: can existing MHD models verify that the main scenarios behind the SOC theory are valid?

These SOC (and associated cellular automata: CA) models identify dissipation occurring at many spatially separated regions. This is illustrated in Figure 1 taken from Vlahos *et al.* (2004). Figure 1a shows a coronal field geometry, reconstructed from photospheric magnetograms, in a volume with characteristic dimensions of 10^9 cm. Figure 1b shows isosurfaces of current in a small part of a coronal volume as yielded by the MHD-consistent CA model of Isliker *et al.* (2001), while Figure 1c shows the local sub-structure not evident in Figure 1b. Finally, a snapshot of the locations where the currents are in excess of the threshold for dissipation in the CA model (referred to as Unstable Current Sheets: UCS) is shown in Figure 1d. Energy release at this time thus occurs at a number of locations.

To assess particle acceleration in such a coronal geometry, Vlahos *et al.* (2004) developed a model in which the particles move between the UCS, gaining (or losing) energy at each one, and undergoing ballistic motion in a background magnetic field between the UCS. The main ideas behind these models are the following. Using the SOC hypotheses we can identify the structure of the spatial distribution of the UCS inside a complex active region. The spatial distribution of the UCS exhibits a specific fractal structure (McIntosh *et al.*, 2002). The knowledge of the spatial distribution of the UCS can help us to reconstruct the probability distribution of the distances between the UCS and as such the probability of a particle colliding with the UCS (Isliker and Vlahos, 2003). An aspect that remains open is the statistical properties of the energy gain or loss for a particle colliding with an UCS, but this is easy to study using three dimensional MHD codes (see Section 4). Therefore we can start from a global reconstruction of the distribution of UCS and recover the kinetic properties of the particles assuming that the SOC theory is valid.

This is illustrated in Figure 2 where a particle first interacts with site i, escapes along the magnetic field, encounters site $i + 1$, escapes again, and so on. The energy gain thus occurs in many increments, rather than at a single location, as

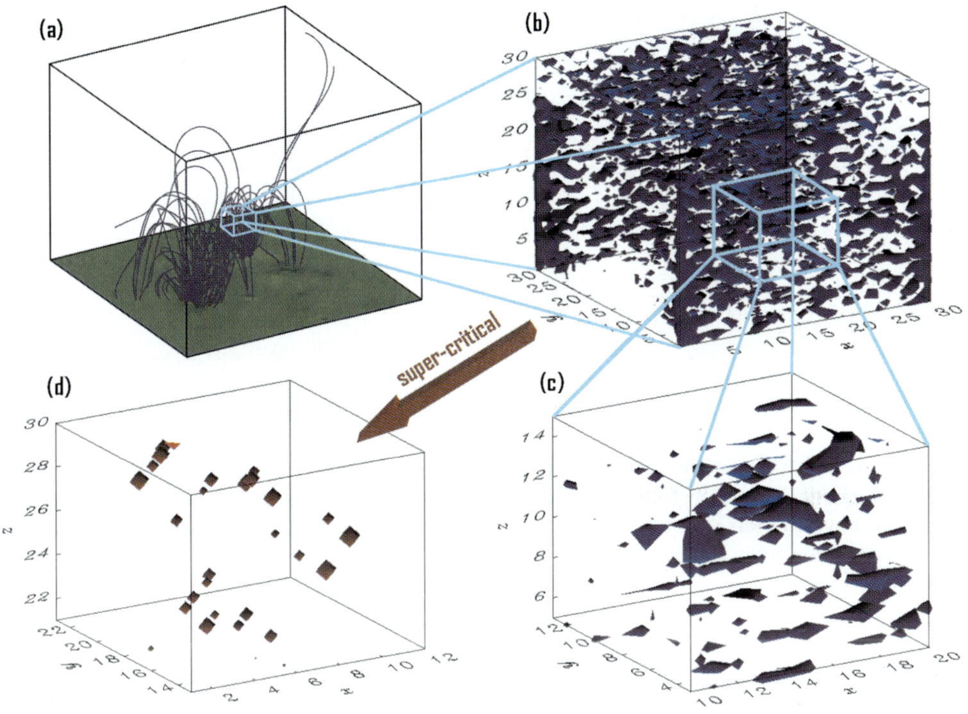

Figure 1. (a) A simulated magnetogram of a photospheric active region and associated coronal force-free magnetic field lines (generated by the model of Fragos *et al.*, 2004). (b) Sub-critical current iso-surfaces in space determined by a X-CA model. (c) An expanded view of panel (b). (d) A temporal snapshot of the X-CA model during a flare, showing the spatial distribution of the UCS (supercritical current iso-surfaces) inside the complex active region. [From Vlahos *et al.*, 2004].

in the monolithic current sheet approach. The acceleration is determined by three probability densities: one defining the distance between a pair of accelerators, one the electric field strength at each accelerator, and a third either the time spent in the UCS, or the length of the UCS. The first two are assumed to be power laws, and the third is taken to be a Gaussian. Note that this model only addresses acceleration by direct electric fields since that can be simply parameterised at the dissipation sites. We return to this point in Section 4. In the example shown in Figure 3, the distance between accelerators ranges from 10^4 to 10^{10} cm, the particle moves in a box of size 10^{10} cm, but the magnetic fields may be such that they are partially trapped and the trajectory is complex. The upper limit of the free travel distances is set to a value larger than the coronal volume in order to allow particles to leave the coronal region without undergoing any interaction with electric fields, besides the initial one (note that for the maximum time of 1 sec for which the system is monitored these escaping particles will move at most a distance 3×10^{10} cm, which is of the order of the length of the coronal volume). The electric field varies between the

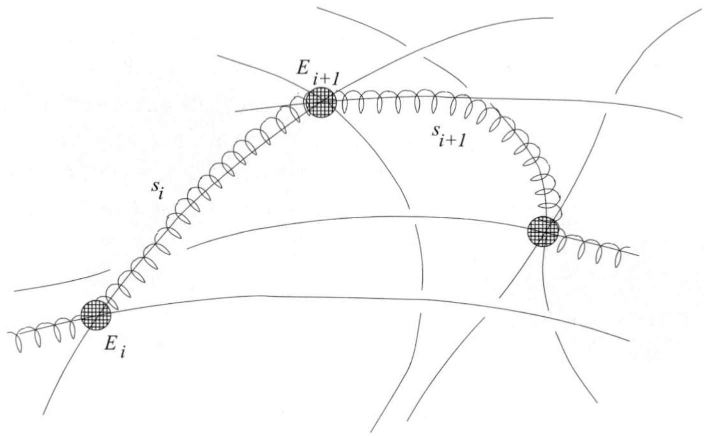

Figure 2. A sketch of particle motion in the coronal X-CA model of Vlahos *et al.* (2004). A particle (the spiraling line) essentially follows the magnetic field lines (solid lines), although also undergoing drifts, and travels in this way freely a distance s_i, until it enters a UCS (filled circles), where it is accelerated by the associated effective DC electric field E_{i+1}. The particle then moves freely again until it meets a new UCS.

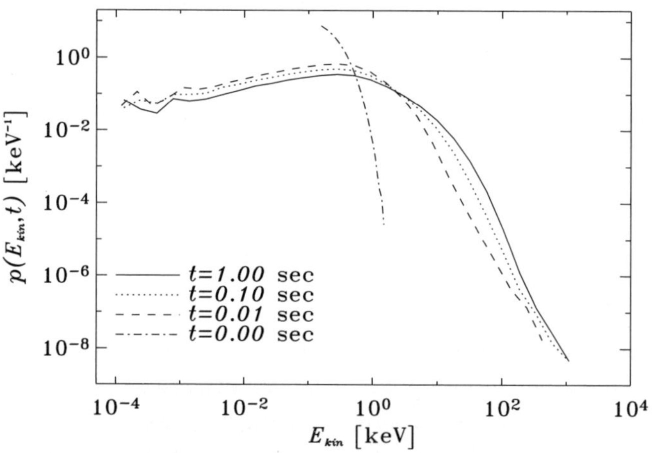

Figure 3. The kinetic energy distributions $p(E_{\mathrm{kin}}, t)$ (probability density function, normalized to 1) at times $t = 0, 0.01, 0.1, 1$ s, for the case in which the acceleration times are prescribed in the X-CA acceleration model of Vlahos *et al.* (2004).

Dreicer field and 10^8 times larger, and the acceleration time has a mean value of 2×10^{-3} s.

A large number of test particles are then tracked through this coronal geometry using the relativistic equation of motion. The details of the acceleration can be found in Vlahos *et al.* (2004), but we note here that most of the particles undergo multiple interactions with UCS, with acceleration taking place very rapidly ($\ll$1 s).

Figure 3 shows the distribution function of electrons at four different times. The initial distribution is a Maxwellian with a temperature of 100 eV (dash-dot line). We see that (a) prompt acceleration to energies well in excess of 100 keV takes place, (b) just under half of the particles end up in the high energy tail and (c) the tail is a power law with approximate slope of −4. When the test particles are protons, it is found necessary to allow for much longer acceleration times to attain similar energies when the acceleration time is prescribed.

4. Acceleration in Self-Consistent MHD Models

The CA models are simple to run, and do not have the limitations that often make MHD models difficult to run (e.g. timestep, number of grid points, numerical stability, artificial Reynolds number). On the other hand, when modelling the large-scale corona, the MHD approach is likely to be valid, and so it is important to address particle acceleration in such a framework. There is considerable evidence from 2.5 dimensional MHD simulations that driving the corona leads to complex magnetic geometries with multiple dissipation sites, as is found in the SOC/CA models (e.g. Buchlin *et al.*, 2003). Although the extension of these results to 3-D is difficult, primarily for computational reasons, it is essential that this be carried out to both conduct a comparison with the CA models and to correctly address loss of accelerated particles from the corona to the photosphere.

Recently Turkmani *et al.* (2005, 2006) have addressed the issue of particle acceleration in a 3-D coronal MHD model. The MHD model was developed by Galsgaard and Nordlund (1996) and Galsgaard (2002), and considers the response of the corona to random photospheric driving. The initial magnetic field extends between the two photospheric ends of a "straightened" coronal loop, and a turbulent cascade develops as energy is injected. This leads to the formation of multiple current sheets in the corona, which appear and disappear over short times, but overall give rise to a continually turbulent coronal state. Figure 4 shows the resistive coronal electric field distribution at one time during such a simulation. In this case the MHD model has a scale of 1.5×10^9 cm between the photospheric ends, and a transverse dimension a factor 20 less. It is clear that current sheets are distributed throughout the corona. The distribution of current sheets will change when different snapshots are considered, giving rise to the possibility of very bursty particle acceleration.

To study particle acceleration, the MHD simulation is "frozen" in time, and test particles are followed through the electromagnetic fields using the relativistic equations of motion. [Since the acceleration time is short compared to the MHD time, we are able to look at a fixed field profile.] A range of particle motions are seen to occur. Some particles undergo rapid and systematic energy gain, and leave the computational box with relativistic energies. Other particles undergo trapping in the corona, gaining some energy, and some do not interact with the current sheets, gaining no energy at all. The model also allows particles to be lost from the

Figure 4. Snapshots of the resistive electric field configurations within the coronal volume, as calculated from the global MHD model. The blue and red regions represent electric field regions that point towards the left and right foot points respectively. [From Turkmani *et al.*, 2005].

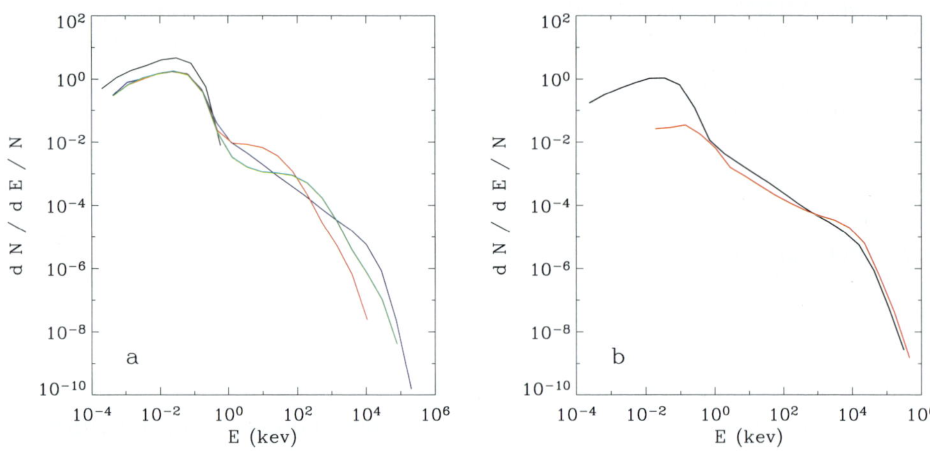

Figure 5. Distribution function obtained by running 40,000 electrons in the MHD model for a coronal volume with $L = 1.5 \times 10^9$ cm. (a) The time evolution of the distribution functions taken at $t = 0$ (black; the curve terminates at $E = 1$ keV), $t_1 = 0.0125$ s (red), $t_2 = 0.025$ s (green), and $t_3 = 0.05$ s (blue). (b) The final distribution function at $t = 0.1$ s from the above model when we consider only the resistive component of the electric field (black) compared to that of a similar model that considers both the resistive and the inductive components on the electric field (red).

computational domain, hence limiting their energy gain. The results show that (a) the resistive electric field is the principal means of energisation: the inductive field plays little role; (b) the particles gain (and lose) energy through multiple interactions with different current sheets; (c) acceleration to relativistic energies is very rapid for both electrons and ions and (d) a large fraction of the particles are accelerated. Figure 5 shows the distribution function of electrons at a number of times during a model run. The distribution has a clear two-part power law structure, with a thermal component at low energies.

It is interesting to briefly compare the two classes of models. The MHD one gives higher particle energies in somewhat faster times than the CA one when the acceleration time is prescribed. This is likely to be due to different choices in the basic parameters, especially the magnetic field strength and density. The same MHD snapshot accelerates both electrons and protons without any need to adjust the electric fields, contrary to the CA based model. It is unclear why this difference arises. The two-part power law does not arise in the CA model, and may be due to the loss of particles in the MHD one. But despite these differences, both approaches show that a fragmented coronal environment is an excellent particle accelerator.

5. Discussion

In this paper we have discussed some of the problems in understanding particle acceleration in the corona, and some ways that these problems may be addressed. In particular, we have demonstrated that a turbulent corona with multiple current sheets is an excellent environment for particle acceleration (see also Dmitruk *et al.*, 2003, 2004). But it is clear that we are still a long way from solving "the flare problem". In particular the following points need to be noted:

- The fact that a large fraction of the flare energy goes into energetic particles suggests that the test particle approach is unlikely to be valid at all times. Instead, feedback needs to be considered on the evolution of the coronal magnetic fields. Also, the influence of energetic particles on the spreading of dissipation may be of importance.
- The present generation of test particle models in coronal fields consider only direct electric field acceleration. Other mechanisms need to be considered.
- Are simplified approaches to coronal magnetic field dynamics such as SOC valid, and what are their limitations?
- Particle transport in the corona, and from corona to chromosphere, is believed to be influenced by scattering, probably due to small-scale turbulence. This is of importance in determining distribution functions entering the chromosphere.

We now address these points in turn. Regarding point (1), theories for current sheet structure do not consider what happens if there is significant particle acceleration. For example, if much of the plasma in the vicinity of the diffusion region is accelerated, and leaves the current sheet rapidly, does this have an effect on the reconnection rate? What happens if beams of charged particles from one acceleration site pass through a current sheet that is not undergoing dissipation? Can such an interaction lead to destabilisation of another current sheet, and so trigger a spreading of dissipation? We have not even the simplest of possible answers to such questions.

Point (2) requires the development of suitable parameterisations of the acceleration processes. This is relatively simple with a DC electric field, since that is a quantity that is "outputted" from an MHD code, but even in this case collective effects influencing the distribution function are ignored. Acceleration by MHD turbulence has been well-studied in isolation (e.g. Miller *et al.*, 1997), but the relationship with coronal dissipation has never been properly established. The difficulty is relating the presence of the required level of turbulence to coronal reconnection. One can make conjectures though. It is well known that magnetic reconnection leads to plasma jets with a velocity at approximately the Alfvén speed based on the reconnecting field components. One can argue that such a jet will be unstable to hydromagnetic instabilities, leading to the generation of plasma eddies, and ultimately a turbulent cascade which can then accelerate particles.

How can one include such a process into a corona with multiple dissipation sites? Clearly it is essential to parameterise the turbulent acceleration, in particular the magnitude of the field fluctuations, in terms of the properties of the reconnection site. One then needs to parameterise the energy gain attained when a particle of given energy interacts with this region of turbulence. These are not unattainable goals, but they do require considerable thought. Similar thoughts apply to diffusive shock acceleration, and we note here that some ideas for drift acceleration have already been considered by Anastasiadis and Vlahos (1991).

The viability of simplified models is an open question. Future work must address in an interactive way the relationship between MHD simulations (including current sheet particle dynamics) and the recovery from MHD codes of the basic rules of the SOC theory (e.g. Isliker *et al.*, 2000, 2001). Moving away from two-dimensional MHD is essential here: many simulations of 3-D MHD processes show turbulent small-scale structures forming that are not present in simulations with reduced dimensionality. It seems essential to abandon once and for all the concept of the monolithic current sheet.

The issue of transport is in many ways not quantifiable given the many possible wave modes that could be responsible for particle scattering. But this is an issue that could be addressed either by introducing a random scattering operator into the equation of motion, or another parameterisation, and so this cannot be regarded as a problem of the same magnitude as the other points.

The material presented above, and in the rest of this paper, offers an alternative approach to solar flares that abandons the monolithic current sheet approach in favour of distributed energy release as suggested by many models of the corona. In fact, the two approaches are related when one realises that the current sheets in the fragmented models are no different from the monolithic ones, but the presence of multiples sheets introduces new dissipation and acceleration scenarios. But the outstanding problem of linking kinetic physics with large-scale MHD in a consistent way with feedback remains of key importance.

Acknowledgements

PC acknowledges the support of a PPARC Senior Research Fellowship and thanks ISSI for their organisation of another excellent workshop. KG was supported by the Carlsberg Foundation in the form of a fellowship. This work was supported by the European Commission under research training network HPRN-CT-2001-00310.

References

Anastasiadis, A., and Vlahos, L.: 1991, *Astron. Astrophys.* **245**, 271.
Arzner, K., and Vlahos, L.: 2004, *Astrophys. J.* **605**, L69.
Buchlin, E., Aletti, V., Galtier, S., Velli, M., Einaudi, G., and Vial, J.-C.: 2003, *Astron. Astrophys.* **406**, 1061.
Dmitruk, P., Matthaeus, W. H., Seenu, N., and Brown, M. R.: 2003, *Astrophys. J.* **597**, L81.
Dmitruk, P., Matthaeus, W. H., and Seenu, N.: 2004, *Astrophys. J.* **617**, 667.
Fragos, T., Rantsiou, E., and Vlahos, L.: 2004, *Astron. Astrophys.* **420**, 719.
Galsgaard, K.: 2002, in *IAU Coll.* 188, *ESA SP* **505**, 269.
Galsgaard, K., and Nordlund, A.: 1996, *J. Geophys. Res.* **101**, 13445.
Isliker, H., Anastadiadis, A., and Vlahos, L.: 2000, *Astron. Astrophys.* **363**, 1134.
Isliker, H., Anastadiadis, A., and Vlahos, L.: 2001, *Astron. Astrophys.* **377**, 1068.
Isliker, H., and Vlahos, L.: 2003, *Phys. Rev. E* **67**, 026413.
Lin, R. P., *et al.*: 2003, *Astrophys. J.* **595**, L69.
Lu, E. T., and Hamilton, R. J.: 1991, *Astrophys. J.* **380**, L89.
McIntosh, S. W., Charbonneau, P., Bogdan, T. J., Liu, H.-L., and Norman, J. P.: 2002, *Phys. Rev. E* **65**, 46125.
Miller, J. A., *et al.*: 1997, *J. Geophys. Res.* **102**, 14631.
Saint-Hilaire, P., and Benz, A. O.: 2005, *Astron. Astrophys.* **435**, 743.
Shibata, K.: 1999, *Astrophys. Space Sci.* **264**, 129.
Sui, L., Holman, G. D., and Dennis, B. R.: 2005, *Astrophys. J.* **626**, 1102.
Turkmani, R., Vlahos, L., Galsgaard, K., Cargill, P. J., and Isliker, H.: 2005, *Astrophys. J.* **620**, L59.
Turkmani, R., Cargill, P. J., Galsgaard, K., Vlahos, L., and Isliker, H.: 2006, *Astron. Astrophys.* **449**, 749.
Vassiliadis, D., Anastasiadis, A., Georgoulis, M., and Vlahos, L.: 1998, *Astrophys. J.* **509**, L53.
Vlahos, L., Georgoulis, M., Kluiving, R., and Paschos, P.: 1995, *Astron. Astrophys.* **299**, 897.
Vlahos, L., Isliker, H., and Lepreti, F.: 2004, *Astrophys. J.* **608**, 540.
Winglee, R. M., Dulk, G. A., Bornmann, P. L., and Brown, J. C.: 1991, *Astrophys. J.* **375**, 382.

THE SEED POPULATION FOR ENERGETIC PARTICLES ACCELERATED BY CME-DRIVEN SHOCKS

M. I. DESAI[1,*], G. M. MASON[2], J. E. MAZUR[3] and J. R. DWYER[4]

[1] *Southwest Research Institute, 6220 Culebra Road, San Antonio, TX 78238-5166, USA*

[2] *Johns Hopkins University, Applied Physics Laboratory, Johns Hopkins Road, Laurel, MD 20723-6099, USA*

[3] *The Aerospace Corporation, Space Sciences Department-Chantilly, 15049 Conference Center Drive, CH3/210 Chantilly, VA 20151, USA*

[4] *Florida Institute of Technology, 150 West University Boulevard, Melbourne, FL 32901-6975, USA*

*(*Author for correspondence: E-mail: mdesai@swri.edu)*

(Received 19 July 2005; Accepted in final form 20 October 2005)

Abstract. Understanding properties of solar energetic particle (SEP) events associated with coronal mass ejections has been identified as a key problem in solar-terrestrial physics. Although recent CME shock acceleration models are highly promising, detailed agreement between theoretical predictions and observations has remained elusive. Recent observations from ACE have shown substantial enrichments in the abundances of ^{3}He and He$^+$ ions which are extremely rare in the thermal solar wind plasma. Consequently, these ions act as tracers of their source material, *i.e.,* ^{3}He ions are flare suprathermals and He$^+$ ions are interstellar pickup ions. The average heavy ion composition also exhibits unsystematic differences when compared with the solar wind values, but correlates significantly with the ambient suprathermal material abundances. Taken together these results provide compelling evidence that CME-driven shocks draw their source material from the ubiquitous but largely unexplored suprathermal tail rather than from the more abundant solar wind peak. However, the suprathermal energy regime has many more contributors and exhibits much larger variability than the solar wind, and as such needs to be investigated more thoroughly. Answers to fundamental new questions regarding the preferred injection of the suprathermal ions, the spatial and temporal dependence of the various sources, and the causes of their variability and their effects on the SEP properties are needed to improve agreement between the simulations and observations.

Keywords: Sun: energetic particles, Sun: coronal mass ejections, Sun: flares

1. Introduction

The earliest observations of solar energetic particle or SEP events extending up to GeV energies were made with ionization chambers and neutron monitors (Forbush, 1946; Meyer *et al.*, 1956). Since such events, also known as ground level events or GLEs, were closely associated with Hα flares on the Sun, it was presumed that there was a causal relationship between the flare and the energetic particles observed at 1 AU. However, on the basis of a close association between the SEP events and slow-drifting Type II and various kinds of Type IV radio bursts, Wild *et al.* (1963) proposed that the energetic particles might be accelerated at magnetohydrodynamic

DOI: 10.1007/s11214-006-9109-7

shock waves that typically accompanied the flares. Later, Lin (1970) reported close associations between 'pure' electron events and flares that only exhibited metric Type III emission on the one hand, and between 'mixed' events with protons and relativistic electrons and flares with TypeII/IV radio events on the other hand, and proposed a 'two-phase' acceleration process for the SEP events observed in space.

Despite these results, a two class paradigm for SEP events was not generally accepted until the mid-1990s. The close association between coronal mass ejections (CMEs) observed on Skylab and large solar proton events led Kahler *et al.* (1978) to suggest an important role for the CME either in creating open field lines for flare particles to escape into the interplanetary medium or for the protons to be accelerated near a region above or around the outward moving ejecta far above the flare site. Subsequently, detailed analyses of flare durations, longitudinal distributions from multi-spacecraft observations, high resolution ionic charge state and elemental composition measurements, and clearer associations with radio bursts led most researchers to accept the view that the SEP events observed at 1 AU belong to two distinct classes, namely impulsive and gradual (*e.g.,* Kahler *et al.*, 1978; Cliver *et al.*, 1982; Kocharov, 1983; Kahler *et al.*, 1984; Luhn *et al.*, 1984; Mason *et al.*, 1984; Cane *et al.*, 1986; Reames, 1988).

In the two-class paradigm, the gradual events occur as a result of diffusive acceleration at CME-driven coronal and interplanetary (IP) shocks, while the impulsive events are attributed to stochastic acceleration during magnetic reconnection in solar flares (*e.g.,* Reames, 1999). The gradual or CME-related events typically lasted several days and had larger fluences, while the impulsive or flare-related events lasted a few hours and had smaller fluences. Impulsive events are typically observed when the observer is magnetically connected to the flare site, while ions accelerated at the expanding large-scale CME-driven shocks can populate magnetic field lines over a significantly broad range of longitudes (Cane *et al.*, 1988). The distinction between impulsive and gradual SEP events was further justified on the basis of the energetic particle composition and radio observations (e.g., Cane *et al.*, 1986). For instance, the flare-related impulsive SEP events were electron-rich and associated with Type III radio bursts. These events also had ^{3}He/^{4}He ratios enhanced between factors of 10^3–10^4 and Fe/O ratios enhanced by up to a factor of 10 over the corresponding solar wind values, had Fe with ionization states up to $\sim$20. In contrast, the gradual events were proton-rich, had average Fe/O ratios of $\sim$0.1 with Fe ionization states of $\sim$14, had no measurable enhancements in the ^{3}He/^{4}He ratio, and were associated with Type II bursts (Reames, 1999; Cliver, 2000).

The fact that IP shocks propagate in the solar wind combined with a lack of detailed composition measurements from the solar wind through the energetic particle ($\sim$1 MeV) energy range led many researchers to propose that the IP shock-associated particle events occurred as a result of diffusive shock acceleration of solar wind ions (e.g., Lee, 1983; Baring *et al.*, 1997). Others pointed out that the suprathermal tail of the solar wind may be the source (e.g., Gosling *et al.*, 1981; Tsurutani and Lin, 1985; Tan *et al.*, 1989). In fact Tsurutani and Lin (1985) and

Tan *et al.* (1989) suggested that the concomitant solar flares might provide the suprathermal seed particles accelerated at the IP shocks. Nonetheless, early theoretical studies based on diffusive shock acceleration theory (e.g., Lee, 1983) were successful in predicting many features of the so-called "Energetic Storm Particle" (ESP) events associated with some IP shocks near 1 AU (e.g., Kennel *et al.*, 1986). However, such detailed agreements between the theory and observations are extremely rare (see also Lario *et al.*, 2005). In contrast, studies involving a large number of IP shocks have shown that the predicted spectral indices for energetic protons and heavy ions near ~100 keV/nuc are significantly different from the observations (van Nes *et al.*, 1984; Desai *et al.*, 2004). Moreover, although the fastest CMEs were typically associated with the largest particle intensity enhancements, CMEs with similar speeds were also found to exhibit huge variations (~3–4 orders of magnitude) in the intensities of 10s of MeV particles at 1 AU (Kahler, 2001).

Understanding the origin of CME-related SEP events has become a top priority for solar-terrestrial physics, since high energy protons in the largest events pose severe radiation hazards for humans and technological systems in space. Recent numerical simulations and analytical models have incorporated particle acceleration at dynamically evolving IP shocks and are highly promising (e.g., Li *et al.*, 2003; Lee, 2005). Nevertheless, much work still needs to be done to achieve closure between theory and observations. One source of uncertainty is the identity of the seed particles and the exact manner in which they are injected into the acceleration process. Since CMEs propagate through interplanetary space, their shocks would presumably accelerate all incident particles that they encounter in the IP medium. However, Gloeckler (2003) pointed out that in addition to the solar wind peak at low energies, the fluences measured at 1 AU also exhibited a continuous presence of a suprathermal tail extending out to cosmic ray energies. Thus, an important question is whether the seed particles originate from the more abundant solar wind peak, or from the suprathermal tail. Although both sources were suggested in the 1980s (e.g., Forman and Webb, 1985; Tsurutani and Lin, 1985; Tan *et al.*, 1989), the instruments flown during that era lacked the sensitivity to measure small compositional differences, and so the question remained unanswered.

2. Observations

Recently, sophisticated mass spectrometers on board the Advanced Composition Explorer (ACE) mission (Stone *et al.*, 1998a) have enabled us to directly compare and distinguish between various particle populations in the heliosphere. In particular, CME-driven IP shocks detected near 1 AU offer a unique opportunity to investigate the seed population and shock acceleration processes during ESP events because the shock properties and the accelerated populations are measured directly. Thus, in-situ ESP events allow us to safely neglect other effects such as propagation, connection to the acceleration region etc., that play a vital role in understanding both the larger CME-related and the smaller flare-related SEP events where the acceleration processes occur near the Sun.

2.1. Ions Associated with CME-Driven IP Shocks at 1 AU

2.1.1. ^{3}*He and* He^{+} *Ions in CME-Driven IP Shocks*

Using measurements from the Ultra Low-Energy Isotope Spectrometer (ULEIS: Mason *et al.*, 1998) on board ACE, Desai *et al.* (2001) surveyed 48 IP shocks between Oct. 1997 and Nov. 2000. These results showed upper limits of ^{3}He in 23 events, while the remaining 25 events had substantial enhancements in the ^{3}He abundance that ranged between factors of ~2–600 times the solar wind value. Figure 1 shows the 0.5–2.0 MeV/nuc He mass histograms in four events where the ^{3}He/^{4}He ratio was over a hundred times more than that in the solar wind. However, such ^{3}He enrichments are routinely present during the smaller flare-related events which are more frequent during periods of high solar activity. The solar flares populate and replenish the interplanetary medium with suprathermal ^{3}He-rich material which subsequently gets re-accelerated by the CME-driven IP shocks whenever they encounter it en route to Earth (Mason *et al.*, 1999). Figure 2 shows a good correspondence between the occurrence frequency of ^{3}He-rich IP shocks and the fraction of time that suprathermal ^{3}He is present at 1 AU (Wiedenbeck *et al.*, 2003), which provides further support for a suprathermal origin for the seed population.

Similarly, He^{+} ions also act as tracers of the interstellar neutral gas that flows unimpeded into the inner solar system (Möbius *et al.*, 1985). Neutral He gets ionized inside 1 AU and is subsequently picked-up by the out-flowing solar wind (Gloeckler

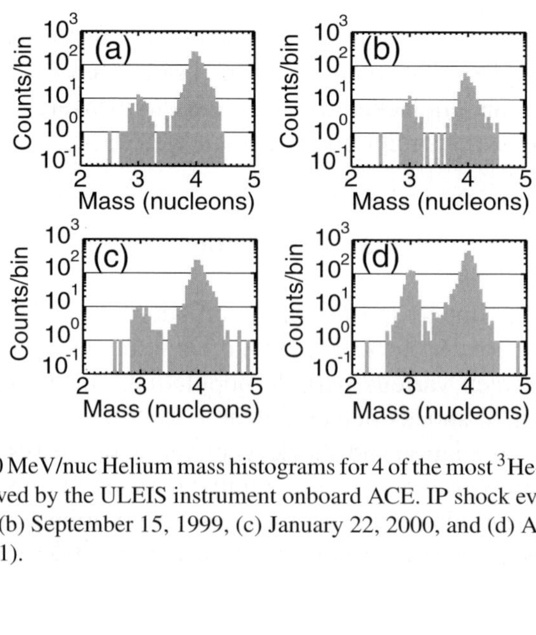

Figure 1. The 0.5–2.0 MeV/nuc Helium mass histograms for 4 of the most ^{3}He-enriched CME-driven IP shock events observed by the ULEIS instrument onboard ACE. IP shock events were observed on (a) January 28, 1998, (b) September 15, 1999, (c) January 22, 2000, and (d) April 24, 2000 (adapted from Desai *et al.*, 2001).

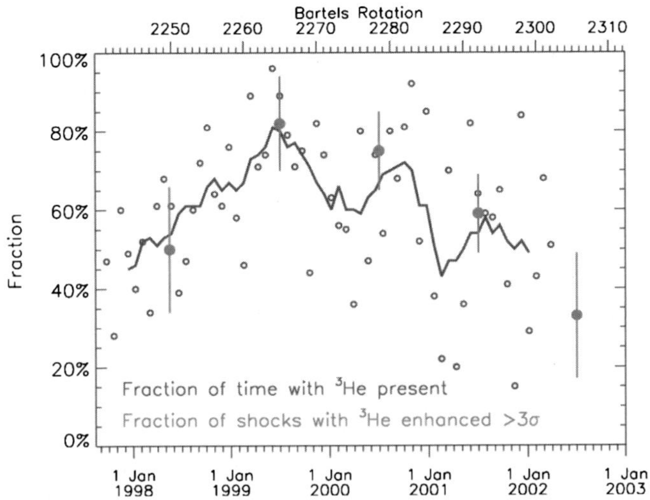

Figure 2. Occurrence frequency of ^{3}He-rich IP shock events at ACE (filled symbols with large error bars) superposed on the fraction of time energetic ^{3}He is present in the IP medium at 1 AU (adapted from Desai *et al.*, 2003 and Wiedenbeck *et al.*, 2003).

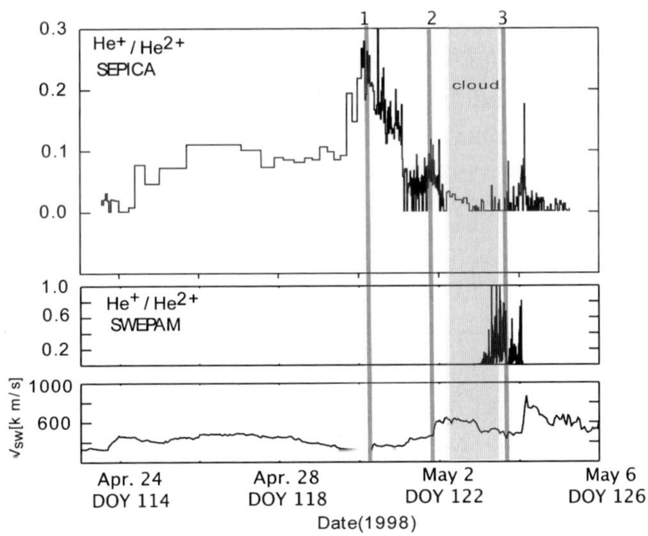

Figure 3. Temporal profiles of 0.25–0.8 MeV/nuc He^+/He^{2+} ratio compared with that measured in the solar wind for 3 CME-driven IP shocks and a magnetic cloud observed at ACE (adapted from Kucharek *et al.*, 2003).

et al., 1993). Near 1 AU, these ions are present in the suprathermal tail with relative abundances of more than 10^3 times the corresponding solar wind value. Figure 3 compares the He^+/He^{2+} ratio at ~0.5 MeV/nuc with that in the thermal solar wind during 3 CME-driven IP shock events (from Kucharek *et al.*, 2003). The energetic ion data were obtained by the Solar Energetic Particle Ionic Compositon Analyzer (SEPICA: Möbius *et al.*, 1998), while the solar wind measurements were obtained by the Solar Wind Electron Proton and Alpha Monitor (SWEPAM: McComas *et al.*, 1998) on board ACE. Notice that the large enhancements in the energetic He^+ abundance at the first 2 IP shocks did not coincide with the one observed later in the thermal plasma. Conversely, even though thermal He^+ ions are relatively more abundant in the "cloud", it appears that the IP shock that followed this cloud did not accelerate as many He^+ ions as the two preceding ones.

Since current shock acceleration models cannot operate solely on a solar wind-like ion composition and produce such enhancements, the mere presence of ^3He and He^+ ions in the energetic particles points to an origin in the suprathermal tail rather than the bulk solar wind. Moreover, since the contribution of pickup ions becomes more prominent near ~twice the solar wind speed (Gloeckler *et al.*, 1994), it appears that the injection process must also become substantially more efficient near or above this energy range (also see Chotoo *et al.*, 2000; Mason *et al.*, 2000).

2.1.2. *Heavy Ion Abundances in IP Shocks*

ACE measurements have also enabled us to investigate whether the more common heavier ions such as C-Fe originate from the bulk solar wind or from the suprathermal tail. Desai *et al.* (2003) compared the average ~1 MeV/nuc ion abundances for 72 IP shocks at ACE with those measured in the solar wind (from von Steiger *et al.*, 2000) and other candidate seed populations. Figure 4a shows the average IP shock abundances normalized to the slow solar wind values and plotted vs. Mass/Charge (M/Q). The ionization states used here are typical of those measured in the slow solar wind (von Steiger *et al.*, 1997). Note that the C/O ratio in IP

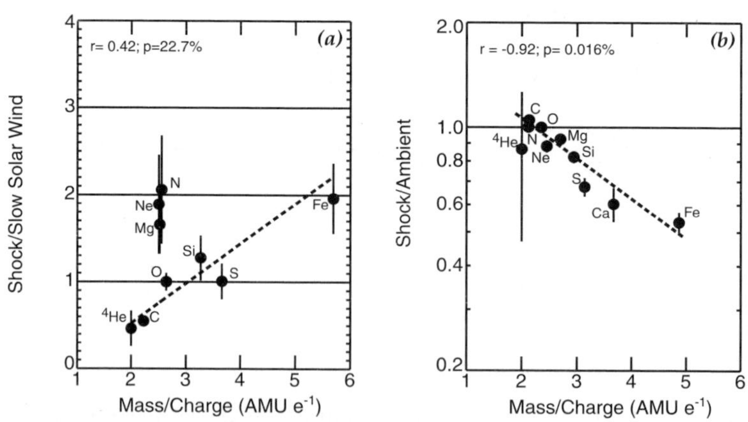

Figure 4. (a) IP shock abundances averaged over 72 events normalized to the slow solar wind values, plotted versus M/Q. (b) Average IP shock abundances compared with those measured in the ambient interplanetary medium prior to the arrival of the shocks (adapted from Desai *et al.*, 2003).

shocks is significantly lower than that measured in the solar wind, even though C and O have essentially the same M/Q ratio. However, since most shock acceleration mechanisms (e.g., Lee, 1983) fractionate ion species according to their M/Q ratios, they cannot distinguish C from O and deplete the C abundance by nearly ~50% after acceleration. In addition, the behavior of elements such as N, O, Ne, and Mg with similar M/Q ratios is highly unsystematic, which results in a poor correlation.

In contrast, Figure 4b shows that the IP shock abundances were well correlated with the average abundances measured in the interplanetary medium prior to the arrival of the IP shocks at ACE (Desai *et al.*, 2003). In particular, elements with higher M/Q ratios are systematically depleted, which is consistent with shock acceleration models wherein ions with higher M/Q ratios are accelerated less efficiently than those with lower M/Q values (*e.g.*, Lee, 2005). In addition to the correlations between the average quantities, Desai *et al.* (2003) also found significant correlations between the IP shock abundances (*e.g.*, ~1 MeV/nuc Fe/O ratio) and those measured in the ambient suprathermal ions for individual events. Since the ambient heavy ion population comprised ~30% material from flare-related SEP events and the remainder from large CME-related SEP events, these results indicate that the heavy ions from C-Fe also originate from a suprathermal tail that is essentially dominated by ions accelerated in prior SEP events (Desai *et al.*, 2003).

2.1.3. *Spectral Properties at IP Shocks*

Desai *et al.* (2004) investigated spectral properties of heavy ions in the ~0.1–100 MeV/nuc energy range during the same 72 CME-driven IP shock events surveyed by Desai *et al.* (2003). They found that the power-law spectral indices for 0.1–0.5 MeV/nuc O ions were poorly correlated with the values predicted by simple one-dimensional steady state models. Likewise, poor agreement between the theoretical predictions and the low-energy ($\lesssim 0.5$ MeV) proton spectral indices were also found earlier by van Nes *et al.* (1984), and more recently by Ho *et al.* (2003). Desai *et al.* (2004) also found that the characteristic *e*-folding or roll-over energy of the O spectra at IP shocks was uncorrelated with shock parameters such as shock normal angle or shock speed.

In contrast, Figure 5 shows that both the O spectral index and the energy-dependent behavior of the Fe/O ratio in IP shocks are very well correlated with the corresponding quantities measured in the ambient population. It is worthwhile noting that such strong correspondence between the accelerated ions and the ambient suprathermal ions are not predicted at all by simple shock acceleration theory.

2.2. CME-Related Solar Energetic Particle Events

2.2.1. ^{3}*He Enhancements*

Understanding large SEP events is somewhat more difficult since the acceleration processes occur near the Sun, and other effects (*e.g.*, propagation to 1 AU) have to be considered. Nevertheless, here too the mere presence of tracer ions like ^{3}He can

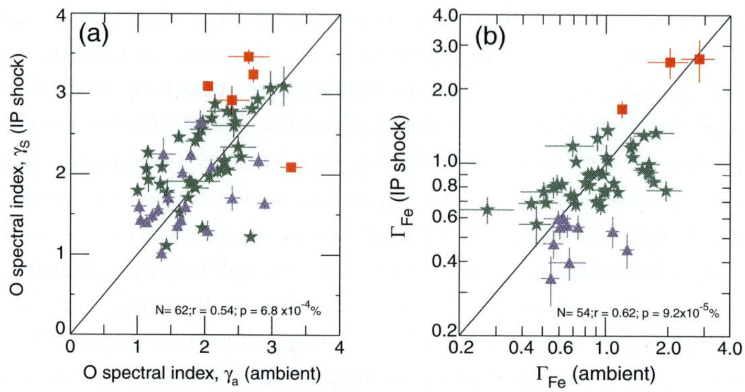

Figure 5. (a) Power-law spectral index of O in IP shocks versus that measured in the ambient material prior to the arrival of the shocks. (b) Energy-dependent behavior of Fe/O ratio at IP shocks vs. that measured in the ambient medium. The energy-dependent behavior of the Fe/O ratio is determined by dividing the 0.62 MeV/nuc Fe/O ratio by that measured at 0.22 MeV/nuc (taken from Desai *et al.*, 2004). Events with different energy-dependent behavior in the Fe/O ratio are identified as follows: red squares: Fe/O increases with increasing energy; green asterisks: Fe/O remains constant with energy; blue triangles: Fe/O decreases with increasing energy.

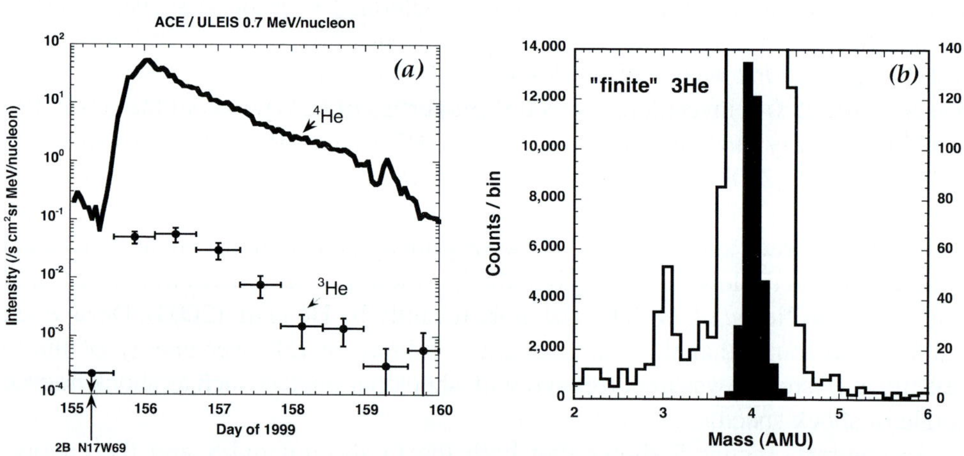

Figure 6. (a) Temporal profiles of ~0.7 MeV/nuc ^{3}He and ^{4}He ions in a large CME-related SEP event. (b) 0.5–2.0 MeV/nuc He mass histogram obtained during several large SEP events. The right scale corresponds to the open histogram (taken from Mason *et al.*, 1999).

be used to identify the origin of the seed population. Figure 6a shows time-intensity profiles for 0.5–2.0 MeV/nuc ^{3}He and ^{4}He ions in a large CME-related SEP event that occurred on June 4, 1999 (from Mason *et al.*, 1999). The temporal profiles of the two species are remarkably similar, which indicates that they probably have the same acceleration and transport history. Here the ^{3}He is enriched by a factor of 16 ± 3 while the Fe/O ratio (not shown) is simultaneously enhanced by about a factor of 10 over corresponding solar wind values. Since the M/Q ratio for Fe is larger than that of O while that of ^{3}He is smaller than that of ^{4}He, these results cannot be reconciled with current rigidity-dependent acceleration mechanisms in which the shock operates on a solar wind-like seed population.

The event in Figure 6a was selected from a list of large CME-related NOAA/SEC events that produced significant 10 MeV proton intensity enhancements at 1 AU. A substantial fraction (~50%) of these events had ^{3}He enrichments (*e.g.,* Mason *et al.*, 1999; Wiedenbeck *et al.*, 2000). Figure 6b shows the low energy He mass histogram from several such events. Notice that the ^{3}He is clearly resolved from ^{4}He and the background. Like the ESP events at 1 AU, these enhancements are also attributed to the presence of residual flare-accelerated ^{3}He-rich suprathermal material in the seed population for the CME-driven shocks near the Sun.

2.2.2. *Heavy Ion Abundances*

ACE measurements during large SEP events have allowed us to explore whether the heavier ions originate from the solar wind peak. Mewaldt *et al.* (2002) normalized the average heavy ion abundances measured by the Solar Isotope Spectrometer (SIS: Stone *et al.*, 1998b) above ~5 MeV/nuc in ~40 large SEP events to those measured in the solar wind and plotted them versus the first ionization potential (FIP), as shown in Figure 7. This figure clearly shows that the normalized abundances are not organized in any systematic fashion by the FIP, but are instead scattered randomly about the 1:1 line. They conclude that the material accelerated in large SEP events is quite distinct from that measured in the solar wind peak, and therefore the SEPs could not have been accelerated directly out of the solar wind.

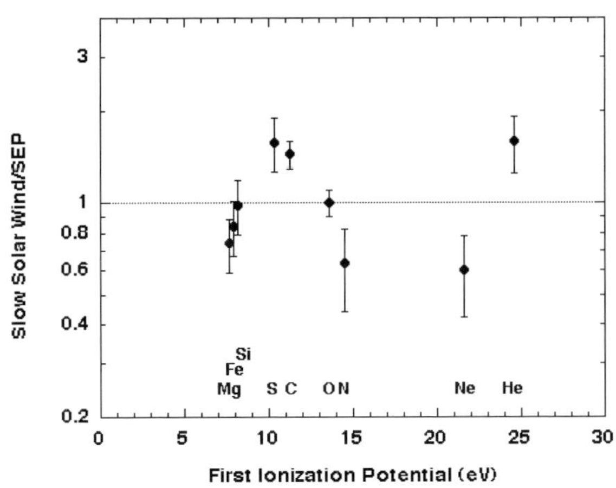

Figure 7. Average heavy ion elemental abundances measured in the slow solar wind divided by that measured above ~5 MeV/nuc in several large CME-related SEP events, and plotted versus the FIP value of each element (taken from Mewaldt *et al.*, 2002).

2.2.3. *Spectral Variability*

One of the outstanding challenges in understanding large SEP events is the inherent event-to-event variability measured in the energy spectra. Figure 8 shows two extreme cases that were associated with flares and CMEs from the west limb, had shocks with similar speeds, and were generally similar in the associated solar and interplanetary signatures (*e.g.*, Cohen *et al.*, 2003; Tylka *et al.*, 2005). Yet the heavy ion spectral behavior was remarkably different. In particular, the Fe/O ratio in the April 21, 2002 event decreases with increasing energy, which is qualitatively consistent with shock acceleration models wherein Fe with its higher M/Q ratio is accelerated less efficiently than O. However, such a model cannot account for the increase in Fe/O with energy observed in the August 22, 2002 event. Tylka *et al.* (2005) attributed the above differences to a quasi-parallel shock preferentially injecting solar wind suprathermals during the April 2002 event, and a quasi-perpendicular shock preferentially injecting flare suprathermals during the August 2002 event (however, see Giacalone *et al.*, 2005).

3. Discussion

Comprehensive surveys of heavy ion abundances and spectra during ESP events associated with CME-driven IP shocks near 1 AU have shown:

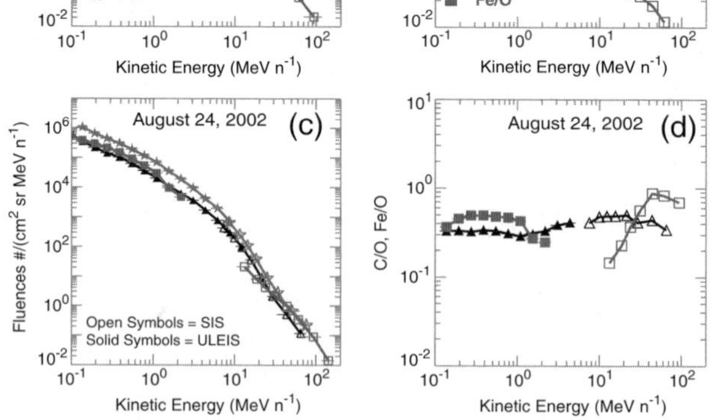

Figure 8. C, O, and Fe energy spectra, and C/O and Fe/O ratios measured by ULEIS (closed symbols) and SIS (open symbols) during the April 21 and August 24 2002 SEP events.

- substantial enrichments of tracer ion species like ^3He and He^+,
- average IP shock abundances that are uncorrelated with solar wind values,
- significant correlation between the spectral properties of the accelerated ions and those of the suprathermal ambient population preceding the shocks.

The acceleration of ^3He and He^+ ions taken together with the highly significant correlations between the heavy ion abundances, the energy spectra, and the energy-dependent behavior of the Fe/O ratio in IP shocks and in the ambient material provide strong evidence that CME-driven IP shocks accelerate ions originating from the suprathermal tail rather than from the solar wind peak. Furthermore, the poor correlation between the average IP shock and the solar wind abundances indicates that the suprathermal tail was dominated by sources such as interstellar pickup ions and ions from prior flare- and CME-related SEP events rather than the heated solar wind (Desai *et al.*, 2001, 2003; Kucharek *et al.*, 2003).

Preliminary results from surveys of large SEP events from ACE show that they too exhibit similar properties as the ESP events. While the inherent variations in the seed population alone could account for the event-to-event variability observed in ESP events (Desai *et al.*, 2004; Channok *et al.*, 2005), such an explanation cannot account for all the puzzling observations during the CME-related SEP events. For instance, on the basis of observations of complex Type III radio bursts during many large SEP events (see Cane and Erickson, 2003), Cane *et al.* (2003) have suggested that events in which the Fe/O ratio increases with increasing energy (*e.g.*, see Figure 8b) are due to the presence of a direct flare component above ~12 MeV/nuc. Thus, even though it is generally accepted that CME-driven shocks produce the largest intensity enhancements at 1 AU up to GeV energies, future in-situ measurements in conjunction with remote sensing observations could yet yield some surprises.

Nonetheless, presently it does appear that the suprathermal tail rather than the solar wind peak is the main source of He and heavy ions in the CME-related ESP and SEP events. Given the fact that SEP events occur more frequently during active solar periods, it is not surprising that the suprathermal tail over the last 6–7 years has been dominated by ions accelerated previously in both flares and CME-related SEP events. Numerical simulations have shown that pickup ions are preferentially injected at shocks due to their shell-like or isotropic velocity distributions (*e.g.*, Scholer and Kucharek, 1999). Since the suprathermal ions from SEP events take several hours to a few days to get to 1 AU, they are also likely to be isotropic. Thus, for injection purposes, the suprathermal ions have at least two distinct advantages when compared with the thermal solar wind ions, namely, (1) they have speeds above the theoretical injection threshold, and (2) they have isotropic velocity distributions (*e.g.*, Giacalone *et al.*, 1994; Ellison *et al.*, 1999).

It is therefore extremely important that we characterize key properties of the suprathermal ion population. Figure 9a shows the O fluence measured at 1 AU, and identifies various ion sources at different energies (Mewaldt *et al.*, 2001) over

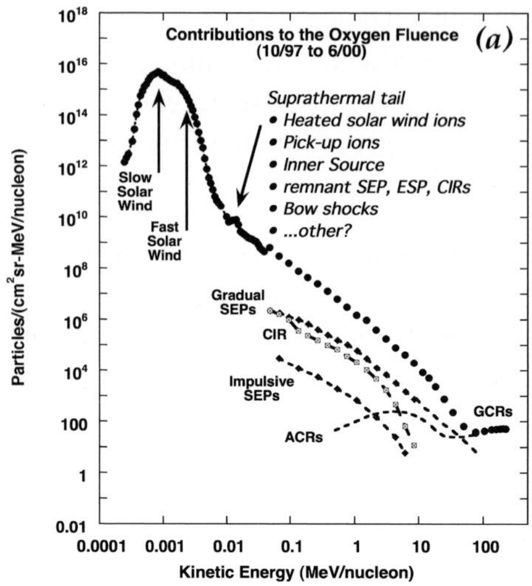

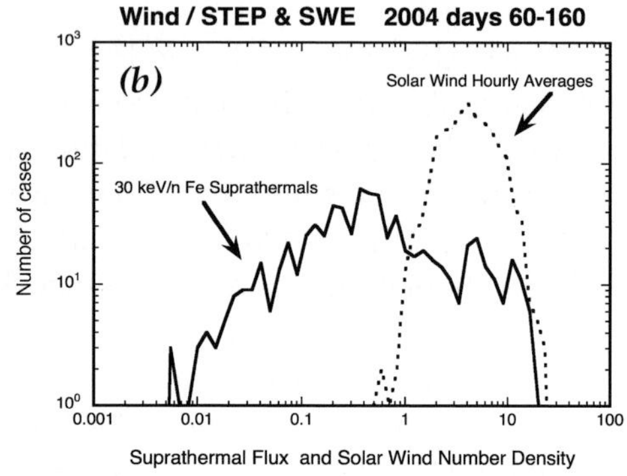

Figure 9. (a) Contributions to the O fluence from solar wind to cosmic ray energies during a 3-year period (taken from Mewaldt *et al.*, 2001). (b) Comparison between the hourly averaged ~30 keV/nuc Fe intensity (solid trace) and solar wind Fe density (dotted curve) measured over a 100-day interval in 2004 at 1 AU (both figures taken from Mason *et al.*, 2005).

a 3-year period. The relative contributions of these sources probably varies with solar activity and spatial distance, but have remained unexplored due to instrumental limitations. Figure 9b shows that the temporal variations in the intensities of ~30 keV/nuc suprathermal Fe are significantly larger (nearly ~3 orders of magnitude) when compared with the solar wind density which varies only by about a factor of 10. It is presently unclear whether such large variations play a critical role in producing the large range in peak proton intensities for CMEs with similar speeds (Kahler, 2001).

Understanding the exact role played by variations in the suprathermal seed population during SEP events is necessary to improve agreement between theoretical predictions and observations. This issue is also critical for understanding particle injection and acceleration at collisonless shocks throughout the Universe. Given the close association between fast CMEs and large SEP events, it is imperative that we develop realistic shock acceleration models that can routinely inject and accelerate suprathermal seed populations from a variety of sources (*e.g.,* see Kocharov and Torsti, 2003; Li and Zank, 2005). Input for the seed population could come from instruments on board spacecraft located well inside 1 AU. Such models should be incorporated into global Sun-to-Earth models of CME initiation, evolution, and propagation through interplanetary space. We can then begin to imagine a future where the end-to-end models in the Sun-Earth chain are constrained and driven by in-situ and remote observations, and can accurately predict the temporal, spatial, and spectral characteristics, and the associated radiation dosages during large

CME-related SEP events near Earth or for that matter, anywhere in the solar system where humans and robots are exploring.

Acknowledgements

We thank ISSI, the organizing committee, and all the participants for a stimulating workshop. A significant portion of the material in this article also appears in Mason *et al.* (2005). Figure 9b used Solar Wind data from the Wind/SWE experiment, kindly provided by K. Ogilvie and the SWE team. This work was partially supported by NASA under Caltech grant 44A1055749 at the University of Maryland.

References

Baring, M. G., Ogilvie, K. W., Ellison, D. C., and Forsyth, R. J.: 1997, *Astrophys. J.* **476**, 889.

Cane, H. V., and Erickson, W. C.: 2003, *J. Geophys. Res. (Space Physics)* **108**, A5, doi: 10.1029/2002JA009488.

Cane, H. V., von Rosenvinge, T. T., Cohen, C. M. S., and Mewaldt, R. A.: 2003, *Geophys. Res. Lett.* **30**(12), 8017, doi: 10.1029/2002GL016580.

Cane, H. V., Reames, D. V., and von Rosenvinge, T. T.: 1988, *Astrophys. J.* **93**, 9555.

Cane, H. V., McGuire, R. E., and von Rosenvinge, T. T.: 1986, *Astrophys. J.* **301**, 448.

Channok, C., Rufollo, D., Desai, M. I., and Mason, G. M.: 2005, Accepted for publication in *Astrophys. J. Lett.*

Chotoo, K., Schwadron, N. A., Mason, G. M., Zurbuchen, T. H., Gloeckler, G., Posner, A., *et al.*: 2000, *J. Geophys. Res.* **105**, 23107.

Cliver, E. W.: 2000, In: Mewaldt, R. A., Jokipii, J. R., Lee, M. A., Möbius, E., and Zurbuchen, T. H. (eds.), *Acceleration and Transport of Energetic Particles Observed in the Heliosphere*, Vol. 528 of *AIP Conference Proceedings*. Melville, NY, pp. 21–31.

Cliver, E. W., Kahler, S. W., Shea, M. A., Smart, D. F.: 1982, *Astrophys. J.* **260**, 362.

Cohen, C. M. S., Mewaldt, R. A., Cummings, A. C., Leske, R. A., Stone, E. C., von Rosenvinge, T. T., *et al.*: 2003, *Adv. Space Res.* **32**, 2649.

Desai, M. I., Mason, G. M., Wiedenbeck, M. E., Cohen, C. M. S., Dwyer, J. R., Mazur, J. E., *et al.*: 2004, *Astrophys. J.* **611**, 1156.

Desai, M. I., Mason, G. M., Dwyer, J. R., Mazur, J. E., Gold, R. E., Krimigis, S. M., *et al.*: 2003, *Astrophys. J.* **558**, 1149.

Desai, M. I., Mason, G. M., Dwyer, J. R., Mazur, J. E., Smith, C. W., and Skoug, R. M.: 2001, *Astrophys. J.* **553**, L89.

Ellison, D. C., Jones, F. C., and Baring, M. G.: 1999, *Astrophys. J.* **512**, 403.

Forbush, S. E.: 1946, *Phys. Rev.* **70**, 771.

Forman, M. A., and Webb, G. M.: 1985, In *Collisonless shocks in the heliosphere,* Vol. 35 of *American Geophysical Union, Geophysical Monograph Series,* Washington, D.C., pp. 91–114.

Giacalone, J.: 2005, *Astrophys. J.* **624**, 765.

Giacalone, J., Jokipii, J. R., and Kota, J.: 1994, *J. Geophys. Res.* **99**, 19351.

Gloeckler, G.: 2003, In: Velli, M., Bruno, R., Malara, F., and Bucci, P. B. (eds.), *Solar Wind Ten*, Vol. 679 of *AIP Conference Proceedings*. Melville, NY, pp. 583–588.

Gloeckler, G, Geiss, J., Roelof, E. C., Fisk, L. A., Ipavich, F. M., Ogilvie, K. W., *et al.*: 1994, *J. Geophys. Res.* **99**, 17637.

Gloeckler, G., Geiss, J., Balsiger, H., Fisk, L. A., Galvin, A. B., Ipavich, F. M., *et al.*: 1993, *Science* **261**, 70.

Gosling, J. T., Asbridge, J. R., Bame, S. J., Feldman, W. C., Zwickl, R. D., Paschmann, G., *et al.*: 1981, *J. Geophys. Res.* **86**, 547.

Ho, G. C., Lario, D., Decker, R. B., Roelof, E. C., Desai, M. I., and Smith, C. W.: 2003, In *Proceedings of the 28th International Cosmic Ray Conference* **28**, 3689.

Kahler, S. W.: 2001, *J. Geophys. Res.* **106**, 20947.

Kahler, S. W., Sheeley, Jr., N. R., Howard, R. A., Michels, D. J., Koomen, M. J., McGuire, R. E., *et al.*: 1984, *J. Geophys. Res.* **89**, 9683.

Kahler, S. W., Hildner, E., and Van Hollebeke, M. A. I.: 1978, *Solar Phys.* **57**, 429.

Kennel, C. F., Coroniti, F. V., Scarf, L. F., Livesey, W. A., Russell, C. T., and Smith, E. J.: 1986, *J. Geophys. Res.* **91**, 11917.

Kocharov, G. E.: 1983, *Proceedings of the 18th International Cosmic Ray Conference, Invited and Rapporteur Papers* **12**, 235.

Kocharov, L., and Torsti, J.: 2003, *Astrophys. J.* **586**, 1430.

Kucharek, H., Möbius, E., Li, W., Farrugia, C. J., Popecki, M. A., Galvin, A. B., Klecker, B., *et al.*: 2003, *J. Geophys. Res. (Space Physics)* **108**, A10, doi: 10.1029/2003JA009938.

Lario, D., Decker, R. B., Ho, G. C., Hu, Q., Smith, C. W., Desai, M. I. *et al.*: 2005, In: Li, G., Zank, G. P., and Russell, C. T. (eds.) *The Physics of Collisionless Shocks* Vol 781 of *AIP Conference Proceedings*. Melville, NY, pp. 180–184.

Lee, M. A.: 2005, *Astrophys. J.* **158**, 38.

Lee, M. A.: 1983, *J. Geophys. Res.* **88**, 6109.

Li, G., and Zank, G. P.: 2005, *Geophys. Res. Lett.* **32**, 2101.

Li, G., Zank, G. P., and Rice, W. K. M.: 2003, *J. Geophys. Res. (Space Physics)* **108**, A2, doi: 10.1029/2002JA009666.

Lin, R. P.: 1970, *Solar Phys.* **12**, 266.

Luhn, A., Klecker, B., Hovestadt, D., Gloeckler, G., Ipavich, F. M., Scholer, M., *et al.*: 1984, *Adv. Space Res.* **4**, 161.

Mason, G. M., Desai, M. I., Mazur, J. E., and Dwyer, J. R.: 2005, In: Li, G., Zank, G. P., and Russell, C. T. (eds.), *The Physics of Collisionless Shocks*, Vol. 781 of *AIP Conference Proceedings*. Melville, NY, pp. 219–226.

Mason, G. M.: 2000, Mewaldt, R. A., Jokipii, J. R., Lee, M. A., Möbius, E., and Zurbuchen, T. H. (eds.): *Acceleration and Transport of Energetic Particles Observed in the Heliosphere*, Vol. 528 of *AIP Conference Proceedings*. Melville, NY, pp. 234–239.

Mason, G. M., Mazur, J. E., and Dwyer, J. R.: 1999, *Astrophys. J. Lett.* **525**, L133.

Mason, G. M., Gold, R. E., Krimigis, S. M., Mazur, J. E., Andrews, G. B., Daley, K. A., *et al.*: 1998, *Space Sci. Rev.* **86**, 409.

Mason, G. M., Gloeckler, G., and Hovestadt, D.: 1984, *Astrophys. J.* **280**, 902.

McComas, D. J., Bame, S. J., Barker, P., Feldman, W., Phillips, J. L., Riley, P., *et al.*: 1998, *Space Sci. Rev.* **86**, 563.

Mewaldt, R. A., Cohen, C. M. S., Leske, R. A., Christian, E. R., Cummings, A. C., Stone, E. C., *et al.*: 2002, *Adv. Space Res.* **30**, 79.

Mewaldt, R. A., Mason, G. M., Gloeckler, G., Christian, E. R., Cohen, C. M. S., Cummings, A. C., *et al.*: 2001, In: Wimmer-Schweingruber, R. F. (ed.), *Solar and Galactic Composition* Vol 598 of *AIP Conference Proceedings*. Melville, NY, pp. 165–169.

Meyer, P., Parker, E. N., and Simpson, J. A.: 1956, *Phys. Rev.* **104**, 768.

Möbius, E., Kistler, L. M., Popecki, M. A., Crocker, K. N., Granoff, M., Turco, S., *et al.*, 1998: *Space Sci. Rev.* **86**, 449.

Möbius, E., Hovestadt, D., Klecker, B., Scholer, M., and Gloeckler, G.: 1985, *Nature* **318**, 426.

van Nes, P., Reinhard, R., Sanderson, T. R., Wenzel, K.-P., and Zwickl, R. D.: 1984, *J. Geophys. Res.* **89**, 2122.

Reames, D. V.: 1999, *Space Sci. Rev.* **90**, 413.

Reames, D. V.: 1988, *Astrophys. J.* **220**, L71.

Scholer, M., and Kucharek, H.: 1999, *Geophys. Res. Lett.* **26**, 29.

von Steiger, R., Schwadron, N. A., Fisk, L. A., Geiss, J., Gloeckler, G., Hefti, S., *et al.*: 2000, *J. Geophys. Res.* **105**, 27217.

von Steiger, R., Geiss, J., and Gloeckler, G.: 1997, Jokipii, J. R., Sonett, C. P., and Giampapa, M. S. (eds.), *Cosmic Winds and the Heliosphere*, Tucson, pp. 581–616.

Stone, E. C., Frandsen, A. M., Mewaldt, R. A., Christian, E. R., Margolies, D., Ormes, J. F., *et al.*: 1998a: *Space Sci. Rev.* **86**, 1.

Stone, E. C., Cohen, C. M. S., Cook, W. R., Cummings, A. C., Gauld, B., Kecman, B., *et al.*: 1998b: *Space Sci. Rev.* **86**, 285.

Tan, L. C., Mason, G. M., Klecker, B., and Hovestadt, D.: 1989, *Astrophys. J.* **345**, 572.

Tsurutani, B. T., and Lin, R. P.: 1985, *J. Geophys. Res.* **90**, 1.

Tylka, A. J., Cohen, C. M. S., Dietrich, W. F., Lee, M. A., Maclennan, C. G., Mewaldt, R. A., *et al.*: 2005, *Astrophys. J.* **625**, 474.

Wiedenbeck, M. E., Mason, G. M., Christian, E. R., Cohen, C. M. S., Cummings, A. C., Dwyer, J. R., *et al.*: 2003, In: Velli, M., Bruno, R., Malara, F., and Bucci, P. B. (eds.), *Solar Wind Ten*, Vol. 679 of *AIP Conference Proceedings*. Melville, NY, pp. 652–655.

Wiedenbeck, M. E., Christian, E. R., Cohen, C. M. S., Cummings, A. C., Leske, R. A., Mewaldt, R. A., *et al.*: 2000, In: Mewaldt, R. A., Jokipii, J. R., Lee, M. A., Möbius, E., and Zurbuchen, T. H. (eds.), *Acceleration and Transport of Energetic Particles Observed in the Heliosphere*, Vol. 528 of *AIP Conference Proceedings*. Melville, NY, pp. 107–110.

Wild, J. P., Smerd, S. F., and Weiss, A. A.: 1963, *Annu. Rev. Astron. Astrophys.* **1**, 291.

ACCELERATION OF SOLAR-ENERGETIC PARTICLES BY SHOCKS

JOE GIACALONE* and JÓZSEF KÓTA
Lunar & Planetary Laboratory, University of Arizona, Tucson, AZ 85721, USA
*(*Author for correspondence: E-mail: giacalon@lpl.arizona.edu)*

(Received 22 July 2005; Accepted in final form 20 October 2006)

Abstract. Our current understanding of the acceleration of solar-energetic particles is reviewed. The emphasis in this paper is on analytic theory and numerical modeling of the physics of diffusive shock acceleration. This mechanism naturally produces an energy spectrum that is a power law over a given energy interval that is below a characteristic energy where the spectrum has a break, or a rollover. This power law is a common feature in the observations of *all* types of solar-energetic particles, and not necessarily just those associated with shock waves (e.g. events associated with impulsive solar flares which are often described in terms of resonant stochastic acceleration). Moreover, the spectral index is observed to have remarkably little variability from one event to the next (about 50%). Any successful acceleration mechanism must be able to produce this feature naturally and have a resulting power-law index that does not depend on physical parameters that are expected to vary considerably. Currently, only diffusive shock acceleration does this.

Keywords: solar energetic particles, particle acceleration, solar magnetic fields, coronal mass ejections

1. Introduction

If we are to understand the space radiation environment in the near vicinity of Earth we require a detailed description of the origin, acceleration, and propagation of high-energy nuclei from the Sun. This is an important, unsolved problem that has far-reaching consequences for a wide variety of astrophysical applications. The problem is particularly difficult because of a lack of *in situ* measurements at acceleration sites near the Sun. It is clear that the largest fluxes of high-energy charged particles result from violent solar explosions that are related to intense magnetic activity on the Sun. It is believed that the largest events are associated with fast coronal mass ejections (CMEs) that drive turbulent shock waves that, in turn, accelerate charged particles to high energies. These particles arrive at Earth within the first minutes to hours after the event.

The fluxes of solar-energetic particles (SEPs) certainly vary markedly from one event to the next. Some events can even be observed by ground-level detectors. However, interestingly, SEP events share one common feature: the observed differential intensity has a power-law dependence on energy below a certain critical energy. This critical energy, where the spectrum rolls over (or "breaks"), also varies considerably from event to event and gives rise to complicated compositional variations

Space Science Reviews (2006) 124: 277–288
DOI: 10.1007/s11214-006-9110-1

at high energies. However, it is important to note that below this energy the spectral index does not vary by much (only about 50%). This is an important starting point for identifying the acceleration mechanism, which must be able to reproduce this feature naturally. Below, we discuss possible acceleration mechanisms and then briefly summarize our current understanding of SEPs and suggest some important outstanding questions and open issues.

2. Mechanisms of Charged-Particle Acceleration

2.1. Swann's Mechanism

Swann (1933) was the first to address the issue of particle acceleration associated with solar magnetic activity. He showed that a gradually increasing magnetic field can accelerate particles because of the effect of the induced electric field. In his picture, which was applied primarily to electron acceleration in sunspots, the magnetic field was assumed to be radial and the electric field was normal to it. Thus, the charged particles gyrate around the field and gain energy as the magnetic field increases with time. Note that this is not the same as acceleration by a parallel electric field. He determined that electrons could reach energies of 10 GeV in one second (assuming a magnetic field increased to 1000 Gauss in 1 day after the formation of a sunspot). Swann's model was, however, not complete because he did not discuss what would happen when the field eventually decreased and it was not discussed how the particles escaped (the gyroradii of the particles decreased in his case making escape more difficult).

2.2. Second-Order Fermi Acceleration

Fermi (1949) suggested a mechanism similar to that proposed by Swann except that he included the effect of particle scattering (see also Alfvén, 1950). In this mechanism, fast-moving charged particles scatter off of slower-moving random magnetic scattering centers. An individual particle either gains or loses enery depending on whether the collision is head on or retreating. A net energy gain results because of the higher probability for head-on collisions (energy gaining).

Second-order Fermi acceleration is conceptually very straightforward. The evolution of the distribution function can be represented as a momentum-diffusion equation which, in steady state, is given by:

$$\frac{1}{p^2}\frac{\partial}{\partial p}\left(p^2 D_{pp}\frac{\partial f}{\partial p}\right) - \frac{f}{\tau_{\rm loss}} = 0 \tag{1}$$

where $\tau_{\rm loss}$ is the loss time (e.g. as in the classic "leaky box" approximation). The usual assumption is $D_{pp} = \langle \Delta p^2 \rangle/(2\Delta t) \approx (v_A/w)^2 p^2/\tau_{\rm scat}$, where v_A is the Alfvén speed, w is the particle speed, and $\tau_{\rm scat}$ is the scattering time.

In the unlikely event that both τ_{scat} and τ_{loss} are independent of momentum, (1) is readily solved giving a power-law dependence of f on p. Even in this highly idealized problem, the resulting power-law index depends on τ_{loss}, τ_{scat}, and v_A, which all vary considerably depending on the strength of the magnetic field and the level of magnetic turbulence. When more reasonable forms of the scattering and loss times are used, the resulting f is not, necessarily, a power law in momentum (J.R. Jokipii, private communication). Another problem with this mechanism is the relatively slow acceleration. Therefore, this mechanism is unconvincing for the acceleration of SEPs (see also Syrovatsky, 1960).

2.3. Resonant Stochastic Acceleration

This mechanism relies on the same basic physics as second-order Fermi acceleration. However, instead of scattering off of random magnetic "blobs", electromagnetic turbulence is assumed to be distributed uniformly in space. The fluctuations are typically assumed to be of small amplitude ($\delta B \ll B$) and are related to the background plasma, e.g plasma waves. The mechanism relies on a particular resonance condition that involves the frequency and wavelength of the wave. D_{pp} in (1) is a sum over all plasma-wave type (e.g. Alfvèn waves, electromagnetic ion-cyclotron waves, etc.) with the resonance condition that is appropriate to that wave type. This mechanism is often applied to the acceleration of ions associated with solar flares (Petrosian and Liu, 2004; Miller, 2000; Emslie *et al.*, 2004). One particular success of this mechanism is the explanation of the enhancement of $^3He^{++}$ relative to $^4He^{++}$ seen to be associated with impulsive solar flares (Fisk, 1978).

2.4. Reconnection Electric Fields

In this mechanism, particles are accelerated by the electric field at the X point of a steady-state reconnection site (e.g. Litvinenko, 1996). A power-law comes about simply because of the geometry. Those particles that are capable of drifting into the X point are accelerated by an amount equal to the full potential difference associated with the electric field. These are the highest-energy particles, whereas the energy gain of other particles depends on how close they drift in to this point. This mechanism is mostly relevant to reconnecting current sheets believed to be associated with solar flares. The role of turbulence and the effect of the accelerated high-energy particles on the dynamics of magnetic reconnection remain open issues.

2.5. Diffusive Shock Acceleration

Charged particles can be accelerated by collisionless shocks provided they can scatter back and forth across the shock many times. They gain energy because the scattering centers are embedded in converging flows (Krymsky, 1977; Axford *et al.*,

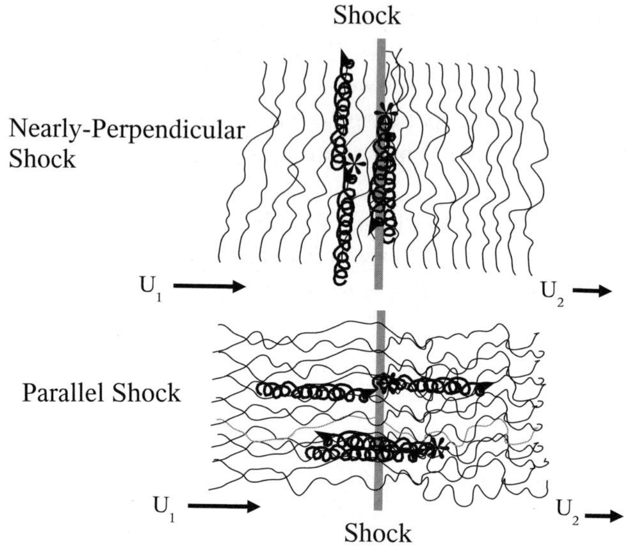

Figure 1. Illustration of particle acceleration at parallel and perpendicular shocks. The gray lines indicate magnetic lines of force.

1978; Bell, 1978; Blandford and Ostriker, 1978). The electric field that is responsible for the energy gain is the usual convective electric field $\mathbf{E} = -(1/c)\mathbf{U} \times \mathbf{B}$, where $\mathbf{U}$ is the plasma velocity, and $\mathbf{B}$ is the magnetic field that contains both a mean and a fluctuating component. For the case in which the upstream magnetic field is normal to the unit normal to the shock front – a perpendicular shock – particles drift along the shock front (due to the increase of the magnetic field) in the same direction as $\mathbf{E}$. Figure 1 shows a simple illustration of shock acceleration.

For a collection of charged particles moving in the vicinity of a shock, the phase-space distribution, f, can be described by the Parker transport equation (Parker, 1965) given by:

$$\frac{\partial f}{\partial t} = \frac{\partial}{\partial x_i}\left[\kappa_{ij}\frac{\partial f}{\partial x_j}\right] - U_i\frac{\partial f}{\partial x_i} + \frac{1}{3}\frac{\partial U_i}{\partial x_i}\frac{\partial f}{\partial \ln p} + Q \qquad (2)$$

where x_i is the position, p is momentum, t is time, κ_{ij} is the diffusion tensor, U_i is the plasma velocity, and Q is any source. Written in this way, particle drifts are contained in the antisymmetric components of κ_{ij}.

Equation (2) is generally valid for high-energy particles such that $w \gg U$, where U is the plasma velocity. It contains terms that include the four major transport effects: convection, spatial diffusion, drifts, and energy change. The electric field, that gives rise to energy changes, does not appear explicitly in (2). It is contained in the terms containing the flow velocity U_i.

To arrive at a quantitative solution for diffusive shock acceleration, we solve (2) for a shock-like discontinuity. Solutions are obtained separately for the upstream and downstream regions, and are matched at the shock with conservation relations. For a one-dimensional planar shock with particle injection at the shock at a momentum p_0 , it is readily shown that the steady-state solution to (2) is given by:

$$f(p) = Ap^{-3r/(r-1)} H(p - p_0) F(x, p) \tag{3}$$

where A is a normalization constant, r is the ratio of the downstream to upstream plasma density, $H(p)$ is the Heaviside step function and

$$F(x, p) = \begin{cases} \exp\left(U_1 x/\kappa_{xx}(p)\right) & x \leq 0 \quad \text{(upstream)} \\ 1 & x > 0 \quad \text{(downstream)} \end{cases}$$

where U_1 is the plasma velocity upstream of the shock and κ_{xx} is the diffusion coefficient normal to the shock. In terms of θ_{Bn}, the angle between the shock normal and mean magnetic field, $\kappa_{xx} = \kappa_\perp \sin^2\theta_{Bn} + \kappa_\parallel \cos^2\theta_{Bn}$.

We see immediately that the downstream spectrum depends only on the shock compression ratio! In terms of the differential intensity, $dJ/dE = p^2 f \propto E^{-\alpha}$, where $\alpha = -(r+2)/(2r-2)$. Thus, for shock compressions in the range $2 < r < 4$, the resulting differential intensity is a power law with an index in the range $1 < \alpha < 2$. Therefore, acceleration by collisionless shocks in the solar atmosphere and inner heliosphere is an attractive explanation for the rather limited variability in the observed power-laws of solar-energetic particles.

2.5.1. *The Origin of Spectral Breaks in Diffusive Shock Acceleration*

At higher energies, the spectrum of SEPs is observed to deviate from the power law derived in the previous section. The critical energy where this spectral break occurs depends on any number of things including free-escape losses, finite size of the shock (e.g. compared to the particle gyroradii), and time dependence. These losses depend on the particle scattering, and other physical parameters related to the shock and its associated turbulence. These spectral breaks are important with regards to our understanding of compositional abundance variations.

One way to understand the origin of spectral rollovers is to realize that the acceleration to high energies takes time. Thus, we need to solve (2) for particles injected at a momentum p_0 at $t = 0$ and determine the resulting evolution of f. We find that, for a strong shock, the spectrum is a power law from the injection momentum to p_c, the critical momentum. p_c increases with time according to:

$$\frac{1}{p_c}\frac{dp_c}{dt} \simeq \frac{4U_1^2}{\kappa_{xx}(p_c)} = \frac{4U_1^2}{\kappa_\perp \sin^2\theta_{Bn} + \kappa_\parallel \cos^2\theta_{Bn}} \tag{4}$$

Thus, we find that the maximum energy attainable over any given time interval depends on U_1, $\kappa_\perp(p_c)$, $\kappa_\parallel(p_c)$, and θ_{Bn}. These quantities vary considerably from one SEP event to the next because of the large variability in magnetic turbulence, global orientation, and shock speeds. Moreover, because these parameters depend on the particle charge and mass, there will be atomic and ionic abundance variations.

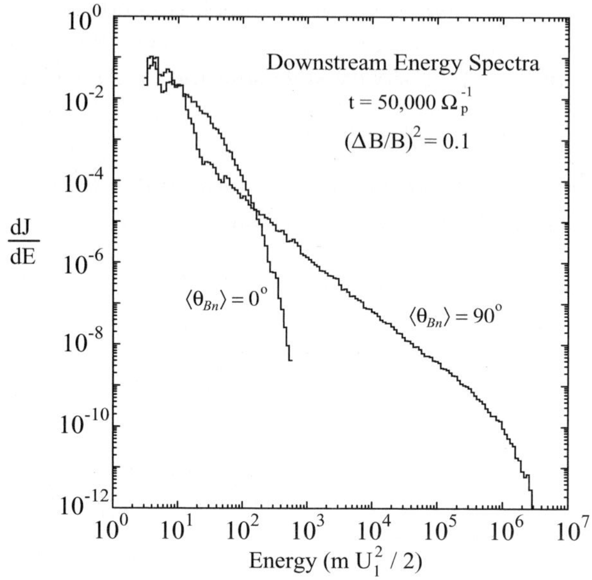

Figure 2. Downstream energy spectra for test-particle numerical simulations. Shown are results from a time-dependent simulation for two different shock-normal angles and rather weak turbulence. This figure is from Giacalone (2005a).

Because $\kappa_\perp \ll \kappa_\parallel$ (except for highly turbulent plasmas), inspection of (4) indicates that quasi-perpendicular shocks, in general, accelerate particles faster than quasi-parallel shocks (Jokipii, 1982, 1987). Therefore, quasi-perpendicular shocks produce the highest energy particles. Recent test-particle simulations by Giacalone (2005a) confirm this. Figure 2 shows one result from this study (Figure 7 from this paper). For the case of a coronal mass ejection expanding into the interplanetary medium, it is reasonable to expect that the highest energy particles originate at the flanks of the shock where the geometry is more favorable (see Figure 4).

2.5.2. *The Limit of Diffusive Shock Acceleration and the Problem of Injection*

The theory of diffusive shock acceleration is only a good approximation to first order in U/w. Therefore, the acceleration of low-energy particles is not described by the theory explicitly. This problem has been discussed mostly for perpendicular shocks because, until recently, it has been believed, that particles are tied to magnetic field lines that convect with the plasma through the shock. If this were true, then the particles would not be able to encounter the shock several times in order to be accelerated (the particles are said to be "de-trapped" from the shock). However, recent studies have shown that this is not the case (Giacalone, 2003, 2005a,b). In fact, the injection efficiency does not depend strongly on the shock-normal angle (provided there is sufficient turbulence of the order $\delta B/B \sim 0.3$ or greater). This is because of the effect of the large-scale magnetic field that leads to meandering of the lines of force in space. This enhances the particle motion normal to the shock

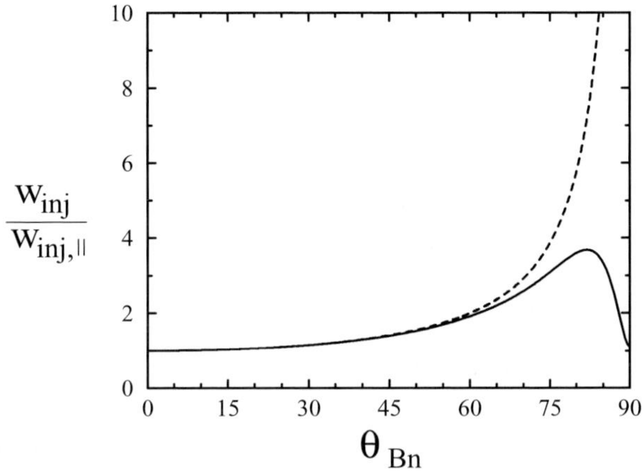

Figure 3. The injection velocity derived from the diffusive streaming anisotropy for the case of field-line random walk (solid line) normalized to that at a parallel shock. The dashed curve assumes the scatter-free approximation. See text for details.

and gives rise to efficient injection and acceleration of low-energy particles, even thermal particles (Giacalone, 2005b).

Strictly speaking, diffusive shock acceleration is applicable if the pitch-angle distribution is nearly isotropic. The anisotropy in the particle distribution is straightforward to determine from the diffusive shock theory. By requiring it to be small, an injection velocity, w_{inj}, can be derived (c.f. Giacalone and Jokipii, 1999). For an anistropy of one, we have:

$$w_{\text{inj}} = 3U_1\left[1 + \frac{\kappa_A^2 \sin^2\theta_{Bn} + (\kappa_\parallel - \kappa_\perp)^2 \sin^2\theta_{Bn}\cos^2\theta_{Bn}}{(\kappa_\perp \sin^2\theta_{Bn} + \kappa_\parallel \cos^2\theta_{Bn})^2}\right]^{1/2} \tag{5}$$

where $\kappa_A = wr_g/3$ is the antisymmetric component of the diffusion tensor and $r_g = w/\Omega_i$, where Ω_i is the cyclotron frequency.

Numerical simulations of charged-particles moving in large-scale (larger than the particle gyroradii) magnetic fluctuations indicate that $\kappa_\perp/\kappa_\parallel$ is independent of energy (Giacalone and Jokipii, 1999). Taking $\varepsilon = \kappa_\perp/\kappa_\parallel \ll 1$ and $\eta = \lambda_\parallel/r_g$, where $\lambda_\parallel$ is the parallel mean-free path (5) becomes:

$$w_{\text{inj}} = w_{inj,\parallel}\left[1 + \frac{(1/\eta)^2 \sin^2\theta_{Bn} + \sin^2\theta_{Bn}\cos^2\theta_{Bn}}{(\varepsilon \sin^2\theta_{Bn} + \cos^2\theta_{Bn})^2}\right]^{1/2} \tag{6}$$

where $w_{\text{inj},\parallel} = 3U_1$ is the injection velocity for a parallel shock.

The solution to (6) for $\eta = 100$ and $\epsilon = 0.02$ is shown in Figure 3. The dashed curve is $\sec\theta_{Bn}$, which is the scatter-free approximation. We find that the injection velocity at a perpendicular shock is similar to that obtained for a parallel shock (Giacalone, 2003)! Moreover, recent self-consistent plasma simulations have shown that thermal ions are efficiently accelerated at a perpendicular shock (Giacalone,

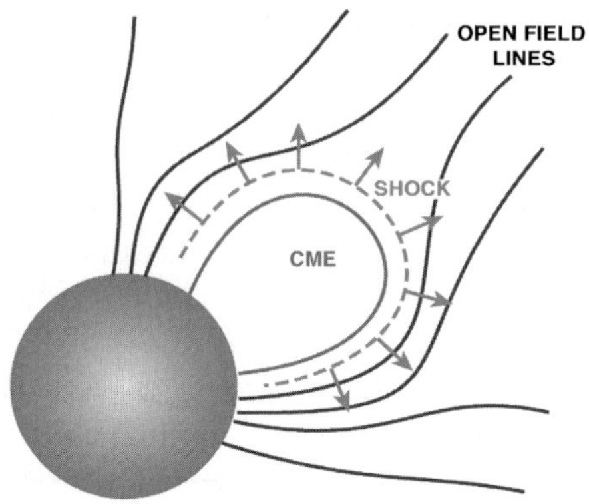

Figure 4. Cartoon illustration of an expanding CME and associated shock (courtesy A. Tylka).

2005b). It was shown that some of the incident thermal ions are specularly reflected by the shock and move upstream, away from the shock, along large-scale magnetic lines of force that are connected elsewhere along the shock giving rise to multiple shock encounters and efficient acceleration.

Thus, we conclude that the injection velocity at a perpendicular shock is the same as at a parallel shock. The acceleration efficiency may, in fact, be independent of the shock-normal angle. At perpendicular shocks, field-line random walk enhances the particle motion normal to the *mean* magnetic field. Yet, because $\kappa_\perp \ll \kappa_\|$, the acceleration *rate* is much higher for perpendicular shocks compared to parallel shocks. We conclude, therefore, that perpendicular shocks are the most important for understanding the highest-energy solar-energetic particles. Figure 4 (courtesy Allan Tylka) gives a simple illustration of an expanding CME showing the locations of perpendicular shocks at the flanks where we suggest that the highest-energy particles originate. Other possible locations of perpendicular shocks should also be explored, such as the environment around explosive magnetic-reconnection sites and near expanding coronal loops.

It is not presently clear whether shock acceleration can account for the origin of SEPs from impulsive solar flares. Any successful application of the theory to particle acceleration by solar flares must be able to explain the enormous enhancement of $^3He^{++}$ relative to $^4He^{++}$ observed at 1 AU, as well as explaining why electrons are accelerated more efficiently than protons. Currently we are aware of no such study in the literature. It is also important to understand how shocks can form low down in the corona. Because of the intense activity associated with active regions on the Sun, it is possible that multiple shocks can exist. This deserves further examination because of the simple fact that SEPs associated with impulsive solar flares are also usually power laws having little variability in the spectral exponent!

3. Solar Energetic Particles

When the first observations of large solar-cosmic ray events were made, largely by balloon flights, or by ground-based neutron monitors, early theories proposed that the particles originated from solar flares. The particles arrived at Earth by moving along the interplanetary magnetic field. Later, this picture was revised to include two separate classes of events: (1) those that are accelerated by solar flares (as in the original picture), and (2) those that are accelerated by a fast-moving shock wave driven by a coronal mass ejection. This is nicely reviewed by Reames (1999) and has been known for some time as the two-class paradigm of SEPs.

More recently, however, this picture is being revisited. More-sensitive spacecraft measurements by ACE, WIND, SoHO, etc. have shown that the simple classification of events into one or the other category is not correct. For example, in Figure 1 of Reames (1988), several events in the 1.9–2.8 MeV/nuc energy range occurring over an 8.5-year period were summed to create a histogram of the ratio, Fe/O, of the abundances of iron and oxygen. The figure suggests a bimodal distribution supporting the claim of two classes of events. However, more recent data from ACE/ULEIS at lower energies (0.5–0.7 MeV/nuc) and ACE/SIS at higher energies (10–36 MeV/nuc) do not give a bimodal distribution (D. Mewaldt and G. Mason, private communications).

Thus, clearly a combination of effects are taking place and the source of energetic particles is uncertain. This is perhaps the most important issue for space weather! Because the acceleration is not observed *in situ*, it is difficult to deconvolve the observations to arrive at an understanding of what the particle sources are, and what the acceleration mechanism is. This makes space-weather forecasting a daunting task.

4. Global Modeling

Recent large-scale models of particle acceleration by shock waves associated with CMEs have been performed in order to help resolve some of these difficulties (Zank *et al.*, 2000; Ng *et al.*, 2003; Li and Zank, 2005; Kocharov and Torsti, 2003). Very recently, an effort to incorporate the global nature of the CME by utilizing the results from magnetohydrodynamic simulations has been used (Kóta *et al.*, 2006).

While it is important to study relatively simple test cases in order to identify and understand the key physical processes, it is equally important to realize that fields around realistic CMEs have more complex structure, which evolves as the CME expands. Descriptions with a single parallel shock at the nose of the CME may be insufficient. The shock is likely to change from an initially perpendicular shock to a more parallel one (Tylka *et al.*, 2005; see also Figure 4). There may also exist multiple shocks and/or compressions that can accelerate particles. Recent 3-D MHD/CME simulations have also revealed that the downstream region has a structure that can contribute to both the acceleration and injection (Manchester

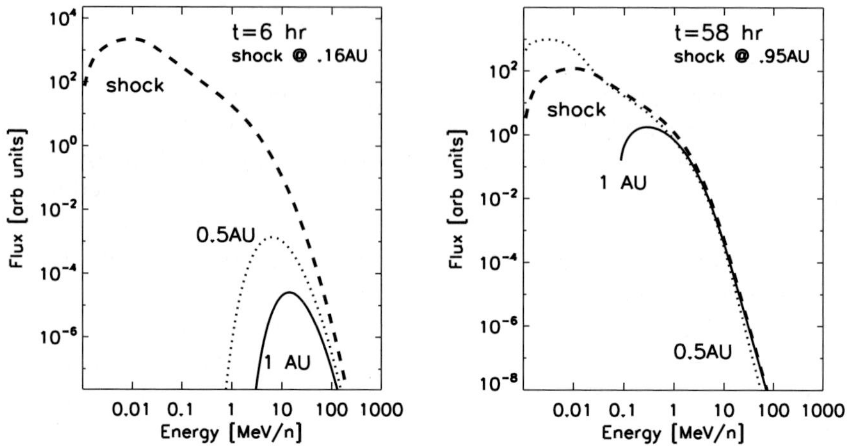

Figure 5. Simulated proton spectra shown at 0.5 AU (dotted) and 1 AU (solid) from the Sun at an early (6 hr – left panel) and a later (58 hr – right panel) stage of the event. The dashed lines represent the spectra at the shock, which is at $r = 0.16$ AU and 0.9 AU, respectively.

et al., 2005; Kóta *et al.*, 2006). These works emphasize the importance of the abrupt strengthening of the magnetic field behind the shock.

Currently there are ongoing efforts to couple particle acceleration with realistic CME simulations, using the constantly changing CME parameters as input into energetic-particle models. These integrated models follow the field lines as they evolve and consider field-aligned transport in either the diffusive approximation (Sokolov *et al.*, 2004) or the more-sophisticated focused transport equation (Kóta *et al.*, 2006). Figure 5 depicts energy spectra obtained from such a global model at different radial distances in an early and a later phase of the event. The underlying reason of the double power-law nature with a breaking point around a few MeV, shown by these spectra is not yet fully explored.

The results from these models seem to be converging on some of the reported features of the observations. Thus, these new models are a promising next step to further our understanding of large gradual SEP events.

5. Summary and Open Issues

We have given a brief overview of solar energetic particles and possible acceleration mechanisms. We have placed most of the emphasis on diffusive shock acceleration. This is because this mechanism provides a natural explanation for the observed power laws seen in nearly *all* solar-energetic particle observations. The origin of spectral rollovers at higher energies, which lead to observed abundance variations, is likely due to finite acceleration time, propagation effects, finite system size, and escaping particles (e.g. Lee, 2005). The critical energy where the spectral break occurs depends strongly on largely unknown quantities and is expected to vary considerably from one SEP event to the next. The new global models combining MHD simulations of CMEs and particle-transport models are an important next step to further our understanding of SEPs and the general problem of the near-Earth radiation environment.

Here we suggest some of the more compelling questions relating to solar energetic particles.

1. What are the particle sources? This is perhaps the most important question for understanding the radiation environment near Earth.
2. How do the shock-accelerated intensity and spectra evolve in time from the acceleration site to Earth? Clearly, the observations at 1 AU represent an integral. The global models may help answer this question because they can work backward in time from the observation point and determine the essential physics.
3. What is the shock geometry? Particularly important is the geometry back at the Sun, but also important is the geometry at 1 AU that is relevant to understanding the local acceleration (i.e. high-energy particles may be coming from different places at the shock). As we have shown, the physics of particle acceleration depends on the shock geometry.
4. What is the nature of the particle transport? In order to answer this question, we need to know the form of the magnetic-field power spectrum and its evolution from the Sun to the Earth.

Acknowledgements

We thank Randy Jokipii for useful discussions related to this work. This work was supported in part by NASA under grants NAG5-7793, NAG5-12919, and NNG04GA79G, and by NSF under grants ATM0327773 and ATM0447354.

References

Alfvén, H.: 1950, *Phys. Rev.* **77**, 375.

Axford, W. I., Leer, E., and Skadron, G.: 1978, *Proceedings of the 15th International Cosmic Ray Conference*, Plovdiv, Bulgaria, vol. 11, p. 132.

Bell, A. R.: 1978, *Mon. Not. R. Astron. Soc.* **182**, 147.

Blandford, R., and Ostriker, G.: 1978, *Astrophys. J.* **221**, L29.

Emslie, A. G., Miller, J. A., and Brown, J. C.: 2004, *Astrophys. J.* **602**, L69.

Fermi, E.: 1949, *Phys. Rev.* **75**, 1169.

Fisk, L. A.: 1978, *Astrophys. J.* **224**, 1048.

Giacalone, J.: 2003, *Planet. Space Sci.* **51**, 659.

Giacalone, J.: 2005a, *Astrophys. J.* **624**, 765.

Giacalone, J.: 2005b, *Astrophys. J.* **628**, L37.

Giacalone, J., and Jokipii, J. R.: 1999, *Astrophys J.* **520**, 204.

Jokipii, J. R.: 1982, *Astrophys. J.* **255**, 716.

Jokipii, J. R.: 1987, *Astrophys. J.* **313**, 842.
Kocharov, L., and Torsti, J.: 2003, *Astrophys. J.* **586**, 1430.
Kóta, J., Manchester, W. B., De Zeeuw, D. L., Jokipii, J. R., and Gombosi, T. I.: 2006, *Astrophys. J.* submitted.
Krymsky, G. F.: 1977, *Dokl. Akad. Nauk SSSR* **234**, 1306.
Lee, M. A.: 2005, *Astrophys. J. Suppl.* **158**, 38.
Li, G., and Zank, G. P.: 2005, *Geophys. Res. Lett.* **32**, L02101, doi:10.1029/2004GL021250.
Litvinenko, Y. E.: 1996, *Astrophys. J.* **462**, 997.
Manchester, W. B., IV, Gombosi, T. I., De Zeeuw, D. L., Sokolov, I. V., Roussev, I. I., Powell, K. G., *et al.*: 2005, *Astrophys. J.* **622**, 1225.
Miller, J. A.: 2000, *ASP Conf. Ser.* **206**, 145.
Ng, C. K., Reames, D. V., and Tylka, A. J.: 2003, *Astrophys. J.* **591**, 461.
Parker, E. N.: 1965, *Planet. Space Sci.* **13**, 9.
Petrosian, V., and Liu, S.: 2004, *Astrophys. J.* **610**, 550.
Reames, D. V.: 1988, *Astrophys. J.* **330**, L71.
Reames, D. V.: 1999, *Sp. Sci. Rev.* **90**, 413.
Sokolov, I. V., Roussev, I. I., Gombosi, T. I., Lee, M. A., Kóta, J., Forbes, T. G., *et al.*: 2004, *Astrophys. J.* **616**, 171.
Swann, W. G. F.: 1933, *Phys. Rev.* **43**, 217.
Syrovatsky, S. J.: 1960, *JETP* **13**, 1257.
Tylka, A. J., Cohen, C. M. S., Dietrich, W. F., Lee, M. A., Maclennan, C. G., Mewaldt, R. A., *et al.*: 2005, *Astrophys. J.* **625**, 474.
Zank, G. P., Rice, W. K. M., and Wu, C. C.: 2000, *J. Geophys. Res.* **105**, 25,079.

SOLAR ENERGETIC PARTICLE CHARGE STATES: AN OVERVIEW

B. KLECKER[1,*], E. MÖBIUS[2] and M. A. POPECKI[2]
[1]*Max-Planck-Institut für extraterrestrische Physik, Postfach 1312, D-85741 Garching, Germany*
[2]*University of New Hampshire, Durham, NH 03824, USA*
*(*Author for correspondence: E-mail: berndt.klecker@mpe.mpg.de)*

(Received 15 December 2005; Accepted in final form 9 June 2006)

Abstract. The ionic charge distributions of solar energetic particles (SEP) as observed in interplanetary space provide fundamental information about the origin of these particles, and the acceleration and propagation processes at the Sun and in interplanetary space. In this paper we review the measurements of ionic charge states of energetic particles in interplanetary space and discuss their implication for our understanding of SEP sources, and acceleration and propagation processes.

Keywords: solar energetic particles, ionic charge composition, energetic particle acceleration, corotating interaction regions

1. Introduction

The ionic charge composition, energy spectra, elemental, and isotopic abundances of solar energetic particles (SEP) carry fundamental information on the source region and their acceleration and propagation processes. High-energy particles originating at the Sun were first reported by Forbush (1946). At that time there was little doubt that these particles were closely related to contemporary solar flares. Later it became clear that acceleration at interplanetary (IP) shocks is also an efficient mechanism for particle acceleration (e.g. Bryant *et al.*, 1962). First ionic charge measurements at energies of ~0.04 – 1 MeV/e showed that C, O, and Fe were not fully stripped (Gloeckler *et al.*, 1976; Sciambi *et al.*, 1977). In the early 70 s also a new type of event was discovered that showed enhanced ^{3}He abundances (Hsieh and Simpson, 1970) with ^{3}He/^{4}He-ratios > 1 (Balasubrahmanyan and Serlemitsos, 1974), while the corresponding ratio in the solar corona and solar wind is 5×10^{-4}. Such events were later found to exhibit enhancements of heavy ions by about an order of magnitude (e.g. Hurford *et al.*, 1975; Mason *et al.*, 1986) relative to coronal abundances, and significantly higher charge states of heavy ions. Whereas in large, IP shock related events the mean ionic charge of heavy ions was compatible with coronal temperatures in the range $1–2 \times 10^6$ K (Q_{Fe} ~10–14, Hovestadt *et al.*, 1981; Luhn *et al.*, 1984), in ^{3}He-rich events $Q_{Fe} \sim 20$ was observed (Klecker *et al.*, 1984; Luhn *et al.*, 1987). Because of this large difference the ionic charge was used as one of several defining parameters for the classification of SEPs as *impulsive* and *gradual*, following a classification of flares based on the length of soft X-ray emission (Pallavicini *et al.*, 1977). In this scenario *impulsive* SEP events were related

DOI: 10.1007/s11214-006-9111-0

to flares, and the large ionic charge of heavy ions was interpreted as being due to high temperatures of $\sim 10^7$ K at the flare site. The *gradual* SEP events were related to coronal mass ejection (CME) driven coronal and interplanetary (IP) shocks with elemental abundances and ionic charge states similar (although not identical) to those in the solar wind (e.g. Reames, 1999; Kallenrode, 2003).

However, new results with advanced instrumentation from several missions (e.g. Wind, SAMPEX, SOHO, ACE) have shown that this picture was oversimplified. One of the key accomplishments with the new generation of instruments was the extension of ionic charge and compositional measurements over a wide energy range and the much improved sensitivity of the instrumentation. These new measurements show that enrichments in ^{3}He relative to ^{4}He are also common in interplanetary shock accelerated populations (Desai *et al.*, 2001) and that enrichments of heavy ions, in particular of Fe, are often observed in large events at high energies (e.g. Cohen *et al.*, 1999b; Cane *et al.*, 2003). In many SEP events also an energy dependence of the ionic charge states is observed, with a large event-to-event variability (Oetliker *et al.*, 1997; Mazur *et al.*, 1999), implying that the ionic charge is not (only) determined by temperature as assumed before.

Therefore, the classification into two distinct types of events is presently in question and the relative contributions of flares and coronal / interplanetary shocks are under debate. The main scenarios discussed are (1) contributions from a suprathermal population from previous *impulsive* events (Mason *et al.*, 1999a) to particles accelerated at interplanetary shocks, (2) systematic differences in the acceleration efficiency by the interplay of shock geometry and different seed populations (solar wind and flare suprathermals, Tylka *et al.*, 2005), and (3) direct injection from the flare acceleration process (e.g. Klein and Trottet, 2001, and references therein; Cane *et al.*, 2003), with or without further acceleration by a coronal shock. In all these models the ionic charge of the heavy ions plays a key role.

The ionic charge determination is also an ideal tool to investigate the relative contribution of different sources to the particle population accelerated at interplanetary shocks and in corotating interaction regions (CIRs), i.e. of pickup ions (e.g. He^+, O^+, Ne^+) and of highly charged ions of solar wind origin.

In this paper we will review the techniques used to determine ionic charge distributions, summarize new results on *gradual* and *impulsive* SEP events and CIRs, and discuss the mechanisms that could cause the observed variation of the ionic charge states with energy in SEP events and their implications.

2. Measurement Techniques

2.1. Direct Determination of the Ionic Charge

Several methods have been developed over the last $\sim$30 years to determine the ionic charge of the solar wind and of energetic ions (e.g. Popecki *et al.*, 2000). The first ionic charge determinations were based on the measurement of energy per charge (*E/Q*) by electrostatic deflection and of the kinetic energy (E), using compositional information derived independently from another sensor (e.g. Gloeckler *et al.*, 1976, and references therein). In order to unambiguously determine the three essential particle parameters kinetic energy, mass (M), and ionic charge (Q), three independent measurements are needed. In the energy range of the solar wind up to suprathermal energies of ~100 keV/e the 3 measured quantities are usually *E/Q* (determined by electrostatic deflection), E (determined by the measurement of the total energy using solid state detectors), and *E/M* (i.e. velocity), determined by a time-of-flight measurement (see e.g. Wüest (1998) for a review of time-of-flight measurements). At higher energies, the mass (or the nuclear charge Z) of the ions can be determined by the dE/dx method utilizing, for example, proportional counters (PC) for dE/dx determination. This technique was first used for ionic charge determination in the MeV/nuc energy range onboard ISEE-1/3 (Hovestadt *et al.*, 1978), and later with much improved sensitivity and resolution by the SEPICA instrument (Möbius *et al.*, 1998) onboard the Advanced Composition Explorer (ACE) spacecraft. These direct measurements provide the most accurate determination of the ionic charge distribution. However, due to the technical limitatons of the high voltages used for the electrostatic deflection, they are limited to energies below a few MeV/nuc.

2.2. Charge Determination Using the Magnetic Field of the Earth

One of the prime scientific objectives of the SAMPEX (Solar Anomalous and Magnetospheric Particle Explorer) mission (Baker *et al.*, 1993) was the determination of ionic charge states of the Anomalous Component of Cosmic Rays (ACR) and of SEPs using the Earth's magnetic field as a *M/Q* spectrometer. SAMPEX provided for the first time ionic charge measurements for many elements in the range C to Fe over the extended energy range of 0.3–70 MeV/nuc (for Fe). This method combines the determination of E and M of energetic ions by instruments of high sensitivity with the determination of the rigidity cutoff in the low altitude polar orbit of SAMPEX (Mason *et al.*, 1995). Time variations of the magnetic field of the Earth during SEP events are taken into account by determining the cutoff variations on an orbit-by-orbit (~90 min) basis, using particles of known ionic charge (i.e. H^+, He^{2+}). Figure 1 illustrates the method: if the particle flux is plotted for ions in the same energy range (in energy/nuc), the different cutoff latitudes directly refer to different *M/Q* values of the ions. These measurements provided for the first time the average ionic charge of heavy ions as a function of energy for the extended energy range of 0.3–70 MeV/nuc (for Fe).

2.3. Indirect Methods

Indirect methods use the rigidity dependence of the acceleration and propagation processes to infer average ionic charge states of heavy ions. These methods assume,

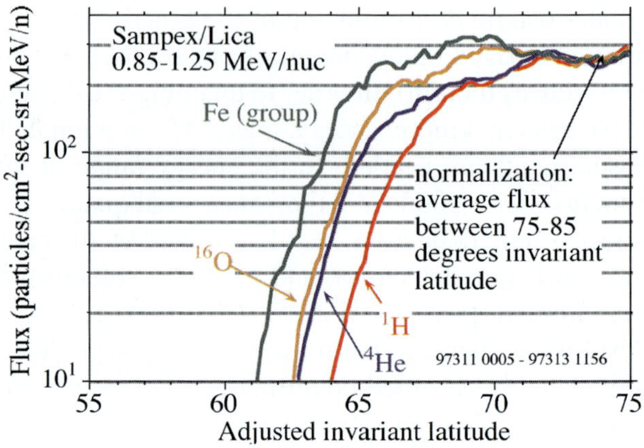

Figure 1. Flux of H, He, O, and Fe in the energy range 0.85–1.25 MeV/nuc as a function of adjusted invariant latitude during the 7 November 1997 SEP event (from Mazur *et al.*, 1999).

for example, diffusive propagation to infer average ionic charge states from the time-to-maximum (O'Gallagher *et al.*, 1976; Dietrich and Tylka, 2003) or from the time profile in the decay phase (Sollitt *et al.*, 2003) of SEP events. Furthermore, the *M/Q*-dependent characteristic 'break' or roll-over often observed in SEP spectra at high energies (Tylka *et al.*, 2000), and elemental fractionation depending on *M/Q* (Cohen *et al.*, 1999a) have been successfully used to infer mean ionic charge states of heavy ions. Although these indirect methods are limited to the determination of average charge states and rely on the assumptions of the models used, they provide a valuable tool if direct measurements are not available.

3. Ionic Charge Composition in Gradual Events

The first measurements of heavy ion ionic charge states over the extended energy range of ~0.3–70 MeV/nuc as obtained with SAMPEX in the October/November 1992 SEP events showed, surprisingly, a significant increase of the mean ionic charge at energies above ~10 MeV/nuc, in particular for Fe (Leske *et al.*, 1995; Oetliker *et al.*, 1997). Subsequent measurements of the mean ionic charge in large IP shock related events with SAMPEX, SOHO and ACE, covering the energy range from solar wind to ~70MeV/nuc, showed that the mean ionic charge as a function of energy is very variable. Figure 2 shows as an example the ionic charge of Si and Fe in two large events as observed with SAMPEX and ACE, in the energy range ~0.3–~70MeV/nuc (Mazur *et al.*, 1999 and references therein).

At suprathermal energies ≤ 250 keV/nuc the mean ionic charge of Fe ($Q_{Fe} \sim 10$) is mostly compatible with typical solar wind charge states (Bogdanov *et al.*, 2000; Klecker *et al.*, 2000). At somewhat higher energies (0.2–0.6 MeV/nuc) the mean

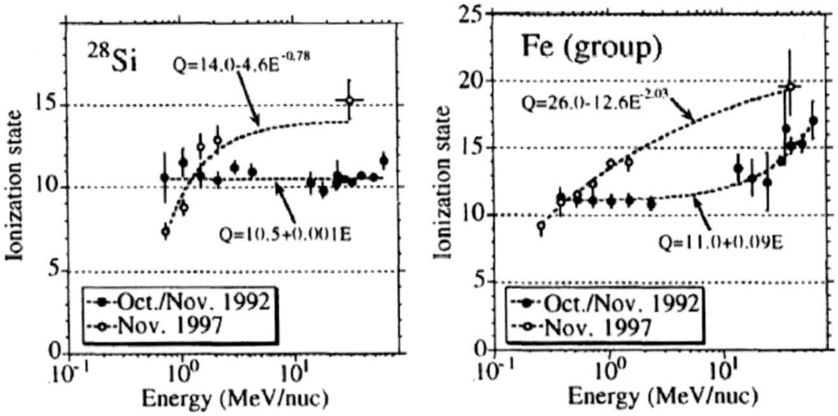

Figure 2. Variation of the mean ionic charge of Si and Fe in the Oct/Nov 1992 and Nov 1997 SEP events (Mazur *et al.*, 1999).

ionic charge shows a large event-to-event variability. It is either constant or increases with energy, in some cases by several charge units (Möbius *et al.*, 1999). At higher energies, however, the mean ionic charge is often observed to be significantly larger than at low energies, with $Q_{\rm Fe} \sim 20$ at energies greater than $\sim$ 10 MeV/nuc (Leske *et al.*, 1995; Oetliker *et al.*, 1997; Leske *et al.*, 2001; Labrador *et al.*, 2003; Popecki *et al.*, 2003).

4. Ionic Charge Composition in Impulsive Events

The early measurements showed that the mean ionic charge of Si and Fe at $\sim$ 1 MeV/nuc was significantly higher in ^{3}He- and heavy-ion-rich events ($Q_{\rm Si} \sim 14$, $Q_{\rm Fe} \sim 20$) than in gradual events (Klecker *et al.*, 1984; Luhn *et al.*, 1987). This was interpreted as being indicative of a high temperature of $\sim 10^7$ K in the source region. However, recent measurements of the mean ionic charge of Fe in six impulsive events (Möbius *et al.*, 2003) showed a systematic increase of $Q_{\rm Fe}$ from 14–16 at 180–250 keV/nuc to 16–20 at 350–550 keV/nuc. Figure 3 shows four out of these six events, together with the average charge of Fe obtained for events 2–4 with STOF onboard SOHO at lower energies (Klecker *et al.*, 2006). This shows that the high charge states of Fe as previously observed at $\sim$1 MeV/nuc were the high energy tail of an energy dependent ionic charge distribution of heavy ions in *impulsive* events. This energy dependence was actually already suggested by the early measurements of Gloeckler *et al.* (1976) reporting $Q_{\rm Fe} \sim 10 - 14$ at energies of 25–50 keV/nuc for Fe-rich events in May 1974, similar to the results from SOHO shown in Figure 3. However, because of limited sensitivity, a systematic study was not possible at those times.

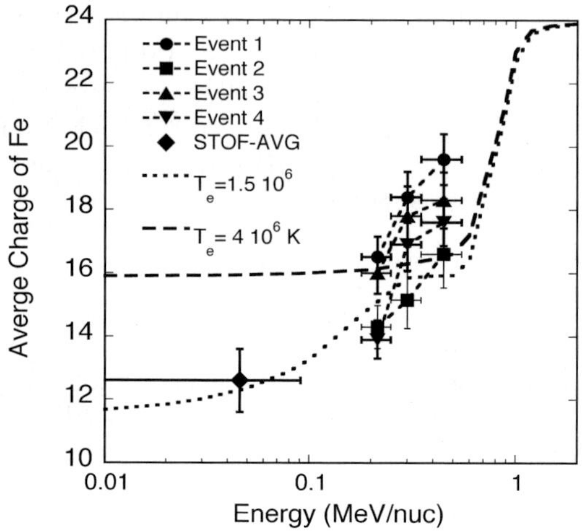

Figure 3. Ionic charge states of Fe in 4 impulsive SEP events as measured with SEPICA/ACE and STOF/SOHO (adopted from Klecker *et al.*, 2006).

5. Energy Dependent Charge States – Mechanisms

5.1. Charge-Changing Effects due to Impact Ionization by Electrons and Ions

A large increase of the mean ionic charge of heavy ions in the small energy range of ~0.18 – 0.55 MeV/nuc can be explained in terms of impact ionization by protons and electrons in a dense environment in the corona. Models including the effect of stochastic acceleration, coulomb energy loss, charge changing processes such as impact ionization and recombination with ambient electrons have been investigated by e.g. Ostryakov *et al.* (2000) and Kartavykh *et al.* (2002). Similar models which employed shock acceleration and might be suitable for gradual events were proposed by Ostryakov and Stovpyuk (1999), Barghouty and Mewaldt (1999), and Lytova and Kocharov (2005). Figure 3 shows as an example the equilibrium charge states of Fe computed in the energy range 0.01 to 2 MeV/nuc with a simple model, including the effects of impact ionization by electrons, protons, and helium. The model is following the approach of Kocharov *et al.* (2000), using updated values of ionization cross sections (Klecker *et al.*, 2006, and references therein). The dashed and dotted lines in Figure 3 show the equilibrium charge states computed for electron temperatures of $T_e = 1.5 \times 10^6$ K and 4.0×10^6 K, respectively. Figure 3 demonstrates that, qualitatively, a strong increase of the mean ionic charge of Fe at energies $\leq$1 MeV/nuc is consistent with the charge stripping effect. Note also that in non-steady state models, at any given energy, the mean ionic charge will be below the equilibrium values shown in Figure 3 (e.g. Kocharov *et al.*, 2000).

Figure 3 also shows that for 3 out of the 4 events studied, the increase of Q_{Fe} with energy is considerably steeper than the results of the charge equilibrium calculation.

As a possible reason for this discrepancy Kartavykh *et al.* (2005) discussed the effect of energy losses by adiabatic deceleration in the solar wind between the Sun and 1 AU and showed that a model combining acceleration, transport and charge stripping in the corona with interplanetary transport including adiabatic deceleration can indeed reproduce the charge dependence on energy as measured at 1 AU.

5.2. Energy Dependence of *Q* due to *M/Q*-Dependent Energy Spectra

The energy spectra in large gradual SEP events can often be described by a power-law with an exponential roll-over at high energies, where the e-folding energy, E_0, depends on *M/Q* (Tylka *et al.*, 2000):

$$J(E) \sim E^{-\gamma} \exp(-E/E_0), \tag{1}$$

with

$$E_0 = E_{0p}(Q/M)^{\alpha} \tag{2}$$

If this functional form of the average differential energy spectra is applied to individual charge states of heavy ions, then the mean ionic charge above an energy $\sim E_0$ will increase (decrease) for positive (negative) α with energy (Klecker *et al.*, 2001). Whether this effect is significant depends on the mass per charge dependence of E_0 (i.e. on α) and can be verified by the investigation of the energy spectra of different elements with different *M/Q* values as, for example O and Fe. Small increases of the mean ionic charge of Fe by 1–2 charge units between solar wind energies and 0.5 MeV/nuc could be explained by this process, assuming a solar wind Fe ionic charge distribution for the source (Klecker *et al.*, 2001).

5.3. Mixing of Sources with Different Ionic Charge Distributions

As an explanation of the increase of the mean ionic charge of Fe at high energies Tylka *et al.* (2001) proposed an admixture of two components: one component with a charge composition similar to the solar wind and another component with high charge states, similar to the ionic charge composition observed in impulsive SEP events. If these two components also have a spectral form as discussed in the last paragraph, the component with high charge states (i.e. large E_0) will dominate at high energies resulting in an increase of the mean ionic charge with energy at energies above $\sim E_0$. In this scenario, the mixing of the two populations could be, for example, at a coronal shock where ambient coronal material and "flare" material with high charge states from an earlier or from the contemporary flare is accelerated.

In a more general treatment of this problem, Tylka *et al.* (2005) proposed as an explanation for the large variability of spectral, compositional, and ionic charge state features at high energies (i.e. above 10 s of MeV/nuc) in large gradual SEP events a model where this variability arises from the interplay of two factors: shock geometry and the mixture of two seed populations with coronal/solar wind composition and "flare" composition, i.e. a composition as observed in ^{3}He- and heavy ion-rich events. In this scenario the shock geometry plays an important role. It is, in particular, assumed that quasi-perpendicular shocks require a higher initial speed of the ions for effective injection and therefore preferentially accelerate suprathermal seed particles from flares, whereas quasi-parallel shocks generally draw their seed particles from the corona/solar wind suprathermals. In this model the shock geometry determines via the injection threshold which of the two components dominates and thus determines spectral shapes, heavy ion abundances and ionic charge states at high energies. However, hybrid simulations of a quasi-perpendicular shock by Giacalone (2005) showed recently that perpendicular shocks can accelerate ions out of the thermal distribution. This implies that the differences in injection speed may not be as large as assumed in the model by Tylka *et al.* (2005), i.e. this scenario needs further investigation.

6. Correlation Between Elemental Abundances and Ionic Charge States

The early measurements that led to the two-class picture of *impulsive* and *gradual* events already showed that, at energies of ~1 MeV/nuc, high abundances of Fe relative to O were correlated with high charge states of Fe. In a statistical survey using 1-day averages of the Fe/O-ratio and the mean ionic charge of Fe, Klecker *et al.* (1983) reported $Q_{\mathrm{Fe}} \sim 20.8 \pm 1.8$ for the average charge of Fe for all days with Fe/O ≥ 1 and ^{3}He/^{4}He ≤ 0.1.

With the new measurements of much improved sensitivity ionic charge measurements of heavy ions in the mass range C–Fe are now available for individual SEP events over a large dynamic range of event intensities. These new results show that there is an almost continuous variation of the Fe abundances between Fe/O ≤ 0.1 to Fe/O ≥ 1, with a clear correlation of Fe and other heavy ion abundances (e.g. Ne/O) with the mean ionic charge of Fe (Möbius *et al.*, 2002a). This correlation is puzzling, because the ions in the mass range C–Mg are mostly fully stripped (i.e. have $Q/M = 0.5$) in Fe-rich events (Möbius *et al.*, 2002a). This is not compatible with *M/Q*-dependent selective heating and acceleration processes that require $Q/M \leq 0.3$ (see e.g. Miller, 1998 for a review). However, the observed energy dependence of the heavy ion charge states in impulsive events may resolve this puzzle, showing at low energies charge states sufficiently low for resonant wave-particle interactions. A study addressing this question by extending the energy range of the ionic charge determination of Fe in individual events to lower energies is presently under way (DiFabio *et al.*, 2006, in preparation).

7. Ionic Charge States in Corotating Interaction Regions

Shocks formed by the interaction of fast and slow solar wind streams, now called corotating interaction regions (CIRs), have been known for a long time to accelerate particles efficiently (e.g. Barnes and Simpson, 1976). The detection of large fluxes of He^+ at energies of ~1 MeV/nuc (Hovestadt *et al.*, 1984) and of pickup He^+ of interstellar origin at 1 AU (Möbius *et al.*, 1985) suggested that, besides ions of solar wind origin, pickup ions of interstellar origin are a significant source in the inner helioshere. It was in fact shown that He^+ on average is the third most abundant species in the suprathermal ion population at 1 AU and at times can even exceed the abundance of He^{2+} (Kucharek *et al.*, 2003). Deviations of the elemental abundances in CIRs from coronal and SEP abundances, in particular for He and C (e.g. Mason *et al.*, 1999b) raised the question whether pickup ions of interstellar origin provide a significant source for CIRs. Indeed, Gloeckler *et al.* (1994) identified He^+ as the major contributor in a CIR at 4.5 AU and Hilchenbach *et al.* (1999) and Chotoo *et al.* (2000) showed that at 1 AU and energies ≤300 keV He^+ contributes about 15 per cent to the energetic particle population. The discovery of inner source heavy pickup ions in the mass range C to Si (Geiss *et al.*, 1994; Gloeckler *et al.*, 2000) raised the question whether these ions also contribute to the accelerated ion population in CIRs. The ideal tool to answer this question is the analysis of the ionic charge distribution that is much different for pickup ions (singly charged) and ions of solar wind origin (e.g. O^{6+}, O^{7+}). The results by Mazur *et al.* (2002) and Möbius *et al.* (2002b) show that the mean ionic charge of ions in the mass range C–Mg in CIRs is similar to IP-shock related events, with an upper limit of ~1 % for the contribution of pickup C^+, O^+, and Mg^+ and 4.7 ± 0.2 % Ne^+ (Möbius *et al.*, 2002b). The apparently much lower acceleration efficiency of inner source heavy pickup ions is possibly due to their different velocity distributions. While the distribution of inner source ions shows a peak just below V_{SW} and falls off steeply from there, the distribution of interstellar pickup He extends up to $2 \times V_{SW}$ (see discussion in Möbius *et al.*, 2002b).

8. Summary

The new measurements of ionic charge states over an extended energy range obtained with instruments onboard SAMPEX, SOHO, and ACE provide new insight into the source location of SEPs in *gradual* and *impulsive* events. The observations show

1. The mean ionic charge of Fe in *impulsive* SEP events increases significantly with energy from 14–16 at 0.18–0.25 MeV/nuc to 16–20 at 350–550 keV/nuc. At lower energies of 10–100 keV/nuc an average value of $Q_{Fe} \sim 12$ was observed for 3 events studied so far. It will be interesting to extend the measurements at

low energies to more events to find out whether low charge states at low energies are a common feature in these events.

2. The mean ionic charge of heavy ions in *gradual* SEP events is highly variable. At low energies of ~100 keV/nuc they are mostly compatible with solar wind charge states. At higher energies a large event-to-event variability is observed, with Q_{Fe} often as high as ~18–20 at energies of 10s of MeV/nuc.
3. The enrichment of heavy ions (e.g. Ne/O, Fe/O) is correlated with a larger mean ionic charge of Fe.
4. The contribution of He^+ ions of interstellar origin to the energetic particle population at interplanetray shocks and CIRs can be significant. However, the contribution of singly charged ions in the mass range C–Mg to the accelerated population (~0.5 MeV/nuc) in CIRs is small ($\leq$1 % for C, O, Mg and 4.8 % for Ne).

The energy dependence of the ionic charge of heavy ions shows that the previous interpretation of the mean ionic charge being determined by the ambient temperature was too simplistic. The strong energy dependence as observed in *impulsive* SEP events is consistent with the combined effects of acceleration, charge stripping in a dense environment, and interplanetary propagation. Thus, the determination of ionic charge states, energy spectra and propagation characteristics can be used to infer temperature, density, and acceleration time scales in the acceleration region.

The mean ionic charge and its energy dependence as observed in *gradual* events provide information on the various possible sources contributing to the accelerated population, i.e. solar wind (or corona), and suprathermal particles with energies $\geq$ 200 kev/nuc from previous (or contemporary) impulsive events with charge states determined by the degree of charge stripping in the low corona.

The correlation of the mean ionic charge with the elemental abundances is difficult to reconcile in the previous scenario with a high temperature of ~10^7 K, where ions in the mass range C–Mg would be fully ionized. However, this puzzle may be resolved if lower ionic charge states at energies of ~10–100 keV/nuc are found to be a consistent feature in impulsive SEP events, as appears to be consistent with the consistently observed energy dependence of the Fe charge states in these events.

Acknowledgements

We thank the International Space Science Institute for their hospitality during the 2005 Workshop on Solar Dynamics. The work was partially supported by NASA under NAG 5-12929.

References

Baker, D. N., Mason, G. M., Figueroa, O., Colon, G., Watzin, J. G., and Aleman, R. M.: 1993, *IEEE Trans. Geosci. Remote Sensing* **31**, 531.

Balasubrahmanyan, V. K., and Serlemitsos, A. T.: 1974, *Nature* **252**, 460.

Barghouty, A. F., and Mewaldt, R. A.: 1999, *Astrophys. J. Lett.* **520**, L127.

Barnes, C. W., and Simpson, J. A.: 1976, *Astrophys. J. Lett.* **210**, L91.

Bogdanov, A. T., Klecker, B., Möbius, E., Hilchenbach, M., Kistler, L. M., Popecki, M. A., *et al.*: 2000, in Mewaldt, R. A. *et al.* (eds.), *Acceleration and Transport of Energetic Particles Observed in the Heliosphere: ACE 2000 Symposium*, AIP Vol. 528, p. 143.

Bryant, D. A., Cline, T. L., Desai, U. D., and McDonald, F. B.: 1962, *J. Geophys. Res.* **67**(16), 4983.

Cane, H. V., von Rosenvinge, T. T., Cohen, C. M. S., and Mewaldt, R. A.: 2003, *Geophys. Res. Lett.* **30**, SEP 5–1, DOI 10.1029/2002GL016580.

Chotoo, K., Schwadron, N. A., Mason, G. M., Zurbuchen, T. H. Gloeckler, G., Posner, A. *et al.*: 2000, *J. Geophys. Res.* **105**(14), 23107.

Cohen, C. M. S., Cummings, A. C., Leske, R. A., Mewaldt, R. A., Stone, E. C., Dougherty, B. L. *et al.*: 1999a, *Geophys. Res. Lett.* **26**, 149.

Cohen, C. M. S., Mewaldt, R. A., Leske, R. A., Cummings, A. C., Stone, E. C., Wiedenbeck, M. E., *et al.*: 1999b, *Geophys. Res. Lett.* **26**, 2697.

Desai, M. I., Mason, G. M., Dwyer, J. R., Mazur, J. E., Smith, C. W., and Skoug, R. M.: 2001, *Astrophys. J. Lett.* **553**, L89.

Dietrich, W., and Tylka, A.: 2003, *Proceedings of the 28th International Cosmic Ray Conference, Tsukuba, Japan*, vol. 6. p. 3291.

Forbush, S. E.: 1946, *Phys. Rev.* **70**, 771.

Geiss, J., Gloeckler, G., Mall, U., von Steiger, R., Galvin, A. B., and Ogilvie, K. W.: 1994, *Astron. Astrophys.* **282**, 924.

Giacalone, J.: 2005, *Astrophys. J. Lett.* **628**, L37.

Gloeckler, G., Sciambi, R. K., Fan, C. Y., and Hovestadt, D.: 1976, *Astrophys. J. Lett.* **209**, L93.

Gloeckler, G., Geiss, J., Roelof, E. C., Fisk, L. A., Ipavich, F. M., Ogilvie, K. W., *et al.*: 1994, *J. Geophys. Res.* **99**(18), 17637.

Gloeckler, G., Fisk, L. A., Geiss, J., Schwadron, N. A., and Zurbuchen, T. H.: 2000, *J. Geophys. Res.* **105**(14), 7459.

Hilchenbach, M., Grünwaldt, H., Kallenbach, R., Klecker, B., Kucharek, H., Ipavich, F. M., *et al.*: 1999, *American Institute of Physics Conference Series*. p. 605.

Hovestadt, D., Klecker, B., Scholer, M., Arbinger, H., Gloeckler, G. Ipavich, F. M., *et al.*: 1978, *IEEE Trans. Geosci. Electron.* **16**, 166.

Hovestadt, D., Höfner, H., Klecker, B., Scholer, M., Gloeckler, G., Ipavich, F. M., *et al.*: 1981, *Adv. Space Res.* **1**, 61.

Hovestadt, D., Klecker, B., Scholer, M., Gloeckler, G., and Ipavich, F. M.: 1984, *Astrophys. J. Lett.* **282**, L39.

Hsieh, K. C., and Simpson, J. A.: 1970, *Astrophys. J. Lett* **162**, L191.

Hurford, G. J., Mewaldt, R. A., Stone, E. C.M., and Vogt, R. E.: 1975, *Astrophys. J. Lett.* **201**, L95.

Kallenrode, M.-B.: 2003, *J. Phys. G Nucl. Phys.* **29**, 965.

Kartavykh, J. J., Wannawichian, S., Ruffolo, D., and Ostryakov, V. M.: 2002, *Adv. Space Res.* **30**, 119.

Kartavykh, J. J., Dröge, W., Kovaltsov, G. A., and Ostryakov, V. M.: 2005, *Solar Phys.* **227**, 123.

Klecker, B., Bogdanov, A. T., Popecki, M., Wimmer-Schweingruber, R. F., Möbius, E., Schaerer, R., *et al.*: 2000, in Mewaldt, R. A., *et al.* (eds.), *Acceleration and Transport of Energetic Particles Observed in the Heliosphere: ACE 2000 Symposium*, p. 135.

Klecker, B., Hovestadt, D., Möobius, E., Scholer, M., Gloeckler, G., and Ipavich, F. M.: 1983, *Proceedings of the 18th International Cosmic Ray Conference, Bangalore, India*, vol. 10. p. 330.

Klecker, B., Hovestadt, D., Scholer, M., Gloeckler, G., Ipavich, F. M., Fan, C. Y., *et al.*: 1984, *Astrophys. J.* **281**, 458.

Klecker, B., Möbius, E., Popecki, M. A., Lee, M. A., and Bogdanov, A. T.: 2001, in Wimmer-Schweingruber, R. F. (ed.), *AIP Conference Proceedings 598: Joint SOHO/ACE Workshop "Solar and Galactic Composition"*, p. 317.

Klecker, B., Möbius, E., Popecki, M., Kistler, L. M., Kucharek, H., and Hilchenbach, M.: 2006, *Adv. Space Res.* **38**, 493.

Klein, K.-L., and Trottet, G.: 2001, *Space Sci. Rev.* **95**, 215.

Kocharov, L., Kovaltsov, G. A., Torsti, J., and Ostryakov, V. M.: 2000, *Astron. Astrophys.* **357**, 716.

Kucharek, H., Möbius, E., Li, W., Farrugia, C. J., Popecki, M. A. Galvin, A. B. *et al.*: 2003, *J. Geophys. Res.* **108**, 15–1–15–13.

Labrador, A., Leske, R., Mewaldt, R., Stone, E., and von Rosenvinge, T.: 2003, *Proceedings of the 28th International Cosmic Ray Conference, Tsukuba, Japan*, vol. 6. p. 3269.

Leske, R. A., Cummings, J. R., Mewaldt, R. A., Stone, E. C., and von Rosenvinge, T. T.: 1995, *Astrophys. J. Lett.* **452**, L149.

Leske, R. A., Mewaldt, R. A., Cummings, A. C., Stone, E. C., and von Rosenvinge, T. T.: 2001, in Wimmer-Schweingruber, R. F. (ed.), *AIP Conf. Proc. 598: Joint SOHO/ACE Workshop "Solar and Galactic Composition"*, p. 171.

Luhn, A., Klecker, B., Hovestadt, D., Scholer, M., Gloeckler, G., Ipavich, F. M. *et al.*: 1984, *Adv. Space Res.* **4**, 161.

Luhn, A., Klecker, B., Hovestadt, D., and Möbius, E.: 1987, *Astrophys. J.* **317**, 951.

Lytova, M., and Kocharov, L.: 2005, *Astrophys. J. Lett.* **620**, L55.

Mason, G. M., Mazur, J. E., and Dwyer, J. R.: 1999a, *Astrophys. J. Lett.* **525**, L133.

Mason, G. M., Mazur, J. E., Looper, M. D., and Mewaldt, R. A.: 1995, *Astrophys. J.* **452**, 901.

Mason, G. M., Reames, D. V., von Rosenvinge, T. T., Klecker, B., and Hovestadt, D.: 1986, *Astrophys. J.* **303**, 849.

Mason, G. M., Von Steiger, R., Decker, R. B., Desai, M. I., Dwyer, J. R. Fisk, L. A., *et al.*: 1999b, *Space Sci. Rev.* **89**, 327.

Mazur, J. E., Mason, G. M., Looper, M. D., Leske, R. A., and Mewaldt, R. A.: 1999, *Geophys. Res. Lett.* **26**, 173.

Mazur, J. E., Mason, G. M., and Mewaldt, R. A.: 2002, *Astrophys. J.* **566**, 555.

Miller, J. A.: 1998, *Space Sci. Rev.* **86**, 79.

Möbius, E., Hovestadt, D., Klecker, B., Scholer, M., and Gloeckler, G.: 1985, *Nature* **318**, 426.

Möbius, E., Kistler, L. M., Popecki, M. A., Crocker, K. N., Granoff, M. Turco, S. *et al.*: 1998, *Space Sci. Rev.* **86**, 449.

Möbius, E., Popecki, M., Klecker, B., Kistler, L. M., Bogdanov, A., Galvin, A. B., *et al.*: 1999, *Geophys. Res. Lett.* **26**, 145.

Möbius, E., Popecki, M., Klecker, B., Kistler, L. M., Bogdanov, A., Galvin, A. B. *et al.*: 2002a, *Adv. Space Res.* **29**, 1501.

Möbius, E., Morris, D., Popecki, M. A., Klecker, B., Kistler, L. M., and Galvin, A. B.: 2002b, *Geophys. Res. Lett.* **29**, 3.1.

Möbius, E., Cao, Y., Popecki, M. A., Kistler, L. M., Kucharek, H., Morris, D., *et al.*: 2003, *Proceedings of the 28th International Cosmic Ray Conf., Tsukuba, Japan*, vol. 6. p. 3273.

Oetliker, M., Klecker, B., Hovestadt, D., Mason, G. M., Mazur, J. E., Mewaldt, R. A. *et al.*: 1997, *Astrophys. J.* **477**, 495.

O'Gallagher, J. J., Hovestadt, D., Klecker, B., Gloeckler, G., and Fan, C. Y.: 1976, *Astrophys. J. Lett.* **209**, L97.

Ostryakov, V. M., and Stovpyuk, M. F.: 1999, *Solar Phys.* **189**, 357.

Ostryakov, V. M., Kartavykh, Y. Y., Ruffolo, D., Kovaltsov, G. A., and Kocharov, L.: 2000, *J. Geophys. Res.* **105**(14), 27315.

Pallavicini, R., Serio, S., and Vaiana, G. S.: 1977, *Astrophys. J.* **216**, 108.

Popecki, M., Möbius, E., Klecker, B., Galvin, A. B., Kistler, L. M., and Bogdanov, A. T. : 2000, in Mewaldt, R. A. *et al.* (eds.), *Acceleration and Transport of Energetic Particles Observed in the Heliosphere: ACE 2000 Symposium*, AIP Vol. 528, p. 63.

Popecki, M. A., Mazur, J. E., Möbius, E., Klecker, B., Bogdanov, A., Mason, G. M. *et al.*: 2003, *Proceedings of the 28th International Cosmic Ray Conference, Tsukuba, Japan*, vol. 6, p. 3283.

Reames, D. V.: 1999, *Space Sci. Rev.* **90**, 413.

Sciambi, R. K., Gloeckler, G., Fan, C. Y., and Hovestadt, D.: 1977, *Astrophys. J.* **214**, 316.

Sollitt, L., Stone, E., Mewaldt, R., Cohen, C., Leske, R., Wiedenbeck, M., *et al.*: 2003, *Proceedings of the 28th International Cosmic Ray Conference, Tsukuba, Japan*, vol. 6. p. 3295.

Tylka, A. J., Boberg, P. R., McGuire, R. E., Ng, C. K., and Reames, D. V.: 2000, in Mewaldt, R. A. *et al.* (eds.), *Acceleration and Transport of Energetic Particles Observed in the Heliosphere: ACE 2000 Symposium*, AIP Vol. 528, p. 147.

Tylka, A. J., Cohen, C. M. S., Dietrich, W. F., Maclennan, C. G., McGuire, R. E., Ng, C. K. *et al.*: 2001, *Astrophys. J. Lett.* **558**, L59.

Tylka, A. J., Cohen, C. M. S., Dietrich, W. F., Lee, M. A., Maclennan, C. G., Mewaldt, R. A., *et al.*: 2005, *Astrophys. J.* **625**, 474.

Wüest, M.: 1998, in Pfaff, R. F. *et al.* (eds.), *Measurement Techniques in Space Plasmas – Particles*, Geophys. Monograph Vol. 102, p. 141.

SOLAR ENERGETIC PARTICLE COMPOSITION, ENERGY SPECTRA, AND SPACE WEATHER

R. A. MEWALDT
California Institute of Technology, Pasadena, CA, USA
(E-mail: rmewaldt@srl.caltech.edu)

(Received 10 June 2006; Accepted in final form 26 June 2006)

Abstract. Recent progress in measuring the composition and energy spectra of solar energetic particles (SEPs) accelerated by CME-driven shocks is reviewed, including a comparison of the observed charge-to-mass dependence of breaks in SEP spectra with model predictions. Also discussed is a comparison of SEP and CME kinetic energies in seventeen large SEP events, and estimates of the SEP radiation dose that astronauts would be subject to once they venture outside the protective cover of Earth's magnetosphere.

Keywords: solar energetic particles, space weather, particle acceleration, coronal mass ejections, space radiation

1. Introduction

Solar cycle 23 has been the best-studied solar cycle in history. This is largely because of the launch of a new generation of instruments that have provided: (1) high-resolution remote-sensing observations of solar eruptions; (2) *in situ* observations of solar wind and solar energetic particle composition spanning decades in energy; (3) continuous solar-wind observations upstream of Earth's magnetosphere; and (4) new geospace instruments that include imaging and multipoint capabilities. As a result, it is now possible to trace the effects of individual solar events continuously from the solar corona, through the interplanetary medium, to the inner magnetosphere and upper atmosphere.

For solar energetic particles (SEPs), the key advances have been the ability to measure composition and energy spectra for the key species from H to Fe, with good statistical accuracy, over an energy range extending from suprathermal energies to >100 MeV/nucleon. When combined with ionic charge-state data, these new multi-instrument data sets make it possible to study the charge-to-mass dependence of acceleration and transport processes, to identify the seed-particle populations accelerated in SEP events, and to test new theoretical models with high-precision composition, spectra, and timing observations.

Recent progress has also been due in part to the Sun's cooperation – solar cycle 23 produced 4 of the 10 largest SEP events of the space era. This activity, and new directions in the space program, have raised the priority of developing improved capabilities to forecast the onset and evolution of SEP events. In this paper we

Space Science Reviews (2006) 124: 303–316
DOI: 10.1007/s11214-006-9091-0

highlight examples of how new SEP observations are testing theoretical models, show a comparison of CME and SEP kinetic energies, and discuss examples of space weather aspects of solar energetic particles.

2. Composition and Spectral Variations in SEP Events

The Halloween 2003 period provided the opportunity to study five large SEP events within a 10-day period, each associated with an X-class flare and very fast (>1500 km/s) CME. In the overview of this period in Figure 1, the two largest events (events 2 and 3), were still accelerating particles to >10 MeV/nuc when the shock reached 1 AU. In each of these five events elements from H to Fe exhibit prominent spectral breaks ranging from ~3 to ~30 MeV/nuc (Cohen *et al.*, 2005; Mewaldt *et al.*, 2005a), similar to those in earlier studies (Tylka *et al.*, 2000; Tylka *et al.*, 2005). Li *et al.* (2005), hereinafter LZR, find similar spectral breaks in their SEP acceleration/transport model, and predict that the break energy/nucleon should be proportional to the square of the charge-to-mass ratio (Q/M), where Q is the ionic charge state and M is the ion's mass number.

This prediction was tested with observations following the arrival of the strong shock from the 28 October 2003 SEP event (Mewaldt *et al.*, 2005b), in which the spectra for nine species have break energies ranging from a few MeV/nuc to ~30 MeV/nuc (see Figure 2a). Note that lighter species have higher break energies

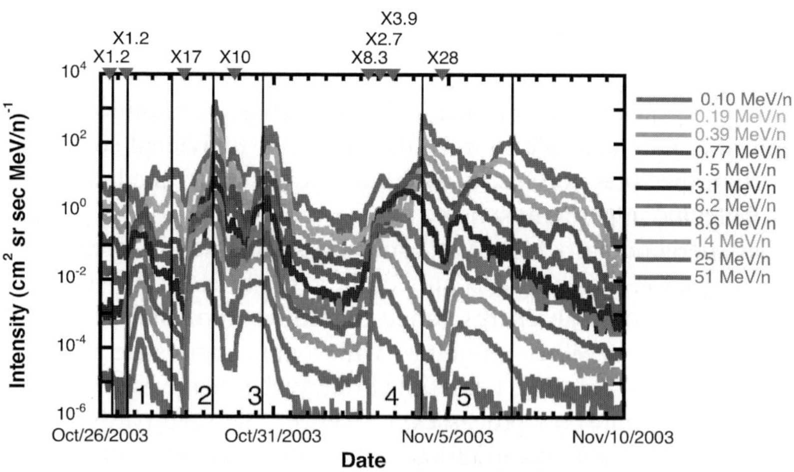

Figure 1. Time history of energetic oxygen nuclei measured by the SIS and ULEIS instruments on ACE during the Halloween series of SEP events (from Cohen *et al.*, 2005). Five large SEP events are evident (numbered 1 to 5). The mean energies range from 0.10 MeV/nuc to 51 MeV/nuc from top to bottom. Interplanetary shocks are indicated by vertical solid lines and X-class flares are shown along the top.

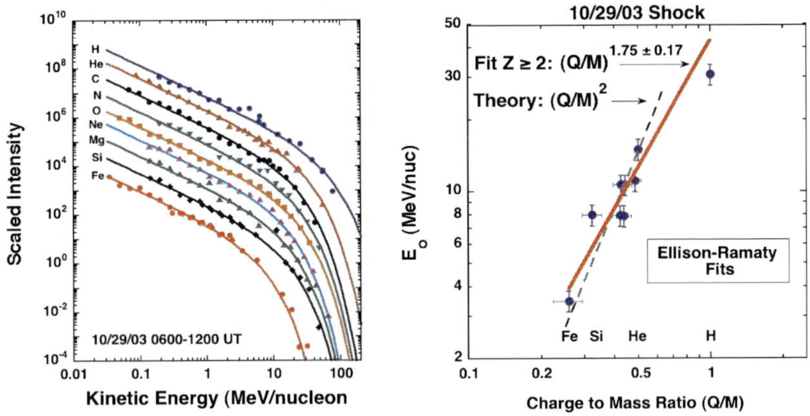

Figure 2. (a) Spectra from the 6-hour period following the shock on 29 October 2003 are fit with the Ellison-Ramaty spectral form with a fixed value of -1.3 for the power-law index. Data are from the SIS, ULEIS and EPAM instruments on ACE, and the EPS sensor on GOES-11 (see Mewaldt *et al.* (2005a, b) for additional details). Each element has been multiplied by a scale factor to separate the spectra. (b) The E_0 values determined from the left panel are plotted versus Q/M. Fits to the $Z \geq 2$ data give a $(Q/M)^{1.75}$ dependence, similar, but somewhat weaker than expected from the theory of LZR (2005).

than heavier species like Fe, as found by Tylka *et al.* (2000). The spectra in Figure 2a are well fit by the Ellison and Ramaty (1985) spectral form:

$$dJ/dE = KE^{-\gamma}\exp(-E/E_0). \tag{1}$$

Here J is the intensity, K, E_0, and γ are constants, with E measured in energy/nucleon. Using this spectrum, which is a power-law at low energies with an exponential roll-over at high energies, the average value of γ for the nine species was 1.3. All spectra were refit using $\gamma = 1.3$ to determine E_0 for each species, which ranged from 31 MeV for protons to 3.5 MeV/nuc for Fe (see Figure 2b).

The *Q/M* dependence of the breaks was investigated using charge-state measurements from the same event by Labrador *et al.* (2005) obtained with the MAST/SAMPEX instrument using the geomagnetic method (see also Klecker et al., this volume) for energy intervals ranging from 15 to 60 MeV/nuc for 0 to 27 to 90 MeV/nuc for Fe. The estimates of the break energies from Figure 2a are plotted against *Q/M* in Figure 2b.

The location of the breaks in Figure 2b is reasonably consistent with the power-law behavior predicted by LZR, although the best-fit slope (1.75 $\pm$ 0.17) is somewhat less than predicted. A weaker *Q/M* dependence of the roll-over in the energy spectra is also predicted by the finite-time shock acceleration model of Channok *et al.* (2005).

In the model of LZR streaming protons escaping upstream from the shock generate enhanced turbulence in the form of Alfvén waves that extends over ~ 2

decades in wave number, k (see also Lee, 1983; Ng *et al.*, 2003). The proton-amplified Alfvén waves play a key role in scattering particles and keeping them near the shock where they can be efficiently accelerated. The break in the power-law spectrum for a given species occurs at the maximum achievable momentum per nucleon (p_{max}) for which there is efficient acceleration, which corresponds to the momentum/nuc that resonates with the minimum k-value (k_{min}) for which there is enhanced turbulence. The sudden decrease in the turbulence level at k values below k_{min} leads to a sudden increase in the diffusion coefficient, thus allowing higher-energy particles to freely escape upstream from the shock. Similar breaks in the wave spectra have been calculated by Ng *et al.* (2003).

Note that the proton break energy is lower than expected from the fit to the heavy-ion data. Li and Zank (personal communication) have pointed out that while heavy ions should act like test particles, protons may not, because protons produce the waves that govern this process (LZR 2005; Ng *et al.*, 2003).

Cohen *et al.* (2005) and Mewaldt *et al.* (2005a) also analyzed fluence spectra for the other Halloween events. These studies support the view that proton-amplified Alfvén waves play a key role in the shock acceleration process, as outlined by Lee (1983). Additional support for this picture comes from the rare observation of proton-amplified Alfvén waves at 1 AU in two large SEP events from solar cycle 23 (Bamert *et al.*, 2004; Kallenbach *et al.*, 2005).

The Ellison-Ramaty form is adequate for fitting the five Halloween events up to ~100 MeV/nuc, but fails to fit proton spectra at higher energies (Mewaldt *et al.*, 2005a). In many cases spectra that extend to >100 MeV/nuc are better fit by a double power-law formula from Band *et al.* (1993). Proton spectra and fits for the five Halloween events are shown in Figure 3, along with the 20 January 2005

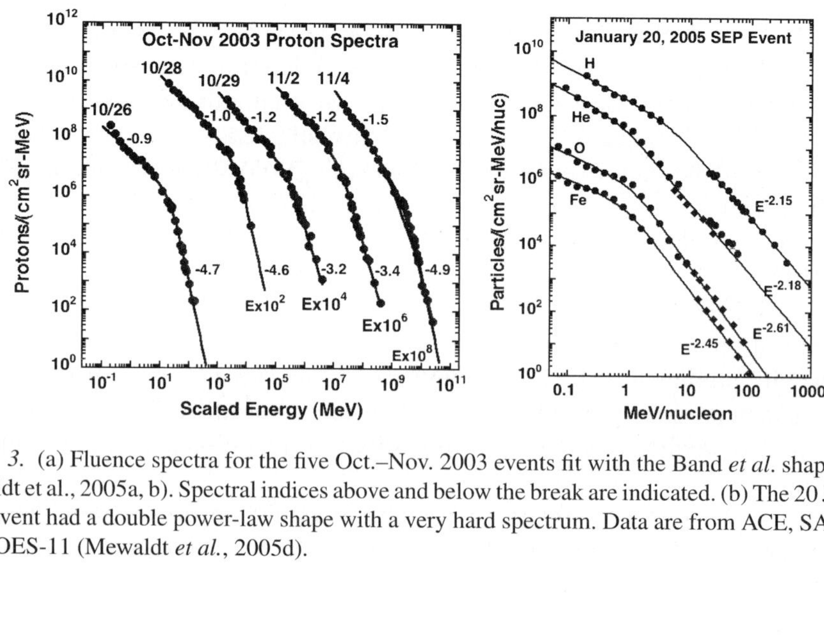

Figure 3. (a) Fluence spectra for the five Oct.–Nov. 2003 events fit with the Band *et al.* shape (from Mewaldt et al., 2005a, b). Spectral indices above and below the break are indicated. (b) The 20 January 2005 event had a double power-law shape with a very hard spectrum. Data are from ACE, SAMPEX and GOES-11 (Mewaldt *et al.*, 2005d).

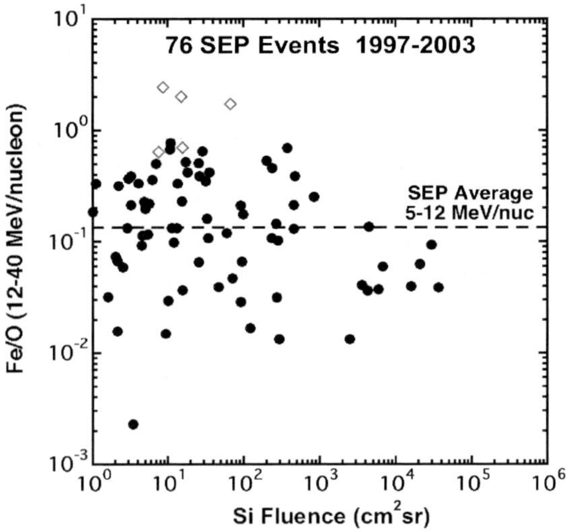

Figure 4. The 12 to 40 MeV/nucleon Fe/O ratio is plotted versus the fluence of Si for 76 large SEP events from late 1997 through 2003 (from Mewaldt *et al.*, 2005c). Data are from the SIS instrument on ACE. The average SEP Fe/O ratio of 0.134 (Reames, 1998) is also indicated. The diamonds are events identified in the literature as "impulsive" (Leske *et al.*, 2003; Mewaldt *et al.*, 2003).

event, the largest ground-level event observed since 1956. In this event the H and He spectra above the breaks extend as power-laws with an index ∼–2.2 for two decades or more in energy/nucleon.

Figure 4 shows a plot of the 12 to 40 MeV/nucleon Fe/O ratio versus the fluence of Si in SEP events. Note that a relatively large number of events have Fe-rich Fe/O ratios with several times the average SEP ratio of 0.134 (Reames, 1998). These Fe-rich events are not the largest events of the solar cycle, and they also tend to be associated with intermediate CME velocities of ∼1000–1500 km/s (Cane *et al.*, 2003). Most of the largest SEP events are Fe-poor, and tend to be associated with the fastest CMEs (e.g., 1500–2500 km/s). The reason that these larger events are Fe-poor above ∼10 MeV/nuc is that the spectra have *Q*/*M*-dependent spectral breaks such as those in Figure 2a, with the break in the Fe spectrum at lower energy/nucleon than that for O. On the other hand, the Fe-rich events tend to have power-law spectra at high energies with a relatively constant Fe/O ratio, as in the January 20, 2005 event (see Figure 3b).

Observations during solar cycle 23 showed that many of the Fe-rich gradual SEP events also included enrichments in ^{3}He (Cohen *et al.*, 1999; Mason *et al.*, 1999), and mean ionic charge states of Fe that increased with energy to values of $\sim Fe^{+20}$ (Mazur *et al.*, 1999; Labrador *et al.*, 2005; see reviews by Popecki (2006) and Klecker *et al.*, this volume (2006)) indicating that these events include a mixture of "impulsive" and coronal material.

These new observations have blurred the observational distinctions between SEP event classes, but it is generally acknowledged that there remains a distinction between the acceleration mechanisms involved. (For discussions of the evolution of the two-class picture see Mewaldt (2000), Klecker (2006), and von Rosenvinge and Cane (2006)). Among the suggested explanations for the presence of impulsive flare material in gradual SEP events associated with CMEs are direct contributions of flare-accelerated material (Cane *et al.*, 2003, 2006) and the acceleration of a suprathermal seed population enriched in ions left over from previous small, impulsive events (Mason *et al.*, 1999; Tylka *et al.*, 2005; Desai *et al.*, 2006a; Desai *et al.*, 2006b), and from earlier gradual events (Mewaldt *et al.*, 2006).

3. SEP and CME Kinetic Energies

In order to estimate the efficiency with which CME-driven shocks accelerate particles it is of interest to compare the kinetic energy of accelerated particles with the CME kinetic energy. CME masses can be estimated from the total excess brightness of the CME and the velocity can be found by fitting the radial profile (Vourlidas *et al.*, 2000). CME masses for large SEP events of 2000–2003 can be found in Gopalswamy *et al.* (2004, 2005) and Emslie *et al.* (2004). Mass and energy estimates are more accurate for events on the limb than for halo CME events.

Several steps are involved in estimating SEP kinetic energies (Mewaldt *et al.*, 2005a, c). (1) Energy spectra from ACE, GOES, and SAMPEX, extending from <0.1 to >100 MeV/nuc, were fit with either the double-power-law form of Band *et al.* (1993) or the model of Ellison and Ramaty (1985), and these fits were integrated from 0.01 to 1000 MeV/nuc to obtain fluences at 1 AU. To obtain the energy/cm^2 escaping from 1 AU the fluences were corrected for the average number of times particles cross 1 AU due to scattering on interplanetary turbulence, which reduces the estimated energy content of accelerated particles by a factor of $\sim$3 to 4 (Mewaldt *et al.*, 2005c). Studies of heavy ions >10 MeV/nuc show that the largest SEP events originate near central meridian. Emslie *et al.* (2004) derived longitudinal e-folding longitudes of $-45°$ for western events and $-25°$ for eastern events, and adopted an e-folding latitude of $-35°$. Using these parameters, the total particle energy escaping through 1 AU can be estimated.

A comparison of CME and SEP kinetic energies for seventeen SEP events is shown in Figure 5. Note that the spread in the SEP kinetic energies is about a factor of 10 greater than the spread in CME kinetic energies. There is a group of eleven events where the SEP kinetic energy ranges from $\sim$3% to 20% of the CME kinetic energy, indicating that shock acceleration can often transform $\sim$10% of the CME kinetic energy into energetic particles. There are also four events where the estimated efficiency is less than 1%. Of course, there are also CMEs for which no SEPs are observed at 1 AU. The events in Figure 5 were selected because the SEP

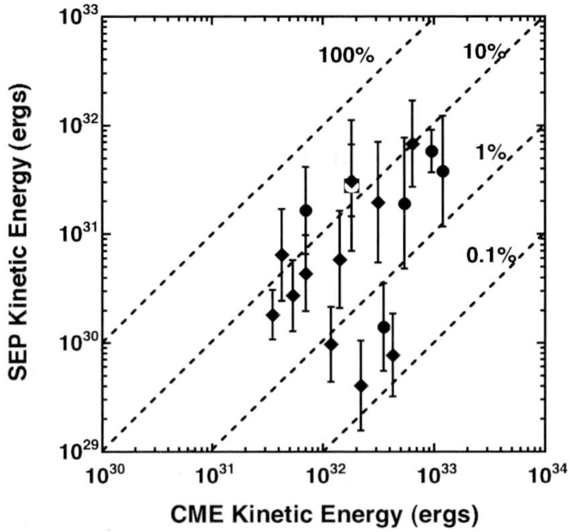

Figure 5. A comparison SEP and CME kinetic energies for 17 SEP events including the 21 April 2002 event (open square), five events from October–November 2003 (circles) and 11 other events observed from 1998–2003 (from Mewaldt *et al.*, 2005c).

intensities were sufficient for spectra to be measured, and because CME kinetic energy estimates were available, so they do not come from a representative sample of CMEs. The SEP kinetic energies could have been underestimated, since adiabatic energy losses have not yet been taken into account (Mewaldt *et al.*, 2005a).

It is interesting that galactic cosmic rays apparently extract ~10% to 30% of the kinetic energy of supernova shocks in order to sustain the cosmic-ray energy-density in the Galaxy (~1.5 eV/cm^3) over the mean cosmic-ray lifetime of ~15 million years (see Ptuskin, 2001). These preliminary comparisons indicate that particle acceleration at CME-driven shocks can also be surprisingly efficient; particles frequently extract ~10% or more of the CME kinetic energy. It remains to be seen why some CME-driven shocks are much more efficient accelerators than others.

4. Radiation Hazards due to Solar Energetic Particles

Although SEP events have been recognized as a radiation hazard to astronauts since the Apollo era, since 1972 human spaceflight has been confined to low-Earth orbit, where Earth's magnetosphere provides an effective shield over most of the orbit of manned platforms such as MIR and the ISS. However, with NASA once again committed to sending astronauts back to the moon, and possibly to Mars, there has been renewed interest in improving our capability to predict the onset of large solar eruptive events, and in forecasting the intensity, energy spectrum, and time history of SEPs resulting from the eruption. Since only a few percent of CMEs produce

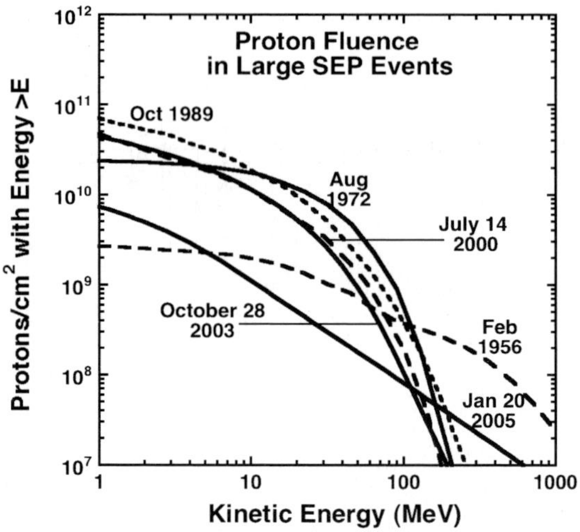

Figure 6. Integral fluence spectra for protons are shown for some of the largest SEP events of the last 50 years (updated from Mewaldt *et al.*, 2005a). Spectra for the events prior to this solar cycle are adapted from Nealy *et al.* (1992). The 14 July 2000 spectrum has been derived from data in Tylka *et al.* (2001), the 28 October 2003 spectrum is from Mewaldt *et al.* (2005a), and the 20 January 2005 spectrum is from Mewaldt *et al.* (2005b).

interplanetary shocks, and only a fraction of these shocks result in a significant SEP event at 1 AU, attention is focused on understanding the conditions under which SEP acceleration is most efficient.

Although heavy ions up through the Fe are the most effective at upsetting space hardware, H and He account for most of the radiation hazard to humans in space because they are the most abundant, and because they penetrate more easily through shielding. Taking into account typical SEP energy spectra and the thickness of available shielding, the most important energy range for protons is from 30 to 100–200 MeV (Turner, 2006).

Figure 6 shows integral energy spectra for protons for some of the largest SEP events of the last 50 years. The August 1972 SEP event, generally recognized as the largest SEP event of the space era, is often used as a standard for assessing requirements for shielding astronauts. Figure 7 indicates the radiation dose to the blood-forming organs (BFO), as a function of the shielding thickness, that would be associated with this event. Although the August 1972 event is the largest, there were four events during solar cycle 23 that had $>$30 MeV proton fluences $\geq$ 60% of the August 1972 event. For reference, if exposed to a BFO dose of 1 Gy in a high exposure-rate situation (roughly equivalent to $\sim$100 cSv for the SEP event in Figure 7), $\sim$5% to $\sim$30% of people would experience nausea and vomiting, while for a dose of 1.5 Gy there would be a $\sim$5% chance of death if there was no treatment (Turner, 2006).

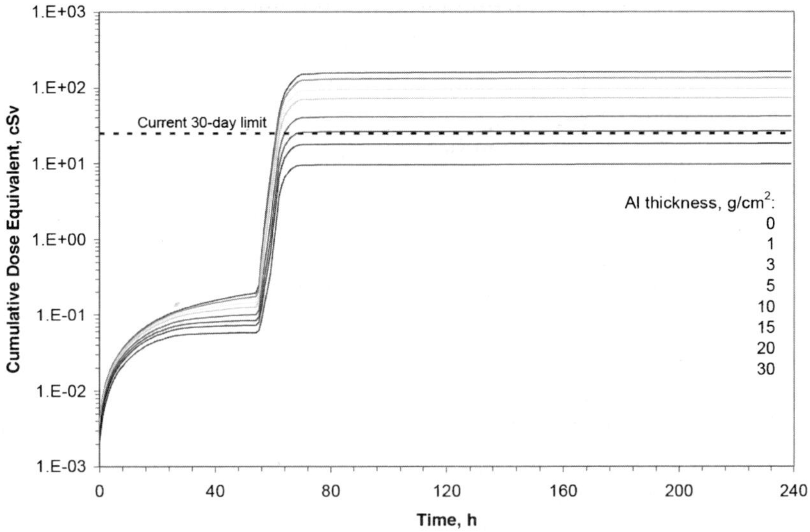

Figure 7. The cumulative dose equivalent is shown for the large August 1972 SEP event for a range of shielding thicknesses measured in g/cm^2 of aluminum (figure from Kim *et al.*, 2005). Also indicated is the current 30-day limit of 0.25 Sv for missions to low-Earth orbit.

It is clear from Figure 7 that shielding can be effective at mitigating the radiation dose from a large SEP event. This also means that it is important to have adequate warning of the onset and evolution of SEP events, along with a close-by safe haven when astronauts are on EVA. A recent LWS report outlines research objectives that can help mitigate the effects of SEPs on humans and hardware in space (Golightly *et al.*, 2005). In addition, Turner (2006) has outlined the interdisciplinary research needed to improve forecasts of SEP events.

With more and more spacecraft exploring the inner solar system, and plans to send the Solar Orbiter, Inner Heliospheric Sentinels, and Solar Probe Missions to explore inside 0.3 AU, it has become important to learn how to characterize and forecast space weather at longitudes away from the Earth-Sun line, and at radial distances ranging from ∼0.3 to 1.5 AU. Studies and models of the dependence of SEP intensities on radial distance from the Sun do not all agree (e.g., Reames and Ng, 1998; Ruzmaikan *et al.*, 2005; Lario *et al.*, 2006) suggesting the need for new measurements by a multi-spacecraft mission such as Inner Heliosphere Sentinels during the next solar maximum.

5. Solar Energetic Particle Impacts on the Geospace Environment

The best known impacts that SEPs have on the geospace environment result from the their effects on communications. During polar cap absorption (PCA) events the sudden ionization of the upper atmosphere that occurs at high latitudes can block short wave communications for periods of hours, affecting airline and ship-to-shore communications. A second, less well-known impact of SEPs is the creation of odd-nitrogen compounds in the upper atmosphere that can deplete ozone. For example, during the October 28–31, 2003 period (see Figure 1) ozone depletions of 5% to 8% were measured in the southern polar stratosphere that lasted days beyond the events (Jackman *et al.*, 2005).

The most intense SEP events at Earth are produced by earthward-directed CMEs in which the shock is still accelerating particles to high energy when it reaches Earth. The largest of these SEP events during the space era were the August 1972 and October 1989 events. Recent examples include the July 14, 2000 "Bastille Day" event and the October 28, 2003 event. Once the CME-driven shock reached 1 AU all these events produced large geomagnetic storms in which the geomagnetic field was severely disturbed and SEP ions had access to much lower latitudes than during quiet times. Leske *et al.* (2001) have measured the geomagnetic cutoff (the minimum latitude that SEPs of a given rigidity can reach) during a number of SEP events during solar cycle 23. They find that the decrease in the cutoff is correlated with geomagnetic parameters such as the Kp and Dst indices For reference, the nominal cutoff rigidity, Rc, is related to the magnetic L-shell, L, by $\text{Rc} = 14.5\ \text{GV/L}^2$. Directly measured geomagnetic cutoffs are somewhat lower than this simple formula for L-shells <2 (Ogliore *et al.*, 2001).

The "conspiracy" between large SEP events and large geomagnetic storms often results in a circumstance where SEP ions and electrons may suddenly have access to the inner magnetosphere, where they can become trapped to form a new radiation belt. The best-measured example of this is the March 1991 storm in which an intense belt of >10 MeV electrons was suddenly created (Blake *et al.*, 1992; Li *et al.*, 1993) and observed to survive for more than two years (Looper *et al.*, 1994).

Mazur *et al.* (2006) have reviewed observations of new radiation belts during solar cycle 23. They observed newly injected heavy ions from He to Fe that could not have originated from other than an SEP source (see example in Figure 8). They find that new radiation belts are created about 1/3 of the time that SEP ions with $\sim$0.5 to 5 MeV/nuc have access to $L = 2$ or below, and discuss some of the conditions that determine when new belts are created, and how long they survive. Large solar events do not only add particles to the radiation belts. According to Looper *et al.* (2005), following the Halloween storms protons with $>$20 MeV at 600 km were almost entirely "scrubbed away" because of the expansion of the upper atmosphere, returning only after a period of months.

6. Space Weather Challenges

As pointed out above, the planned return of astronauts to the Moon has led to increased concern about our ability to predict the onset and evolution of large SEP events. The discussion above highlights several questions that need to be answered

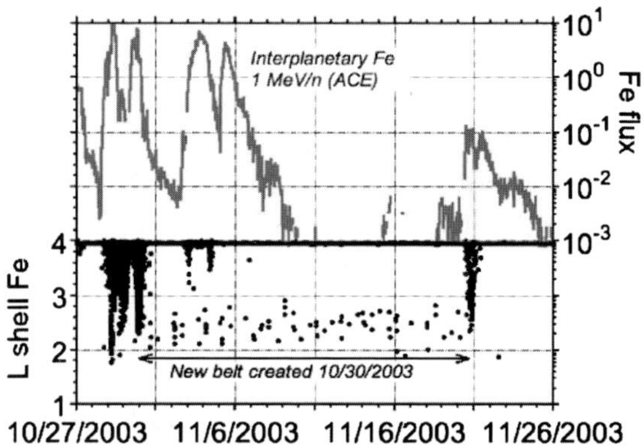

Figure 8. Example showing the creation and disappearance of a new radiation belt coinciding with the occurrence of two SEP events with associated geomagnetic storms. The top trace shows the 1 MeV/nuc Fe intensity at ACE (in units of $cm^{-2}\,sr^{-1}s^{-1}$), upstream of the magnetosphere, while the bottom trace shows SAMPEX measurements of Fe penetrating into $L = 2$, where a fraction of it was trapped for ~3 weeks (from Mazur *et al.*, 2006).

to improve our ability to forecast the near-Earth radiation environment, leading to the following challenges:

1. The results in Figure 5 suggest that the largest SEP events can extract ~10% of the CME kinetic energy. Yet there are clearly many relatively fast CMEs where this does not happen. *What determines when CME-driven shocks accelerate SEPs efficiently and when this efficiency is low or zero?*
2. The radiation hazard to humans in space is mainly due to the fluence of >30 MeV protons, which is determined to a large degree by the location of the break in the proton spectrum, and the spectral index above the break. *What determines the spectral index of particles above the commonly observed break in the spectrum?*
3. As more and more spacecraft leave the protective cover of low Earth orbit and explore the inner solar system, it has become more important to forecast space weather at longitudes away from the Earth-Sun line and at radial distances ranging from 0.3 to 1.5 AU. *How do solar energetic particles vary with longitude and distance from the Sun, and what interplanetary conditions determine how fast their intensity increases?*
4. The past solar maximum has provided a number of examples of how the combined effects of CMEs and SEPs can suddenly alter the radiation belts. *Can recent examples of CME/SEP effects on trapped radiation lead to an understanding of their long-term impact on the formation, structure, and lifetime of the radiation belts, and perhaps result in the ability to forecast these effects?*

The coming generation of spacecraft that are designed to explore the inner heliosphere, including Solar Orbiter, Inner Heliosphere Sentinels, and Solar Probe, should be in an ideal position to address these and other key challenges.

Acknowledgement

I am very grateful to ISSI for their hospitality and for the opportunities for interdisciplinary discussion provided by this workshop. This work was supported by NASA under grants NNG04GB55G, NNG04088G, and NAG5-12929. I also thank the editor, Berndt Klecker, for his patience and assistance, and Jasmin Zanker-Smith for editorial assistance.

References

Bamert, K., Kallenbach, R., Ness, N. F., Smith, C. W., Terasawa, T., Hilchenbach, M., *et al.*: 2004, *Astrophys. J. Lett.* **601**, L99.

Band, D., Matteson, J., Ford, L., Schaefer, B., Palmer, D., Teegarden, B., *et al.*: 1993, *Astrophys. J.* **413**, 281.

Blake, J. B., Kolasinski, W. A., Fillius, R. W., and Mullen, E. G.: 1992, *Geophys. Res. Lett.* **19**, 821.

Cane, H. V., von Rosenvinge, T. T., Cohen, C. M. S., and Mewaldt, R. A.: 2003, *Geophys. Res. Lett.* **30**, SEP 5-1, doi: 10.1029/2002GL016580.

Cane, H. V., von Rosenvinge, T. T., Cohen, C. M. S., and Mewaldt, R. A.: 2006, *J. Geophys. Res.*, A06S90, doi: 10.1029/2005JA011071.

Channok, C., Ruffolo, D., Desai, M. I., and Mason, G. M.: 2005, *Astrophys. J. Lett.* **633**, L53.

Cohen, C. M. S., Mewaldt, R. A., Leske, R. A., Cummings, A. C., Stone, E. C., Wiedenbeck, M. E., *et al.*: 1999, *Geophys. Res. Lett.* **26**, 2697.

Cohen, C. M. S., Stone, E. C., Mewaldt, R. A., Leske, R. A., Cummings, A. C., Mason, G. M., *et al.*: 2005, *J. Geophys. Res.* **110**, doi: 10.1029/2004JA011004.

Desai, M. I., Mason, G. M., Gold, R. E., Krimigis, S. M., Cohen, C. M. S., Mazur, J. E., *et al.*: 2006a, *Astrophys. J.* **649**, 470.

Desai, M. I., Mason, G. M., Mazur, J. E., and Dwyer, J. R.: 2006b, this volume, doi: 10.1007/s11214-006-9109-7.

Ellison, D. C., and Ramaty, R.: 1985, *Astrophys. J.* **298**, 400.

Emslie, A. G., Kucharek, H., Dennis, B. R., Gopalswamy, N., Holman, G. D., Share, G. H., *et al.*: 2004, *J. Geophys. Res.* **109**, A10104, doi:10.1029/2004JA010571.

Golightly, M., *et al.*: 2005, In: *NASA Technical Paper 3211.*

Gopalswamy, N., Yashiro, S., Krucker, S., Stenborg, G., and Howard, R. A.: 2004, *J. Geophys. Res.* **109**, A12105, doi: 1029/2004JA10602.

Gopalswamy, N., Yashiro, S., Liu, Y., Michalek, G., Vourlidas, A., Kaiser, M. L., *et al.*: 2005, *J. Geophys. Res.* **110**, A09S15, doi: 10.1029/2004JA010958.

Jackman, C. H., DeLand, M. T., Labow, G. J., Fleming, E. L., Weisenstein, D. K., Ko, M. K. W., *et al.*: 2005, *J. Geophys. Res.* **110**, A09S27, doi: 10.1029/2004JA010888.

Kallenbach, R., Bamert, K., Hilchenbach, M., and Smith, C. W.: 2005, in: Li, G., Zank, G. P., and Russell, C. T. (eds.), *AIP Conf. Proc. 781: The Physics of Collisionless Shocks: 4th Annual IGPP International Astrophysics Conference*, pp. 129–134.

Kim, M. Y., Hu, X., and Cuccinotta, F. A.: 2005, In: *American Institute of Aeronautics and Astronautics, Space 2005 Conference, Paper AIAA-6653*, pp. 1–16.

Klecker, B.: 2006, In: Gopolswamy, N., Mewaldt, R. A., and Torsti, J. (eds.): *AGU Monograph Series, Vol. 165 "Solar Eruptions and Energetic Particles"*, pp. 147–156.

Klecker, B., Möbius, E., and Popecki, M. A.: 2006, *Space Sci. Rev.*, this volume, doi: 10.1007/s11214-006-9111-0.

Labrador, A., Leske, R., Mewaldt, R., Stone, E., and von Rosenvinge, T.: 2005, In: *Proc. 29th Intern. Cosmic Ray Conf., Pune, India*, Vol. 1, pp. 99–102.

Lario, D., Kallenrode, M.-B., Decker, R. B., Krimigis, S. M., Roelof, E. C., Aran, A., *et al.*: 2006, *Astrophys. J.*, submitted.

Lee, M. A.: 1983, *J. Geophys. Res.* **88**(17), 6109.

Leske, R. A., Mewaldt, R. A., Stone, E. C., and von Rosenvinge, T. T.: 2001, *J. Geophys. Res.* **106**, 30011.

Leske, R. A., *et al.*: 2003, In: *28th Intern. Cosmic Ray Conf.*, pp. 3253–3236.

Li, X., Roth, I., Temerin, M., Wygant, J. R., Hudson, M. K., and Blake, J. B.: 1993, *Geophys. Res. Lett.* **20**, 2423.

Li, G., Zank, G. P., and Rice, W. K. M.: 2005, *J. Geophys. Res.* **110**, doi: 10.1029/2005JA011038.

Looper, M. D., Blake, J. B., Mewaldt, R. A., Cummings, J. R., and Baker, D. N.: 1994, *Geophys. Res. Lett.* **21**, 2079.

Looper, M. D., Blake, J. B., and Mewaldt, R. A.: 2005, *Geophys. Res. Lett.* **32**, doi: 10.1029/2004GL021502.

Mason, G. M., Mazur, J. E., and Dwyer, J. R.: 1999, *Astrophys. J. Lett.* **525**, L133.

Mazur, J. E., Mason, G. M., Looper, M. D., Leske, R. A., and Mewaldt, R. A.: 1999, *Geophys. Res. Lett.* **26**, 173.

Mazur, J. E., Blake, J. B., Slocum, P. L., Hudson, M. K., and Mason, G. M.: 2006, in: Gopolswamy, N., Mewaldt, R. A., and Torsti, J. (eds.), *AGU Monograph Series, Vol. 165 "Solar Eruptions and Energetic Particles"*, pp. 345–352.

Mewaldt, R. A.: 2000, In: Dingus, B. L., Kieda, D. B., and Salamon, M. H. (eds.), *AIP Conf. Proc. 516: 26th Internat. Cosmic Ray Conf.*, pp. 265–270.

Mewaldt, R., Cohen, C., Mason, G., Desai, M., Leske, R., Mazur, J., *et al.*: 2003, in: *Proc. 28th Intern. Cosmic Ray Conf., Tsukuba, Japan*, Vol. 6. pp. 3229–3232.

Mewaldt, R. A., Cohen, C. M. S., Labrador, A. W., Leske, R. A., Mason, G. M., Desai, M. I., *et al.*: 2005a, *J. Geophys. Res.* **110**, 9.

Mewaldt, R. A., Cohen, C. M. S., Mason, G. M., Labrador, A. W., Looper, M. L., Haggerty, D. E., *et al.*: 2005b, in: Li, G., Zank, G. P., and Russell, C. T. (eds.), *AIP Conf. Proc. 781: The Physics of Collisionless Shocks: 4th Annual IGPP International Astrophysics Conference*, pp. 227–232.

Mewaldt, R. A., Cohen, C. M. S., Mason, G. M., *et al.*: 2005c, in: *ESA SP-592: Solar Wind 11/SOHO 16, Connecting Sun and Heliosphere.*

Mewaldt, R. A., Looper, M. D., Cohen, C. M. S., Mason, G. M., Haggerty, D. K., Desai, M. I., *et al.*: 2005d, in: *Proc. 29th Intern. Cosmic Ray Conf., Pune, India*, Vol. 1, pp. 111–114.

Mewaldt, R. A., Cohen, C. M. S., and Mason, G. M.: 2006, in: Gopolswamy, N., Mewaldt, R. A., and Torsti, J. (eds.), *AGU Monograph Series, Vol. 165 "Solar Eruptions and Energetic Particles"*, pp. 115–126.

Nealy, J., Streipe, S., and Simonsen, L.: 1992, in: *NASA Technical Paper 3211.*

Ng, C. K., Reames, D. V., and Tylka, A. J.: 2003, *Astrophys. J.* **591**, 461.

Ogliore, R. C., Mewaldt, R. A., Leske, R. A., Stone, E. C., and von Rosenvinge, T. T.: 2001, in: *27th Intern. Cosmic Ray Conf.*, pp. 4112–4115.

Popecki, M. A.: 2006, in: Gopolswamy, N., Mewaldt, R. A., and Torsti, J. (eds.), *AGU Monograph Series, Vol. 165 "Solar Eruptions and Energetic Particles"*, pp. 127–136.

Ptuskin, V. S.: 2001, *Space Sci. Rev.* **99**, 281.

Reames, D. V.: 1998, *Space Sci. Rev.* **85**, 327.

Reames, D. V., and Ng, C. K.: 1998, *Astrophys. J.* **504**, 1002.

Ruzmaikan, A., Li, G., Feynman, J., and Jun, I.: 2005, in: *ESA SP-592: Solar Wind 11/SOHO 16, Connecting Sun and Heliosphere*, pp. 441–444.

Turner, R. E.: 2006, in: Gopolswamy, N., Mewaldt, R. A., and Torsti, J. (eds.), *AGU Monograph Series, Vol. 165 "Solar Eruptions and Energetic Particles"*, pp. 367–374.

Tylka, A. J., Boberg, P. R., McGuire, R. E., Ng, C. K., and Reames, D. V.: 2000, in: R. A. M. *et al.* (ed.), *Acceleration and Transport of Energetic Particles Observed in the Heliosphere: ACE 2000 Symposium*, pp. 147–152.
Tylka, A. J., Cohen, C. M. S., Dietrich, W. F., Maclennan, C. G., McGuire, R. E., Ng, C. K., *et al.*: 2001, *Astrophys. J. Lett.* **558**, L59.
Tylka, A. J., Cohen, C. M. S., Dietrich, W. F., Lee, M. A., Maclennan, C. G., Mewaldt, R. A., *et al.*: 2005, *Astrophys. J.* **625**, 474.
von Rosenvinge, T. T., and Cane, H. V.: 2006, in: Gopolswamy, N., Mewaldt, R. A., and Torsti, J. (eds.), *AGU Monograph Series, Vol. 165 "Solar Eruptions and Energetic Particles"*, pp. 103–114.
Vourlidas, A., Subramanian, P., Dere, K. P., and Howard, R. A.: 2000, *Astrophys. J.* **534**, 456.

COMMONALITIES BETWEEN IONOSPHERE AND CHROMOSPHERE

GERHARD HAERENDEL
International University Bremen
(E-mail: hae@mpe.mpg.de)

(Received 29 July 2005; Accepted in final form 15 February 2006)

Abstract. Three types of processes, occurring in the weakly ionized plasmas of the Earth's ionosphere as well as in the solar chromosphere, are being compared with each other. The main objective is to elaborate on the differences introduced primarily by the grossly different magnitudes of the densities, both with respect to the neutral and, even more so, to the plasma constituents. This leads to great differences in the momentum coupling from the plasma to the neutral component and becomes clear when considering the direct electric current component transverse to the magnetic field, called "Pedersen current" in the ionosphere, which has no quasi-static counterpart in the chromosphere. The three classes of processes are related to the dynamical response of the two plasmas to energy influx from below and from above. In the first two cases, the energy is carried by waves. The third class concerns plasma erosion or ablation in the two respective regions in reaction to the injection of high Poynting and/or energetic particle fluxes.

Keywords: chromosphere, ionosphere, ion-neutral coupling, plages, cavitation

1. Introductory Remark

It has often been said that the study of plasmas near Earth can help understanding astrophysical plasmas. Although nobody would seriously dispute the validity of this statement and although there are subjects, like the physics of magnetic reconnection, which find application in a wide number of cosmic objects as well as in near-Earth and laboratory plasmas, relatively few papers have been published in which knowledge has been transferred, for instance, from the magnetosphere to other objects. This paper is not a communication of latest data or insights into solar or magnetospheric topics, but is primarily meant to raise interest in and sharpen the eyes for some commonalities and differences between plasmas in the Sun's and Earth's atmospheres. To this end, I have chosen two plasma realms, which, in search for commonalities, have received much less attention, namely the Earth's ionosphere and the solar chromosphere. Some of the subsequently discussed physics is far from being fully understood or contains conjectures of the author and, therefore, may stimulate further exploration. The reader may also discover more differences than commonalities.

On the other hand, the paper does not contain the most often quoted commonality between solar and magnetospheric physics, the role of *reconnection* in flares and substorms. The sole reason for this omission is that this process does not play in the

Space Science Reviews (2006) 124: 317–331
DOI: 10.1007/s11214-006-9092-z

cool and dense plasmas of the arena chosen for this paper, but rather in the hot and more dilute plasmas of corona and plasma sheet, respectively. The reader looking for a comparative treatment of this topic is referred to a review by this author (Haerendel, 2001), in which reconnection is covered in the solar, magnetospheric, and laboratory contexts on equal grounds.

2. Neutral-Ion Coupling

Both regions, ionosphere and chromosphere, are partially ionized. The ion/neutral ratio in the ionosphere varies from 10^{-8} to 10^{-3} between 100 and 300 km. In the chromosphere, the ratio is smaller, ranging from about 10^{-4} to 1 between the temperature minimum (around 400–500 km above the photosphere) and the transition region (at about 2000 km height, on average). In both regions one finds distinct effects owed to ion-neutral coupling. However, the different ratios between the relevant dynamic time-scales and the neutral-ion coupling times create widely different responses. In both cases, we find $\nu_{ni} \ll \Omega_i$, with Ω_i being the gyro-frequency and ν_{ni} the momentum coupling frequency from ions to neutrals. But in the ionosphere, the typical inverse dynamic time-scale, ω, is much higher than ν_{ni}, whereas the opposite holds for the chromosphere. The reason for that lies in the high absolute number density of the ions, around 10^{11} cm^{-3}, in the chromosphere, about six orders of magnitude higher than in the ionosphere. Correspondingly, the neutral component couples to the ion motion within 10 ms, or so, whereas this takes about one hour in the ionosphere. As a consequence much of the plasma dynamics in the ionosphere, at least well above the level of $\nu_{in} \approx \Omega_i$, ν_{in} being the ion-neutral collision frequency, can react to changing electric fields independent of the neutrals. In the chromosphere, the relevant dynamic time-scales in response to waves from below or above are in the range of seconds to minutes, i.e. much longer than the neutral-ion response time (Haerendel, 1992). To lowest order, the neutrals can be regarded as strictly coupled to the ions when subject to changing electromagnetic fields. Piddington (1956) already recognized this fact implying that the inertial mass in Alfvén waves is the sum of ions and neutrals.

In this context let us look at the so-called *Pedersen current*:

$$j_p = \frac{en}{B}\frac{\nu_{in}}{\Omega_i} E_\perp^* \tag{1}$$

with:

$$E_\perp^* = E_\perp + v_n \times B = (v_n - v_i) \times B \tag{2}$$

Since in the ionosphere the neutral and ion velocities can be vastly different, the electric field in the neutral frame can be substantial and thus the Pedersen current. For Alfvén waves in the chromosphere, however, $E_\perp^*$ practically vanishes. No quasi-static Pedersen current can exist. This has an impact on the ohmic dissipation of the respective waves.

To second order, one finds a more subtle result (Haerendel, 1992; De Pontieu and Haerendel, 1998). The high-frequency Pedersen current, still for $\omega \ll \nu_{ni}$, does have a (small) direct current component:

$$j_\perp = \frac{\rho_n + \rho_i}{B^2}\left(i\omega + \frac{\rho_n}{\rho_n + \rho_i}\frac{\omega^2}{\nu_{ni}}\right)E_\perp \tag{3}$$

Whereas the first term in the brackets describes the familiar inertial current, the second term introduces damping. The damping length is:

$$\lambda_{\text{damp}} = 2\frac{\rho_{\text{tot}}}{\rho_n}\frac{\nu_{ni} v_A}{\omega^2} \tag{4}$$

The surprising result is that for $\omega \ll \nu_{ni}$ the damping becomes weaker if the collision frequency increases. The reason for that lies in the reduction of the slippage between neutrals and ions, which lowers the possibility of dissipative energy exchange between the two components. In the following applications the damping is weak, i.e. the wavelength is short compared with damping length. As a quantitative example let us consider an Alfvén wave with a frequency of 0.5 Hz. At a height of 1000 km, we have $\nu_{ni} \approx 70$ s^{-1}, $\mathbf{v}_A = 100$ km/s, $\lambda_{\text{wave}} \approx 50$ km, and $\lambda_{\text{damp}} \approx 1400$ km. Most important is the fact that by being damped an Alfvén wave does not only heat the plasma but also exerts a force along the direction of the magnetic field (Haerendel, 1992).

3. Wave Drive from Below

Our first comparison between similar processes in ionosphere and chromosphere deals with the effect of waves launched from below and pushing matter up the magnetic field lines. The respective phenomena in the chromosphere are the *spicules*, which are omnipresent on the quiet Sun. The most closely related process in the ionosphere is the equally frequent formation of *sporadic E* layers. Both generate inhomogeneous density distributions. Both involve ion-neutral collisions and the Pedersen current. The main distinction between the two is that for spicules it is the $\mathbf{j} \times \mathbf{B}$ force of Alfvén waves, and here the small current component in phase with the transverse electric field (Equation (3)) in cooperation with the transverse magnetic perturbation field, what drives the plasma upward, whereas in the E layer, near $\nu_{in} \approx \Omega_i$, the driver is d.c. transverse electric field.

The formation of *spicules* by the here considered ion-neutral wave damping process was discussed in detail by De Pontieu and Haerendel (1998), and more recently criticized by James *et al.* (2003). The main unsolved question is whether the WKB approximation, employed in the original theory, is valid, or whether accounting for reflection of the upward traveling Alfvén waves at the transition

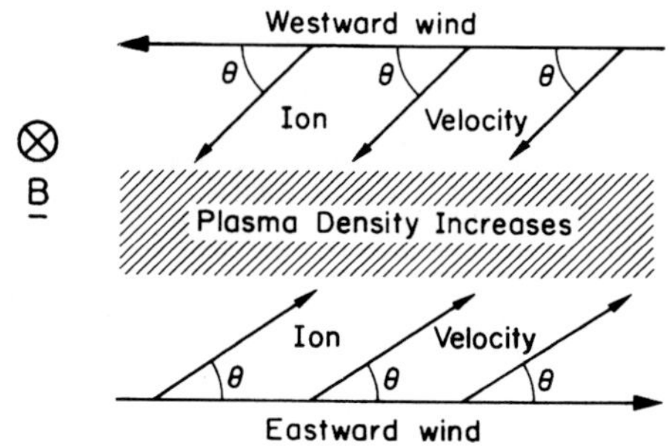

Figure 1. Cartoon demonstrating the increase of plasma density in the low ionosphere (Sporadic E) caused by converging Pedersen currents under wind shear conditions (Hines, 1974).

region does not introduce a more powerful mechanism via slow shock generation. The subject is still open for further clarification by introducing into the modeling the so far neglected effects of thermal conduction and radiative cooling.

The *sporadic E* process coming closest to the spicules was proposed by Hines (1974). The density irregularities, most prominent in the ionosphere between 90 and 110 km after sunset, originate from shears in the zonal neutral wind and their effect on the Pedersen current (s. Equation (2)), creating opposing components of transverse electric field and thus divergences in the accompanying plasma transport where $\nu_{in} \approx \Omega_i$. The sketch contained in Figure 1 illustrates the process.

A slightly different process is operational at somewhat higher altitudes where $\nu_{in} \ll \Omega_i$ (Dungey, 1959). Vertical shears of the neutral wind push the plasma component respectively up or down along the field lines and thus create the density irregularities. Direct collisional momentum exchange creates the effect in this case.

Besides the common aspect of ion-neutral coupling, there may be another reason for considering the commonality between spicules and sporadic E. The ultimate causes for both phenomena seem to be gravity or sound waves launched from below. In the case of the ionosphere, they arise from weather fronts or thunderstorms, and for spicules, from a leakage of energy from p-mode oscillations through the temperature minimum (De Pontieu *et al.*, 2004). Figure 2, taken from the latter paper, exhibits a 5-minute periodicity in the occurrence of fibrils (equivalent to spicules) in active regions. Furthermore, it shows the result of model calculations, involving the formation of slow shocks above the temperature minimum as driving agents for the spicules. The observations are, however, not sufficiently well resolved to establish a one-to-one correlation between local p-mode oscillations and spicule formation.

4. Wave Heating from Above

The now to be discussed commonality of a solar and a magnetospheric process is much more subtle than that of the previous section. On the Sun as well as in

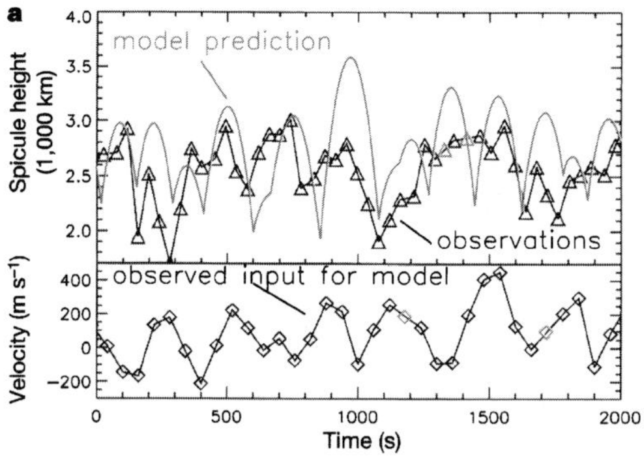

Figure 2. Comparison of EUV brightness variations (171 Å, black line and triangles) as proxy for spicule occurrence with model calculations of spicule height (grey line), based on input from photospheric velocity measurements (lower panel) (De Pontieu *et al.*, 2004).

the Earth's environment, energy is "raining down" into the weakly ionized atmospheric layers in various ways, basically either as energetic particle or Poynting fluxes. The latter occur dominantly in form of Alfvén waves. Whereas the energetic particles deliver their energy "simply" via collisions, wave energy can be converted into thermal energy either by ion-neutral collisions or, in very low-density plasmas, by non-linear wave coupling to high frequency waves and subsequent ion cyclotron heating. We will invoke these two wave heating processes as underlying the origin of plages in the chromosphere and of so-called Alfvénic arcs in the ionosphere.

As we have seen above, the relatively high density of the chromospheric plasma allows collisions to be effective energy exchange processes. High-frequency Alfvén waves, generated by fast plasma processes in the corona, may penetrate through the transition region and be damped in the upper chromosphere by the same neutral-ion coupling processes discussed above. Most effective would be waves in the frequency range from 0.5 to 10 Hz, because of their very low reflectivity (De Pontieu *et al.*, 2001). Existence and origin of such waves at sufficiently high flux level are still a matter of speculation, since the temporal resolution of Doppler measurements in chromospheric spectral lines is not yet sufficient. But one can theoretically pursue the possibility of generating the thermal properties of *plages*, a brightening of the active region chromosphere, best observed in hydrogen and calcium lines, by heating through the damping of such waves entering from above.

Since in the most recent work on this subject (De Pontieu *et al.*, 2001) the subject of plage heating by high-frequency waves is not treated explicitly, I refer in the following to a preceding, still unpublished work by De Pontieu and myself. Here we investigated, in a one-dimensional model, the damping of a flux, F_w, of monochromatic Alfvén waves guided along the magnetic field and calculated the

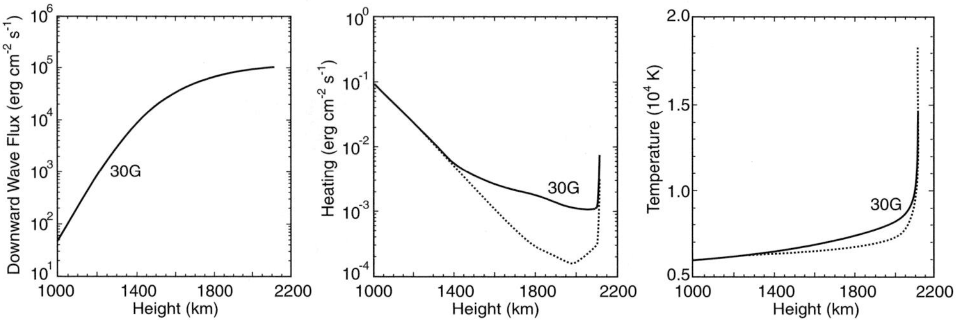

Figure 3. Model calculations of the absorption of a one-dimensional downward flux of Alfvén waves in the chromosphere and the related heating rate and temperature rise (Haerendel and De Pontieu, unpublished).

corresponding heating or pressure increase of the upper chromosphere. The crucial relations, apart from many details, are the hydrostatic equation:

$$\frac{dp}{dz} = -\frac{\mu}{kT}(g_0 + a_w)p \tag{5}$$

and the conservation of wave energy:

$$\frac{dF_w}{dz} = \frac{F_w}{\lambda_{\text{damp}}} \tag{6}$$

with

$$a_w = \frac{F_w}{2\rho_{\text{tot}} v_A \lambda_{\text{damp}}} \tag{7}$$

The downward directed wave force per unit mass, a_w, enhances the effective gravity and lowers the scale height. In addition to the above barometric and energy equations, the heat balance under radiative cooling has been incorporated in the model. Figures 3a-c show a set of results for the height distribution of F_w, dF_w/dz, and $T(z)$ for the following set of parameters: $f = 2$ Hz, $F_{w0} = 10^5$ erg/cm^{2}s and a magnetic field strength of 30 G. Observations show temperature increases in plage regions quite similar to the distribution in Figure 3b.

A possible origin of such fluxes of high-frequency Alfvén waves may be local reconnection processes between adjacent narrow twisted magnetic flux tubes in active coronal regions as exhibited in Figure 4. The jets emerging out of such local reconnection sites with Alfvén velocities of the order 1000 km/s will be braked within seconds or fractions thereof by "collisions" with adjacent strong fields. Their kinetic energy and momentum will be converted into heat and wave excitations. Without getting into details, one can thus easily imagine that such relaxations of a multitude of small-scale reconnection events constitute a powerful source of high-frequency Alfvén waves (see Haerendel, 1994). The demands on the wave energy

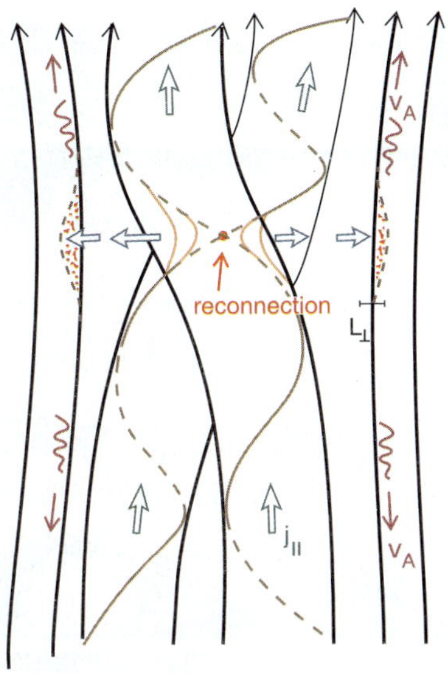

Figure 4. Generation of small-scale, high-frequency Alfvén waves from local reconnection events between adjacent twisted flux-tubes.

flux for the generation of plages by the above process are rather low in comparison with the overall energy flux out of the active corona (Withbroe and Noyes, 1977).

We know of a similar phenomenon in the magnetosphere, called *Alfvénic arcs.* In this type of rather irregular and faint aurora borealis, occurring near the polar cap border of the auroral oval, the origin of the particle fluxes is distinctly different from that underlying the brighter and more common auroral displays. Not yet fully explored, the sequence of events may be the following (Erlandsen *et al.*, 1994). Small-scale inertial Alfvén waves, with frequencies in the range one 1 to 10 Hz, deposit their energy in accelerating cold electrons of the topside ionosphere parallel to **B** to energies of several 100 eV. These highly anisotropic electrons, besides creating the optical emissions, excite electromagnetic ion cyclotron waves, which can heat the ionospheric ions transversely to **B**. Figure 5 shows energy-time spectra and pitch-angle distributions of such an event (Tung *et al.*, 2001). Not only H^+ ions are heated, but quite prominently also O^+ ions. A statistical study of Keiling *et al.* (2003) has shown that a substantial part of the dayside and nightside aurora is indeed powered, in addition to more stationary field-aligned currents, by the energy flux of transient Alfvén waves.

As hinted above, we are still lacking a full understanding of the exact energy transfer mechanisms. But the commonality with the process proposed for the origin of chromospheric plages is obvious. Electromagnetic energy flowing into the topside atmospheres from above, is coupled to the "cool" plasma, which is visibly

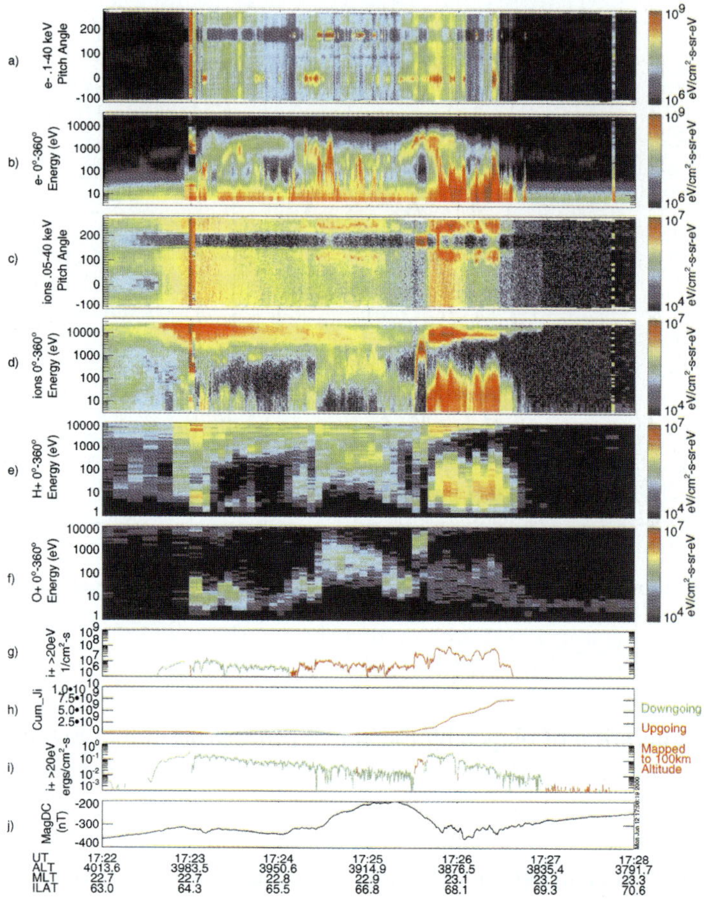

Figure 5. Energy spectra and angular distributions of electrons and ions in Alfvénic arcs, i. e. regions where small-scale Alfvén waves dominate the energy influx into the ionosphere (Tung *et al.*, 2001).

heated. The coupling processes though are different. In the case of the plages, it is ion-neutral collisions, in Alfvénic arcs, due to the much lower ionospheric density a complex nonlinear wave coupling chain takes place. Much more work has to be done on both phenomena.

5. Plasma Erosion and Cavitation

An obvious commonality between solar and magnetospheric phenomena exists in the light emissions caused by electrons energized in the outer plasma realms and precipitating on the denser atmospheric layers, at X-ray energies in the solar flares and in visible and UV light in the aurora borealis. There is another common effect. The deposition of energy from above not only causes heating and light emissions, but also erosion of the ionic component in the upper atmosphere and corresponding mass injections into the regions from which the energy is being derived, i.e. corona and magnetosphere. Post-flare loops, so beautifully appearing in the X-ray images of the TRACE satellite, bear witness thereof in the solar corona. In the magnetosphere, measurements with the Freja and, in particular, the FAST satellites have identified both, regions of upward as well as downward field-aligned currents in the ionosphere as the sources. Owing to the direct accessibility of the source regions in the magnetosphere, we have derived a much more detailed understanding of the underlying physical processes than can be claimed for the plasma injections into the post-flare loops. It is tempting to try to transfer some of the insights obtained near the Earth to the solar phenomena.

Ion outflows from ionosphere into the magnetosphere exist all around the auroral oval. In wintertime, they dominate in the evening sector. An interesting feature is that the velocity distributions of these ion streams are indicative of an energy transfer from the lighter and faster H^+ ions to the He^+ and O^+ ions (Moebius *et al.*, 1998). As a consequence of the plasma outflow, the topside ionosphere is noticeably eroded (Persoon *et al.*, 1988). The persistence of these density holes, well beyond the energy deposition time, has, among others, the consequence that the ionosphere-magnetosphere system develops a "memory" for the sites of energy deposition. This may be the reason for the relatively long existence and often-observed repetitiveness of individual auroral arcs.

Figure 6 displays the energetic particle as well as the background plasma components of such an auroral cavity above a wide region of night-time auroral precipitation with associated ion outflows, as measured by the FAST satellite (Mc Fadden *et al.*, 1999). Three features are remarkable, the energy concentrations of both, ions and electrons, at a few keV, the strong alignments with the direction of **B**, the electrons downward, the ions upward, and the very low total plasma density. All of this is in accordance with the existence of strong field-aligned potential drops and a satellite position somewhere in the middle of it. The upward directed electric field reflects and excludes cold ionospheric electrons from below, while ions are accelerated to form upward beams. Almost all of the plasma density is constituted by the beam components (s. lowest panel of Figure 6).

There are two different ways to interpret the working of the cavitation or erosion process. Ergun *et al.* (2000) created a 1-D numerical model for the interface between the high-density cold plasma of the ionosphere and the dilute hot magnetospheric plasma in a region of strong upward directed field-aligned current. Figure 7 shows the result. The total potential drop being prescribed, the model yields a two-step drop with strong parallel electric fields (two lower panels). The lower drop has been designated as the "electron transition layer", because here the cold electrons are being reflected, while the cold ions are accelerated upward, thus lowering the density. The upper drop has been called "ion transition layer", because here the hot magnetospheric ions are reflected, while the cold ion beams are further accelerated upward. In between the two drops is a very low-density cavity, dominated

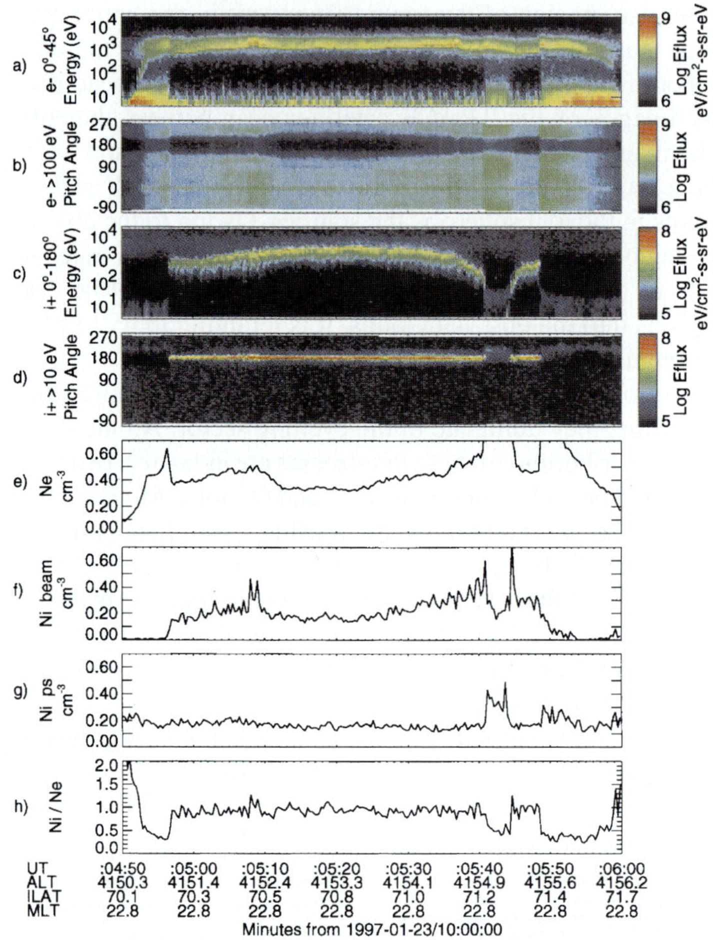

Figure 6. Energy spectra and angular distributions of the energetic particle components as well as the total plasma density inside an auroral cavity in a region of parallel potential drop (McFadden *et al.*, 1999).

by the counter-streaming beams and a trapped component of secondary or scattered electrons maintaining charge neutrality. These model particle distributions, admittedly subject to the choice of some open parameters, match very well the features emerging from the data displayed in Figure 6.

The other interpretation of the way, the concentrated electric potential fields above auroral arcs interact with the ionosphere, was given by the author (Haerendel, 1999). I took into account the U-shaped distribution of the potential, the observation that auroral arcs tend to move with respect to the background plasma (Frey *et al.*, 1996), and the observation of strong field-aligned transversely cold electrons are seen at the edge of many arcs (Boehm *et al.*, 1994). The main point is sketched in Figure 8. A sharp transverse gradient between dense plasma and cavity is postulated.

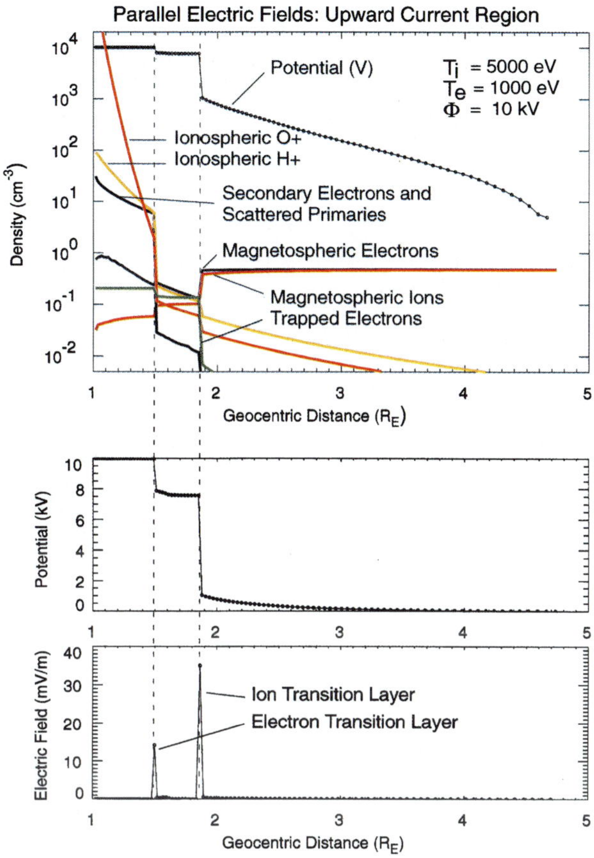

Figure 7. Densities of downward magnetospheric electron and ion fluxes as well as of upward accelerated ionospheric ions plus neutralizing trapped electrons in a 1-D model of field-aligned potential drops in the lower magnetosphere (Ergun *et al.*, 2000).

Here ions are extracted by the strong transverse field component, while the electrons are downward accelerated by the weak parallel component. The transition layer has a thickness of the order of the inertial length of the extracted(!) ions. As current sheet and auroral arc proceed through the dense plasma environment, the cold plasma is being eroded and upward ion beams are being formed. It should be noticed that both interpretations are not mutually exclusive. They address the physics on different transverse scales.

An erosion process must also occur when the chromosphere is strongly heated by the impact of high-energy electrons during solar flares. This is suggested by the amount of hot ions ($>10^6$ K) quickly filling the *flare loops*, which manifest themselves through thermally excited line emissions of certain highly-ionized ions like Fe XVI, Fe X, O VIII, or C IV. Traditionally, the fast filling of magnetic flux-tubes up to heights of tens of thousands of kilometers has been attributed to an overpressure

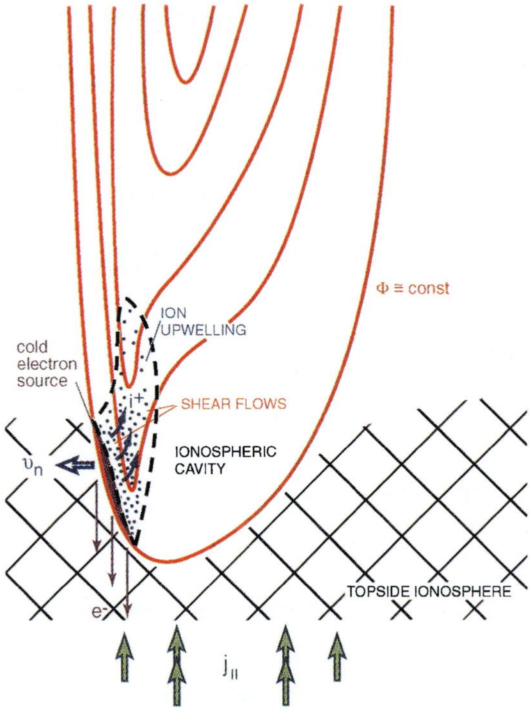

Figure 8. Erosion of the topside ionosphere due to transverse ion extraction and parallel electron acceleration at the leading edge of a progressing auroral acceleration region (U-shaped potential above an auroral arc) (Haerendel, 1999).

of the secondary electrons, which, upon expanding upward, drag the ions with them via ambipolar electric fields (Winglee, 1989). The resulting differential motions of the various ion species are largely wiped out by two-stream instabilities, further accelerating the heavier ion species. De Jager (1985) attributed the plasma erosion to the downward progression of a very thin ablation front heating the chromospheric gas to temperatures of 50 MK, as sketched in Figure 9. The expanding plasma is further heated by the lower energy electron component. The subsequent upward progression of the plasma into the coronal loops is mostly treated as a hydrodynamic outflow (e.g., Antonucci *et al.*, 1999).

While this scenario seems to explain satisfactorily the origin of the post-flare loops, there are a few difficulties arising from the disagreement of the observed temperatures with those needed to explain the upward injection process. Winglee *et al.* (1991) therefore postulated the existence of strong field-aligned currents and associated small-scale quasi-static electric fields aiding the plasma injection. These are pretty much the basic ingredients of the auroral acceleration and ion injection process. The necessary high current densities in the corona are attributed to a current filamentation process implying also the existence of balancing return currents.

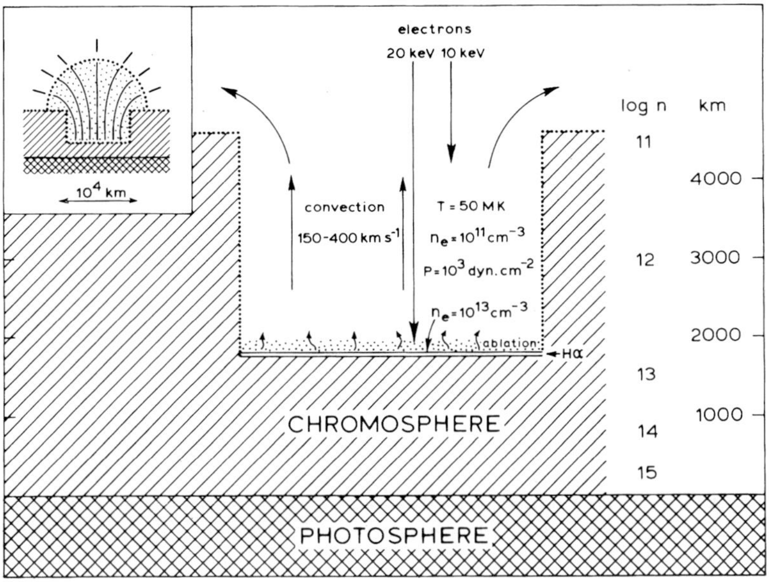

Figure 9. Model of an ablation front in which the chromospheric temperature is raised to 50 MK by the impact of energetic solar flare electrons (de Jager, 1985).

The latter theory is a striking example for the transfer of knowledge from near-Earth plasmas to another cosmic object. The chain of arguments is rather convincing, since it is hard to imagine that the downward progression of the energy, liberated by reconnecting magnetic fields in the corona, will proceed entirely via the energized particle component, whereas the transport of energy as Poynting flux in form of Alfvén waves can be ignored. As suggested in the context of the origin of plages (Section 4), it is highly likely that in flare events the powerful reconnection jets, when braked, serve also as current generators. Filamentation of such currents to smaller and smaller transverse scales and related high densities, for instance owing to multiple reflections, will proceed until some sort of anomalous dissipation sets in. I believe that the further exploration of the complex solar flare process will have to take into account such other routes of energy transfer thereby supporting the validity of the initial statement of this paper.

6. Final Remark

We have discussed three classes of phenomena in ionosphere and chromosphere and found strong commonalities as well as significant differences between them. Mechanical waves penetrating into these regions from below can set up density inhomogeneities by the action of a Pedersen current. However, in the chromosphere it is not the quasi-static current known to exist in the ionosphere, but rather a high-frequency relative. Alfvén waves from above can heat the ionospheric as well as the chromospheric plasmas, but the heating processes are probably very different, owing to the highly different coupling time-scales between neutrals and ions. Our third case is one for which the differing explanations for the erosion of cold plasma, caused by an influx of high energy from above, may in reality not be intrinsic to the different plasma regimes, but rather be an expression of the differing states of exploration. Thus, close comparisons of such seemingly similar, but not identical processes occurring in two different cosmic plasmas, have the potential of being more than an intellectual exercise, but rather serve as a creative tool to further our understanding.

References

Antonucci, E., Alexander, D., Culhane, J. L., de Jager, C., McNeice, P., Somov, B. U., *et al.*: 1999, in K. T., Strong, J. L. R. Saba, B. M. Haisch, and J. T. Schmelz (eds.), *The Many Faces of the Sun. A Summary of the Results From NASA's Solar Maximum Mission*, Springer, Berlin, p. 331.

Boehm, M. H., Paschmann, G., Clemmons, J., Haerendel, G., Eliasson, L., and Lundin, R.: 1994, *Geophys. Res. Lett.* **21**, 1895.

De Pontieu, B., and Haerendel, G.: 1998, *Astron. Astrophys.* **338**, 729.

De Pontieu, B., Martens, P. C. H., and Hudson, H. S.: 2001, *Astrophys. J.* **558**, 859.

De Pontieu, B., Erdélyi, R., and Stewart, S. P.: 2004, *Nature* **430**, 536.

Dungey, J.: 1959, *J. Geophys. Res.* **64**, 2188.

De Jager, C.: 1985, *Sol. Phys.* **98**, 267.

Erlandsen, R. E., Zanetti, L. J., Acuña, M. H., Eriksson, A. I., Eliasson, L., Boehm, M. H., *et al.*: 1994, *Geophys. Res Lett.* **21**, 1855.

Ergun, R. E., Carlson, C. W., McFadden, J. P., Mozer, F. S., and Strangeway, R. J.: 2000, *Geophys. Res. Lett.* **27**, 4053.

Frey, H. U., Haerendel, G., Knudsen, D., Buchert, S., and Bauer, O. H.:1996, *J. Atmosph. Terr. Phys.* **58**, 169.

Haerendel, G.: 1992, *Nature* **360**, 241.

Haerendel, G.: 1994, *Ap. J Supp. Series* **90**, 765.

Haerendel, G.: 1999, *Adv. Space Res.* **23**, 1637.

Haerendel, G.: 2001, in J. A. M., Bleeker, J. Geiss, and M. C. E. Huber (eds.), 'The Century of Space Science', Kluwer Academic Publ., Dordrecht, p. 1007.

Hines, C. O.: 1974, *Geophys. Monograph* **18**, Am. Geophys. Union, Washington D.C.

James, S. P., Erdélyi, R., and De Pontieu, B.: 2003, *Astron. Astrophys.* **406**, 715.

Keiling, A., Wygant, J. R., Cattell, C. A., Mozer, F. S., and Russell, C. T., 2003, *Science* **299**, 383.

McFadden, J. P., Carlson, C. W., Ergun, R. E., Klumpar, D. M., and Moebius, E.: 1999, *J. Geophys. Res.* **104**, 14,671.

Moebius, E., Tang, L., Kistler, L. M., Popecki, M., Lund, E. J., Klumpar, D., *et al.*: 1998, *Geophys. Res. Lett.* **25**, 2029.

Persoon, A. D. G., Peterson, W. K., Waite, J. J. H., Burch, J., and Green, J. L.: 1988, *J. Geophys. Res.* **93**, 1871.

Piddington, J.: 1956, *MNRAS* **116**, 314.

Tung, Y.-K., Carlson, C. W., McFadden, J. P., Klumpar, D. M., Parks, G. K., Peria, W. J., *et al.*: 2001, *J. Geophys. Res.* **106**, 3603.

Winglee, R. M.:1989, *Astrophys. J.* **343**, 511.

Winglee, R. M., Kiplinger, A. L., Zarro, D. M., Dulk, G. A., and Lemen, J. R.: 1991, *Astrophys. J.* **375**, 366.
Withbroe, G.L., and Noyes, R. W.: 1977, *Ann. Rev. Astron. Astrophys.* **15**, 363.

SHOCKS: COMMONALITIES IN SOLAR-TERRESTRIAL CHAINS

STEVEN J. SCHWARTZ
Blackett Laboratory, Imperial College London, South Kensington, London SW7 2BW, United Kingdom
(E-mail: s.schwartz@imperial.ac.uk)

(Received 25 May 2005; Accepted in final form 14 September 2005)

Abstract. Shocks are found throughout the heliosphere, wherever supersonic (or super-magnetosonic) flows encounter obstacles or other, slowly moving, media. Although some of the physical parameters are in different regimes, all shocks heat and decelerate the media incident upon them. Most shocks must propagate in a collisionless plasma, thereby adding importance to the particle interactions with the electromagnetic fields, and enabling some particles to be accelerated to high energies. This paper explores the commonalities, and differences, in shocks throughout the heliosphere, and concentrates on the role of shock microstructure in effecting the shock transition and in governing the resulting energy partition amongst the constituent species. Shocks play a significant role in the solar-terrestrial chain.

Keywords: shock waves, interplanetary medium, plasmas

1. Introduction

Shocks are found throughout the heliosphere, wherever supersonic (or super-magnetosonic) flows encounter obstacles or other, slowly moving, media, or under driven conditions which steepen to effect such a supersonic configuration. Near the Sun, explosive events drive shock waves in the solar corona which then propagate out into the interplanetary medium. Slow solar wind streams, comets, planets, and the interstellar medium all present obstacles to the solar wind flow which result in shock formation and propagation. Since the Earth's magnetosphere responds to dynamic changes in the solar/solar wind input, shocks play a significant role in mediating and influencing the energy and momentum balance. The solar wind that actually hits the magnetosphere has been shocked at least once, at the Earth's bow shock.

Throughout most of the heliosphere, collisions between particles are negligibly rare. At 1 AU, for example, the collisional mean free path of a proton or electron is approximately 1 AU. Thus, we must turn to collisionless processes to investigate the formation, structure, and propagation of heliospheric shock waves. This opens up new avenues for the plasma to (re)distribute energy and momentum, leading to, for example, the acceleration of some particles to high energies, and to the differential heating of different species.

Space Science Reviews (2006) 124: 333–344
DOI: 10.1007/s11214-006-9093-y

In this brief article, I will explore some of the commonalities of shocks in the heliosphere, so that we may apply what we have learned from the very detailed studies of the Earth's bow shock, for example, to shocks deep in the solar corona. We shall need to be careful to address the key differences as well. The remainder of the article then focuses on key, new results.

2. Why Are There Shock Waves?

Shock waves arise whenever information about an obstacle in a flow needs to be communicated to that flow faster than a characteristic information or wave speed (i.e., the sound speed in a simple fluid, the fast magnetosonic speed in an MHD plasma, etc.). The ratio of this required communication speed to the linear wave speed is known as the Mach number, M, and is a key parameter of shock physics. The medium must support some nonlinear steepening or compression, which can be balanced by either dissipative processes (removing and diffusing the energy) or dispersive ones (radiating away energy or otherwise dispersing it away from the steepened profile). Under collisionless regimes, a variety of processes can participate to bring about the shock transition, and can be fit into this rather simple paradigm of steepening vs. dissipation or dispersion.

There are at least four generic circumstances that give rise to shocks in the solar-terrestrial chain:

Transients such as solar flares, Coronal Mass Ejections, and explosive reconnection drive shock waves through the impulsive injection of energy into the surrounding medium.

Fast streams in the solar wind overtake slow streams to form a stream-stream interface, as shown in Figure 1. The impact drives a forward shock ahead of the interface, and a reverse shock may propagate sunward relative to the interface.

Magnetised obstacles (such as most planets) form magnetospheres around which the solar wind must be deflected. The planetary magnetic field prevents (to a good approximation) exchange of mass and launches a bow shock that stands in the upstream flow, as sketched in Figure 2.

Unmagnetised obstacles such as comets may interact with the solar wind through the presence of an atmosphere. Newly ionised particles mass-load the solar wind flow and lead to a deceleration which may be sufficiently rapid to form a steepened structure. In this case, the "information speed" may be ill-defined.

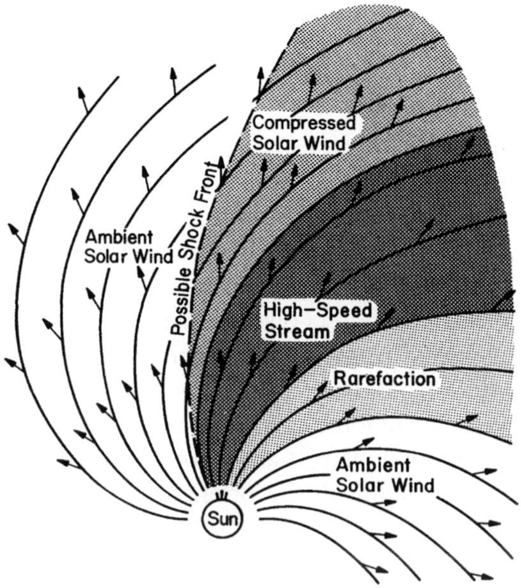

Figure 1. Formation of a stream interface by a high speed solar wind stream. A forward shock is launched ahead of the contact surface, while a reverse shock may propagate sunward (relative to the fast flow) due to the need to slow down that flow (from Hundhausen, 1972).

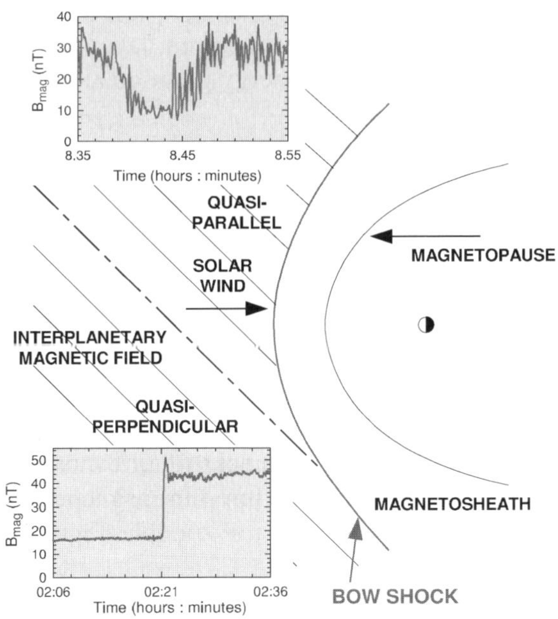

Figure 2. Formation of a bow shock upstream of the Earth. The orientation of the interplanetary magnetic field plays a central role in determining the structure and microphysics at the shock (from Balogh *et al.*, 2005).

3. Commonalities and Differences

3.1. Common Aspects

Many of the shock environments described above share common elements and parameters, making it fruitful to consider them together. For example, outside a few solar radii, the interplanetary medium is practically free of particle collisions, especially on the scale of the shock transition, so that most shocks in the heliosphere are collisionless. In all cases, shocks are important for the overall heating and deceleration of the plasma; that is, they effect overall changes in the bulk parameters of the medium. Being collisionless, however, they are also important "non-thermalisers," resulting in the acceleration of some particles to high energies, and to the partition of energy amongst the different ionic species and electrons. For many applications, this partition is of fundamental interest and importance.

Finally, as sketched in Figure 2, the orientation of the magnetic field in the unshocked medium controls the (collisionless) behaviour of the particles incident on the shock. When the field is nearly perpendicular to the shock normal ("quasi-perpendicular") it serves to bind the particles to within one gyroradius and results in a relatively smooth, laminar shock transition. By contrast, under quasi-parallel conditions, particles are free to traverse the shock in both directions, giving rise to an extended foreshock region filled with shock-modified particles and generally more dis-ordered in structure. Shocks change the direction and magnitude of the magnetic field, and thus are necessarily current layers. Their current-carrying properties are linked to the problem of energy partition and dissipation within the shock.

3.2. Different Aspects

Despite these similarities, not all heliospheric shocks are the same. The Mach number of the shocks range from very weak $M \sim 1$ interplanetary shocks to very strong $M > 10$ CME-driven or outer heliospheric planetary bow shocks. The Mach number controls the dissipation required at the shock. Low Mach number, "sub-critical" shocks can, in theory, be brought into steady state through the action of electrical resistivity alone. Super-critical shocks require more dissipation than can be accomplished by the (current-limited) jump in magnetic field, and so require ion "viscosity" or equivalent process (see, for example, Figure 8 of (Kennel *et al.*, 1985)).

Shocks driven by explosive events (e.g., solar flares or CME eruptions) tend to be stronger near the Sun and weaken with increasing heliocentric distance as they run out of energy. Continuously driven shocks, such as those associated with Corotating Interaction Regions (where fast solar wind streams collide with slower solar wind) and standing shocks at planets, are more stationary.

The plasma β (ratio of thermal to magnetic pressures) also varies from $\ll 1$ in the corona to ~ 1 at 1 AU. In collisionless shocks, the plasma β controls the shock's ability to differentiate different segments of the population, e.g, to reflect a portion of the ions (see below). Low β shocks tend to be unsteady.

Obstacles with a neutral atmosphere can mass load the solar wind through ionisation. This alters significantly the solar wind-obstacle interaction at comets and un-magnetised or weakly magnetised planets.

Finally, the scale and curvature of a shock can influence a variety of processes. Scales comparable to the gyroradius of the various particle species can give rise to new kinetic effects if particle gyration removes (or adds) particles from specific locations. Shock curvature enables particles from one domain, e.g., quasi-perpendicular portions, to contaminate other regions thereby modifying the plasma conditions experienced locally by the shock. Large-scale shocks may be modified by shock-accelerated particles ("cosmic ray mediated shocks") which escape too quickly at smaller shocks to be of any major consequence. Additionally, the relative lack of escape at larger scale shocks, including CME/CIR shocks and the termination shock, means these shocks can accelerate particles to higher energies than their planetary brethren.

4. Shock Microphysics

4.1. Ion Microphysics

In the absence of collisions, shocks effect the transition from fast, cool, unshocked plasma to slow, heated, and deflected shocked plasma by the influence of the electromagnetic fields on the particles. For ions, the most important consequence is the reflection of a fraction of the ion population at the shock front. Under quasi-perpendicular conditions, subsequent particle gyromotion returns these reflected particles back to the shock front where, having picked up energy due to their tangential motion parallel to the upstream $-\mathbf{V} \times \mathbf{B}$ electric field, they then overcome the field barrier and enter the downstream region. As revealed by the sketch in Figure 3, the result in the downstream region, and even in the shock "foot" formed by the reflected ions in the upstream region, is a multiple component ion distribution that has an overall reduced bulk flow speed and an enhanced "thermal" spread relative to that bulk flow. This is precisely what the shock needs to do, and all that remains is for turbulent scattering or some other process to mix these populations. The result is a fairly laminar transition with a relatively broad foot region and steeper ramp as shown in the inset in Figure 2.

Under quasi-parallel conditions (when the angle θ_{Bn} between the shock normal and upstream magnetic field is $<45^\circ$), such reflected particles do not return to the shock front but have guiding centre motion directed into the upstream region. There, the resulting counterstreaming plasma populations give rise to plasma instabilities

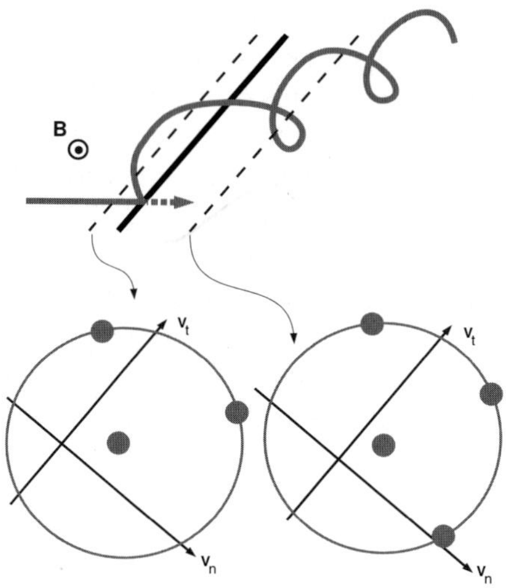

Figure 3. Ions incident at a quasi-perpendicular shock are divided into a directly transmitted component (dotted arrow) and a reflected, gyrating component (top). When viewed in velocity space (bottom), the multiple components have a reduced overall bulk flow and enhanced spread relative to that flow, thereby effectively decelerating and heating the incident plasma. (after Figure 1 of Sckopke *et al.* (1983)).

which scatter the particles and result in large amplitude magnetic fluctuations, as seen in the inset to Figure 2. Further encounters with the shock can lead to an efficient first-order Fermi acceleration of some of these particles. We shall see below in Section 6 that the Earth's bow shock is a very good laboratory for studying this process. The result is an extended foreshock region filled with field fluctuations and energised particles. The actual shock transition is accomplished not across a simple, planar boundary, but by the growth of Short Large Amplitude Magnetic Structures (SLAMS), as sketched in the cartoon of Figure 4 and discussed in more detail below.

4.2. Electron Microphysics

Owing to their small mass, the electron thermal speed is typically larger than the solar wind speed and shock speeds in the heliosphere. Thus electrons are free to traverse the shock in both directions, subjected only to the electromagnetic fields self-consistently established to maintain quasi-neutrality and to ensure that the total changes across the shock in mass-, momentum-, and energy-fluxes are zero. Typically, the electrons account for approximately 25% of the total shock heating (Schwartz *et al.*, 1988). The electrons help to establish a cross-shock electric potential (Feldman *et al.*, 1983; Scudder *et al.*, 1986; Scudder, 1995) that (a) prevents too many heated downstream electrons from escaping upstream (b) contributes to the reflection of incident ions, and (c) inflates the phase space distribution of electrons.

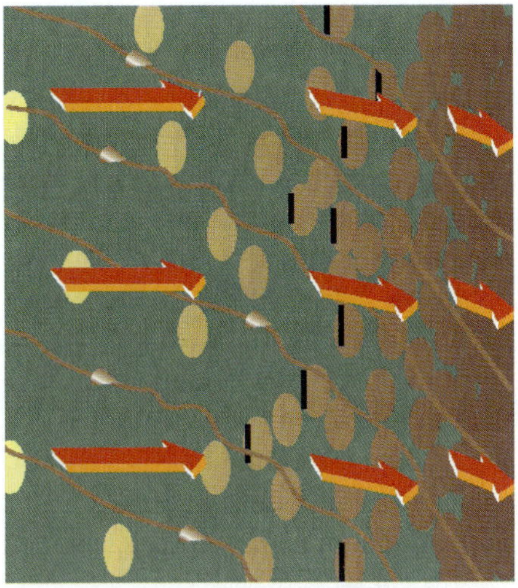

Figure 4. The transition from unshocked to shocked plasma regimes under quasi-parallel field conditions by the growth, convection, and action of Short Large Amplitude Magnetic Structures (SLAMS). (after Schwartz and Burgess (1991)).

If the scales within the shock layer are large enough, this potential can account for essentially all the electron heating (Scudder *et al.*, 1986). However, if the variations in the shock electric field are less than an electron gyroscale, the electrons can become (partially) demagnetised, leading to significantly more electron heating (Lembège *et al.*, 2004; Balikhin *et al.*, 1993). Recent evidence suggests that the shock magnetic ramp scales with ion Larmor radii while electric field spikes within that ramp have scales of a few electron inertial lengths. These results are shown in Figures 5 and 6. This area of shock scales, electron dynamics, and electrostatic potentials is one where future work is needed.

5. Shock Variability

Many applications of shock phenomena treat shocks as steady, one-dimensional discontinuities. It is therefore of some interest to explore how close real shocks are to this idealisation. We have already alluded in Section 3.2 that gross curvature of shock waves can influence the interplay between macroscopic (curvature) and microscopic (particle) processes. Additionally, temporal variability can play a role in a variety of shock processes.

5.1. Quasi-Perpendicular Shocks

High Mach number, quasi-perpendicular shocks are intrinsically unstable, alternating between nearly 100 % reflection of the incident population and none (Hellinger

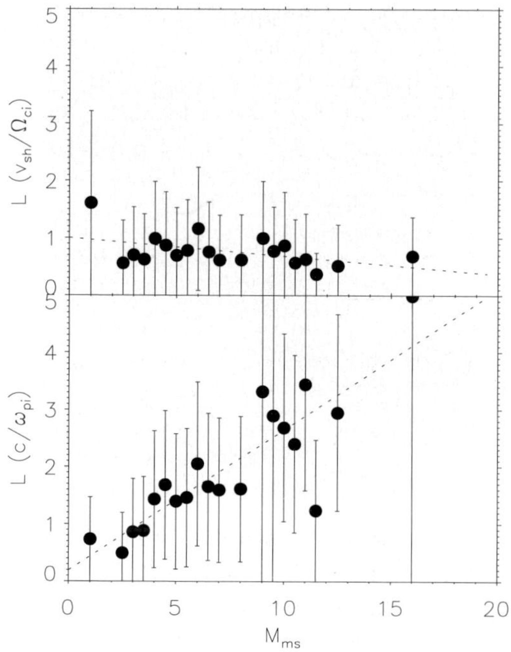

Figure 5. Scale lengths L of hyperbolic fits to the density profile (which mimics the magnetic field profile) of many shocks, as determined by the Cluster spacecraft, from Bale *et al.* (2003). The top panel shows this scale to be roughly constant with the convected gyroscale V_{sh}/Ω_{ci} while increasing relative to the ion inertial length c/ω_{pi}.

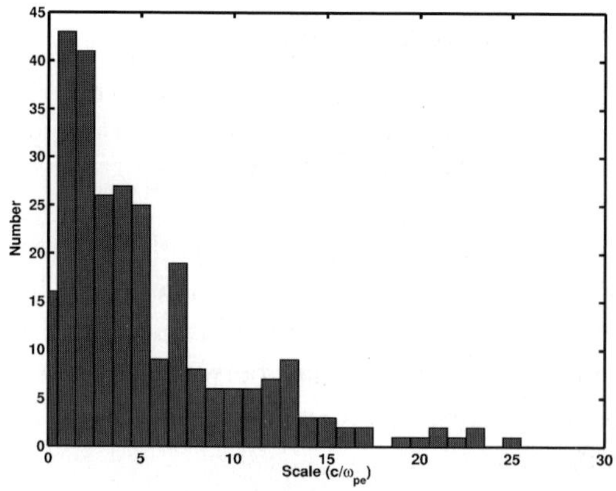

Figure 6. Histogram of the scale sizes of electric spikes within shock transitions as determined using the Cluster spacecraft, from Walker *et al.* (2004). This scale, while not necessarily the same as that of the overall cross-shock potential, is much smaller than the magnetic ramp scale.

et al., 2002) as the shock proceeds through various reformation cycles. Even at lower Mach numbers, quasi-perpendicular shocks can show variability in their profile which is confined largely to the foot region (Horbury *et al.*, 2001). The ramp, "overshoot/undershoot" and downstream oscillations remain coherent.

Another piece of evidence suggesting that the quasi-perpendicular shock is generally well-behaved comes from a study by Horbury *et al.* (2001) of the local orientation of the shock normal, derived directly from 4-spacecraft measurements using Cluster, compared to normals based on historical, empirical models. These agree to within 10° for a large sample of shock crossings, whereas single spacecraft techniques (minimum variance, coplanarity, etc.) are typically much worse. Since the Cluster spacecraft were within a shock foot-distance of one another during this period, this shows that even on small scales, the quasi-perpendicular shock shows little spatial variability, at least in terms of overall orientation.

5.2. Quasi-Parallel Shocks

We know that quasi-parallel shock regions are populated by energised ions, SLAMS, and other turbulence, so it is not surprising here that variability is the norm rather than the exception. This variability appears different in different parameters (e.g., magnetic field and electric potential as shown in Figure 7). Worse, although SLAMS are roughly $1R_e \approx 6400$ km in scale (15 seconds at 300 km/s convection speeds), even over scales one tenth as large (600 km) they appear quite structured, as shown in Figure 7. Lucek *et al.* (2004) argue that this structure is spatial rather than

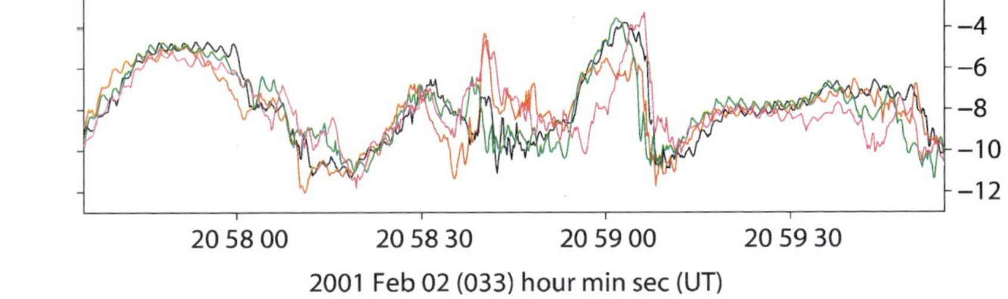

Figure 7. Short Large Amplitude Structures (SLAMS) at a quasi-parallel shock observed by the four Cluster spacecraft with typical separations of ~600 km in magnetic field (top) and electric potential (bottom) data. Even at this small scale, SLAMS show very different features implying that they are quite filamentary despite their overall scale ~6000 km (from Lucek *et al.* (2004)).

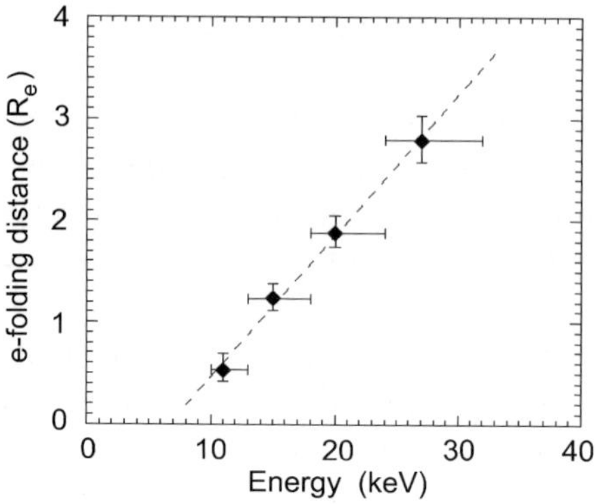

Figure 8. E-folding distance of energised particles in the quasi-parallel foreshock as a function of energy (from Kis *et al.* (2004)).

temporal in nature, implying that SLAMS are quite filamentary in nature. Only at scales $\sim$ 100 km do SLAMS show more spatial coherence.

6. Diffuse Ion Acceleration

Upstream of the quasi-parallel shock is a foreshock region populated by back-streaming, energised particles. Scattering due to convected turbulence upstream and SLAMS or slower downstream turbulence provides a natural environment for first order Fermi acceleration. Recent tests of this scenario have been made using Cluster to measure spatial gradients in the energetic particles by Kis *et al.* (2004). In the Fermi picture, the density of particles of an energy E should fall off exponentially with distance upstream of the shock over a diffusive scale $L(E) = \kappa(E)/V_{\rm solar\ wind}$ where κ is the diffusion coefficient due to the upstream turbulence. Kis *et al.* measured this gradient directly for several energies. They noted firstly that the gradient was almost exactly exponential. The scalelength L for particles with 24–32 keV was approximately $2.4R_e$, somewhat smaller than previous estimates, perhaps due to the unusually large solar wind speed for the period used in their study. Moreover, that scale increased approximately linearly with energy, as shown in Figure 8.

This study shows conclusively that the foreshock processes are diffusive, and by inference that the Fermi mechanism must be present. The energy-dependent scattering length is presumably related to the efficiency in self-exciting the necessary turbulence, although a detailed comparison using the observed turublent fields has not yet been carried out for this event.

7. Conclusions

Shocks are found throughout the heliosphere wherever fast flows collide with other flows or with obstacles. Most heliospheric shocks are collisionless, enabling them to be important for particle acceleration and other kinetic processes in addition to their more traditional role in modifying the bulk flow parameters. In terms of STP chains, shocks participate in coronal processes (heating, solar wind acceleration, etc.), interplanetary processes (further particle acceleration), and in pre-conditioning the solar wind prior to its impact on the magnetosphere (the bow shock). Much of what we know and have learned in recent years is based on detailed studies of the Earth's bow shock.

At the macroscopic level, the bow shock and interplanetary shocks behave much as the idealised fluid models of steady, locally 1-D, discontinuities would anticipate. However, at the microscopic level, and in terms of particle acceleration, shocks are neither steady nor dull. The self-consistent establishment of shock profiles in magnetic and electric fields partitions the incident energy flux amongst the thermal and non-thermal particle populations. While this brief paper has sketched some of the basic properties and new results, there remains much work to be done exploiting numerical simulations, existing datasets, and new missions. Extrapolation to parameter regimes relevant to the corona and, at the other extreme, to extra-heliospheric phenomena, remains a challenge.

Acknowledgements

Much of the material in this paper is drawn from the special issue (Vol. 118:1–4) of Space Science Reviews, published in parallel as Vol. 20 of the Space Science Series of ISSI, devoted to Cluster results at the dayside boundaries. Four chapters of that book are devoted to the bow shock, and I am grateful to all the authors for the material deposited therein. I thank ISSI for their warm hospitality and scientific support.

References

Bale, S. D., Mozer, F. S., and Horbury, T. S.: 2003, *Physical Review Letters* **91**(26), 265004.

Balikhin, M., Gedalin, M., and Petrukovich, A.: 1993, *Physical Review Letters* **70**, 1259.

Balogh, A., Schwartz, S. J., Bale, S. D., Balikhin, M. A., Burgess, D., Horbury, T. S., *et al.*: 2005, in G. Paschmann, S. J. Schwartz, C. P. Escoubet, and S. Haaland (eds.), *Outer Magnetospheric boundaries: Cluster results*, Vol. 20. ISSI, p. 155.

Feldman, W. C., Anderson, R. C.,Bame, S. J., Gary, S. P., Gosling, J. T., McComas, D. J.: 1983, *J. Geophys. Res.* **88**(17), 96.

Hellinger, P., Trávnícek, P., and Matsumoto, H.: 2002, *Geophys. Res. Lett.* **29**, 87.

Horbury, T. S., Cargill, P., Lucek, E. A., Balogh, A., Dunlop, M. W., Oddy, T., *et al.*, *Annales Geophysicae* **19**, 1399.

Hundhausen, A. J.: 1972, *Coronal Expansion and Solar Wind*. New York: Springer.

Kennel, C. F., Edmiston, J. P., and Hada, T.: 1985, in R. G. Stone and B. T. Tsurutani (eds.), *Collisionless Shocks in the Heliosphere: A Tutorial Review*. American Geophysical Union, pp. 1.

Kis, A., Scholer, M., Klecker, B., Möbius, E., Lucek, E., Réme, H., *et al.*: 2004, *Geophys. Res. Lett.* **31**, L20801, doi:10.1029/2004GL020759.

Lembège, B., Giacalone, J., Scholer, M. Hada, T., Hoshino, M., Krasnoselskikh, V., Kucharek, H., *et al.*: 2004, *Space Science Reviews* **110**, 161.

Lucek, E. A., Horbury, T. S., Balogh, A., Dandouras, I., and Rème, H.: 2004, *Annales Geophysicae* **22**, 2309.

Schwartz, S. J., and Burgess, D.: 1991, *Geophys. Res. Lett.* **18**, 373.

Schwartz, S. J., Thomsen, M. F., Bame, S. J., and Stansberry, J.: 1988, *J. Geophys. Res.* **93**(12), 12923.

Sckopke, N., Paschmann, G., Bame, S. J., Gosling, J. T., and Russell, J. T.: 1983, *J. Geophys. Res.* **88**(17), 6121.

Scudder, J. D.: 1995, *Adv. Space Res.* **15**, 181.

Scudder, J. D., Mangeney, A., Lacombe, C., Harvey, C. C., and Wu, C. S.: 1986, *J. Geophys. Res.* **91**(10), 11075.

Walker, S., Alleyne, H., Balikhin, M., Andre, M., and Horbury, T.: 2004, *Annales Geophys.* **22**, 2291.

THEORY AND SIMULATION OF RECONNECTION

In memoriam Harry Petschek

J. BÜCHNER

Max-Planck-Institut für Sonnensystemforschung, Max-Planck Str. 2, 37191 Katlenburg-Lindau, Germany

(E-mail: buechner@mps.mpg.de)

(Received 3 October 2005; Accepted in final form 12 March 2006)

Abstract. Reconnection is a major commonality of solar and magnetospheric physics. It was conjectured by Giovanelli in 1946 to explain particle acceleration in solar flares near magnetic neutral points. Since than it has been broadly applied in space physics including magnetospheric physics. In a special way this is due to Harry Petschek, who in 1994 published his ground breaking solution for a 2D magnetized plasma flow in regions containing singularities of vanishing magnetic field. Petschek's reconnection theory was questioned in endless disputes and arguments, but his work stimulated the further investigation of this phenomenon like no other. However, there are questions left open. We consider two of them – "anomalous" resistivity in collisionless space plasma and the nature of reconnection in three dimensions. The Cluster and SOHO missions address these two aspects of reconnection in a complementary way – the resistivity problem *in situ* in the magnetosphere and the 3D aspect by remote sensing of the Sun. We demonstrate that the search for answers to both questions leads beyond the applicability of analytical theories and that appropriate numerical approaches are necessary to investigate the essentially nonlinear and nonlocal processes involved. Necessary are both micro-physical, kinetic Vlasov-equation based methods of investigation as well as large scale (MHD) simulations to obtain the geometry and topology of the acting fields and flows.

Keywords: solar magnetic activity, magnetic reconnection, acceleration of particles, electrical resistivity, plasma instabilities, magnetohydrodynamics – MHD, plasma waves and turbulence, numerical simulation

1. Introduction

Magnetic reconnection is a major commonality of solar and magnetospheric plasmas. Ronald Gordon Giovanelli (1946) first conjectured that atoms could be excited at neutral points of sunspot magnetic fields to cause the (flare-) emission of photons. Later Jim Dungey (1953) called such process reconnection – another widely used term is "magnetic merging" (Vasyliunas, 1975). Reconnection requires a decoupling of plasma and magnetic field, i.e. the plasma has to be locally non-ideal. In other words, the frozen-in condition, valid in most space plasmas has to be violated locally. In the important first quantitative reconnection models (Parker, 1957; Sweet, 1958) the energy conversion rate, also called "reconnection rate", strongly depends on the electrical resistivity in the non-ideal plasma region. In the hot and dilute plasma of the solar corona, in magnetospheres and in astrophysics, however, the resistivity due to binary particle collisions is small. For

DOI: 10.1007/s11214-006-9094-x

a fluid description the electrical resistivity can be quantified by the dimensionless magnetic Reynolds number $R_m = Lvc^2/\eta\epsilon_o$. The magnetic Reynolds number compares the resistive scale length $\eta c^2 \epsilon_o / v$, where v is a typical plasma flow velocity, η is the electrical resistivity, c the speed of light and ϵ_o the vacuum dielectric constant, with the size of the dissipative region L. If the Alfvén speed $v_A = c\,B\,\sqrt{\epsilon_o / n M_i}$ is taken as the characteristic velocity $R_m(v_A)$ is frequently called also Lundquist number. For the Spitzer (1956)-resistivity due to binary particle collisions the magnetic Reynolds number in space plasmas is usually many orders of magnitudes too large to explain the fast observed time scales by Parker-Sweet reconnection.

In his ground-breaking paper of 1964 Harry Petschek suggested an alternative reconnection model. He suggested that slow and intermediate MHD waves could appreciable decrease the size of the nonideal plasma region, effectively shrinking, this way the Reynolds number by decreasing L. With only a small nonideal region of vanishing magnetic field around an X-point (in 2D) instead of an extended resistive current sheet, the reconnection rate could cease to depend strongly on the resistivity (see Petschek, 1964, 1999). Harry Petschek's theory essentially encouraged the application of the reconnection concept in space physics and the search for its evidences – remotely at sun and *in situ* in magnetospheres. However, despite endless disputes and arguments it is not obvious, whether the fast "Petschek reconnection" at a rate, which practically does not depend on resistivity, really exists in nature. Two-dimensional resistive MHD simulations, for example, reveal elongated current sheets instead of X-points. The formation of current sheets slows the reconnection rate down and makes it strongly dependent on resistivity (Biskamp, 1986). Anyway, though his reconnection rate weakly depends on resistivity Harry Petschek was aware of the problem that binary particle collisions cannot provide sufficient dissipation in the hot and dilute space plasmas.

Beyond the single-fluid theory there are more possibilities to balance the electric field in current carrying systems. This can be seen already in a two-fluid-approach, where Hall currents due to the different mass of electrons and ions, inertial effects due to the finite mass of the current carriers and off-diagonal terms of the pressure tensor due to non-gyrotropic particle orbits come into play. Another way to finite dissipation is the scattering of the current carriers at their own, collectively generated electric field fluctuations. For a weak fluctuation level a quasi-linear theory can describe the resulting wave-particle interaction (Dupree, 1970, see also Galeev and Sagdeev, 1984). Unfortunately it was found that the quasilinear, weak turbulence approach to the two most appropriate plasma instabilities – lower hybrid drift and ion-acoustic – does not provide the amount of turbulent dissipation needed for reconnection in space (Coroniti, 1985). Cluster observations have now shown that the amplitudes of electric field fluctuations can be much higher than the level predicted by the quasi-linear theory. This means that strongly nonlinear effects such as holes in the phase space might take place. Due to their strongly nonlinear character the excitation of such holes and their interaction with the current carriers, leading to enhanced collisionless plasma resistivity, have to be investigated by means of kinetic computer simulations. In Section 2 we discuss the current understanding of collisionless resistivity for reconnection in space plasmas based on numerical simulation results.

A second open question left after Harry Petschek's work is the rate of reconnection in three dimensions (3D). It appeared that before clarifying this point, even more basic questions about reconnection in 3D have to be solved first: those about its three-dimensional geometry and topology. One particular, still open question, is whether the 3D generalization of X-points, nulls, do matter in the real space and if so, to what extent. The complexity of reconnection in three dimensions became especially obvious after the high resolution solar observations by SOHO, TRACE and RHESSI. Also, in contrast to magnetospheric reconnection its solar counterparts are driven from the photosphere, where the footpoints of reconnecting magnetic fluxes are anchored and moved together with the highly resistive photospheric plasma. Numerical simulations are necessary to investigate the potential reconnection configuration, now with respect to its large-scale geometrical and topology aspects. In Section 3 we discuss possible 3D configurations including finite-B-reconnection without nulls and separatrices.

2. Collisionless, or as it is Often Called, "Anomalous" Resistivity

Current energy dissipation due to wave-particle interaction is the normal dissipation mode in collisionless plasmas. By historical reasons it is often called "anomalous" in contrast to dissipation due to binary particle collisions, important in plasmas with a small "plasma parameter" $N_p = n\lambda_D^3 > 1$. Here n is the number density and $\lambda_D = v_t/\omega_{pe}$ the Debye length, $v_t = \sqrt{\kappa T/m_e}$ is the electron thermal velocity, $\omega_{pe} = \sqrt{ne^2/\epsilon_o m_e}$ the electron plasma frequency, m_e, e the electron's mass and charge, respectively. The efficiency of binary collisions ceases in the hot and dilute space plasmas for which $N_p \approx 10^{13}$–10^{19}. Such plasmas are called collisionless, the particles' momentum exchange is controlled by their interaction with turbulence and wave fields. Collisionless plasmas are described by a Boltzmann-equation for distribution functions $f(\mathbf{r}, \mathbf{v}, t)$ with no explicit "collision term" on the right-hand-side. Instead, the resulting Vlasov equation self-consistently considers the action of the mean electromagnetic fields on the particles (Vlasov, 1938):

$$\frac{\partial f_j}{\partial t} + \mathbf{v} \cdot \frac{\partial f_j}{\partial \mathbf{r}} + \frac{e_j}{m_j}(\mathbf{E} + \mathbf{v} \times \mathbf{B}) \cdot \frac{\partial f_j}{\partial \mathbf{v}} = 0. \tag{1}$$

where the subscript j denotes the particle species (electrons, ions). In order to determine the action of the collectively created mean electromagnetic fields ($\mathbf{E}$, $\mathbf{B}$) on the particle distribution self-consistently, it is appropriate to split distribution functions and fields into an ensemble-averaged part (denoted $\langle\rangle$) and the deviations caused by interaction, i.e. $f_j = \langle f_j \rangle + \delta f_j$, $\mathbf{E} = \langle \mathbf{E} \rangle + \delta \mathbf{E}$ and $\mathbf{B} = \langle \mathbf{B} \rangle + \delta \mathbf{B}$.

By definition the ensemble averages of the deviations vanish, i.e. $\langle \delta f_j \rangle = \langle \delta \mathbf{E} \rangle = \langle \delta \mathbf{B} \rangle = 0$. From Equation (1) one obtains the following equation for the ensemble averaged distribution function $\langle f_j \rangle$, with a right hand side that contains second order correlations between the fluctuations:

$$\frac{\partial \langle f_j \rangle}{\partial t} + \mathbf{v} \cdot \frac{\partial \langle f_j \rangle}{\partial \mathbf{r}} + \frac{e_j}{c m_j} (\mathbf{v} \times \langle \mathbf{B} \rangle) \cdot \frac{\partial \langle f_j \rangle}{\partial \mathbf{v}} = \left(\frac{\partial f_j}{\partial t} \right)_{\text{eff}} = - \frac{e_j}{m_j} \left\langle (\delta \mathbf{E} + \mathbf{v} \times \delta \mathbf{B}) \cdot \frac{\partial \delta f_j}{\partial \mathbf{v}} \right\rangle \tag{2}$$

The right-hand-side of Equation (2) can be considered as an effective collision term which has to be evaluated. The r.h.s of Equation (2) does not vanish only if the field and particle fluctuations are correlated. One obtains an effective momentum exchange rate by multiplying Equation (2) with $m_j v_y$ and integrating over $\mathbf{v}$. Assuming that y is the direction of a current flow, Equation (2) yields for the effective collision rate for particles, drifting in the current direction,

$$\nu_{\text{eff,j}} = \frac{1}{n_j m_j \langle v_{y,j} \rangle} \left\langle \frac{\partial}{\partial t} n_j m_j v_{y,j} \right\rangle_{\text{eff}} = \langle \delta E_y \delta \rho_j + [\delta \mathbf{j}_j \times \delta \mathbf{B}]_y \rangle \tag{3}$$

where $\langle v_{y,j} \rangle = v_{dj}$ is the drift velocity of the particles, giving rise to the current. Wave particle scattering causes collisionless current dissipation, i.e. resistivity, if the motion of the current carriers is slowed down. Their kinetic energy can be transferred to the waves which grow and in a second wave-particle interaction may broaden the particle distribution function, i.e. heat the plasma by dissipating the kinetic energy of the drift to a small scale chaotic motion. Such transfer corresponds to positive effective, sometimes called "anomalous", collision rates $\nu_{\text{eff},j}$. Negative ν_{eff} mean that the particles gain momentum and kinetic energy, i.e. particle acceleration. For particles, coherently accelerated and decelerated in a periodic wave field, the net effect could also be zero. Hence, one has to find out whether, in a given situation, the correlated fluctuations of density and electric field (and/or of the currents and magnetic fields in electromagnetic waves) reveal on average positive ν_{eff}, i.e. whether really irreversible wave-particle interaction processes take place. In this case ν_{eff} can be associated with an "anomalous", or better, effective resistivity:

$$\eta_{\text{eff}} = \frac{1}{\epsilon_o \omega_{pe}^2} \nu_{\text{eff}} = \frac{m_e}{n e^2} \langle \delta E_y \delta \rho_j + [\delta \mathbf{j}_j \times \delta \mathbf{B}]_y \rangle \tag{4}$$

It remains to determine the level of positive correlation between electromagnetic field, density and current fluctuations for an appropriate plasma wave turbulence. Obviously, the resulting effective collision rate and collisionless resistivity will depend on the energy of the waves. For sufficiently low wave energies the weak turbulence theory reveals the general expression

$$\nu_{\text{eff}} \approx \frac{\epsilon_o \delta E^2}{2 n m_e v_d} \left[\frac{\partial \omega \epsilon(\omega)}{\partial \omega} \right]_{\max} \left[\frac{k(\omega) \gamma[\omega]}{\omega} \right]_{\max} \approx \frac{\gamma_{\max}}{v_{\text{ph,max}}} \frac{\epsilon_o \delta E^2}{2 m_e n v_d} \tag{5}$$

where $\gamma_{\max}$ and $v_{\mathrm{ph,max}}$ are the linear growth rate and phase velocities taken for the fastest growing unstable wave modes $k_{\max}(\omega)$ (see, e.g., Treumann, 2001). In magnetospheres the wave energy δE^2 can be measured *in situ* while the wave dispersion has to be determined theoretically. For low excitation levels wave amplitudes can be estimated using the quasi-linear (QL) approach. A QL theory uses the linear dispersion obtained for the mean-field particle distribution $\langle f_j \rangle$ to estimate the saturated wave energy. A QL approach neglects the feedback of waves to the particles (see, e.g. Galeev and Sagdeev, 1984). Analyzing the power of waves observed in the Earth's magnetosphere, Coroniti, 1985, came to the conclusion that for the observed wave energies, the resulting anomalous resistivity would be too small to provide the dissipation necessary for fast reconnection. Currently the Cluster spacecraft could verify these observations. Also, advanced numerical simulation techniques became available to go beyond the limits set by the quasi-linear theory. In the following we address the two main plasma instabilities typical for current sheets, the sites of reconnection: the gradient driven lower-hybrid-drift instability (see Section 2.1) and the current driven ion-acoustic instability (see Section 2.2).

2.1. Lower-Hybrid-Drift Waves

Lower hybrid drift (LHD) waves are excited by strong plasma pressure gradients, including those of thin current sheets important for reconnection (Büchner and Daughton, 2006). The frequency of the fastest growing LHD waves is about the lower hybrid frequency Ω_{LH}. For typical space plasmas, where $\omega_{pe} \gg \Omega_{ce}$ ($\Omega_e = eB/m_e$), the lower hybrid frequency is approximately $\Omega_{LH} \approx \sqrt{\Omega_i \Omega_e} = \Omega_e \sqrt{m_e/M_i}$. Effective particle scattering in LHD waves takes place as electric field fluctuations δE perpendicular to the magnetic field create fluctuating $\delta E \times B$-drifts. These fluctuating drifts transport current carriers across the magnetic field. The corresponding diffusion coefficient is roughly $D \approx (\delta E/B)^2\, \tau_{\mathrm{corr}}$, where the correlation time τ_{corr} is approximately the growth time of the waves. For strongly driven LHD waves this is about the inverse lower hybrid frequency Ω_{LH}^{-1}. The diffusion coefficient (D) reveals an effective collision frequency $\nu_{\mathrm{eff}} = D/\rho_e^2$, where $\rho_e^2 = v_{te}/\Omega_e$ is the electron gyro-radius. Hence

$$\nu_{\mathrm{eff}} \approx \frac{\omega_{pe}^2}{\Omega_{LH}} \quad \frac{\epsilon_o \delta E^2}{2n\kappa T_e} \tag{6}$$

A quasi-linear value of LHD wave saturation and, from that an effective collision frequency was obtained by Davidson and Gladd (1975):

$$\nu_{\mathrm{eff}}|_{LHD1} = \sqrt{\frac{\pi}{8}\frac{M_i}{m_e}\frac{T_i}{T_i+T_e}}\,\Omega_{LH}\frac{\epsilon_o \delta E^2}{2n\kappa T_e} \approx \frac{1}{2}\frac{M_i}{m_e}\frac{T_e}{T_i+T_e}\left(\frac{v_d}{v_{te}}\right)^2 \Omega_{LH} \tag{7}$$

Unfortunately, the quasi-linear estimate of Davidson and Gladd (1975) – second expression (7) – has turned to be way too optimistic. From *in situ* wave observations

Coroniti (1985), concluded that the wave power is three orders of magnitude below the level necessary for reconnection in the Earth's magnetotail. More recently Bale *et al.* (2002) re-analyzed Polar spacecraft data obtained at the Earth's magnetopause. With an effective collision rate calculated according to the first expression (7) they obtained a negligible contribution of the effective resistivity, 200 times smaller than the measured ratio of parallel electric field to the parallel current.

Currently Cluster allowed a direct determination of the fluctuations of density and of two components of the electric field fluctuations by using a special mode of the Swedish electric field and wave instrument EFW. This allows an estimate of the electrostatic component of the effective collision rate according to expression (3), i.e. without using the weak turbulence assumption necessary for obtaining expression (5). For example, in an electron-scale thin current layer discovered by André *et al.* (2004) at the Earth's magnetopause, the electric field was $E \approx 20$ mV/m, the plasma density $n_e = 20\ \mathrm{cm}^{-3}$, $\langle \delta N_e \delta E \rangle \approx 10^{-8}\ \mathrm{V/m}^4$. Hence, the electrostatic part of ν_{eff} was $\nu_{\mathrm{es,eff}} \approx 70$ Hz (cf., e.g., Silin *et al.*, 2005). Notice that the local lower hybrid frequency was $f_{LH} = \Omega_{LH}/2\pi \approx 60$ Hz. The corresponding value of anomalous resistivity is $\eta_{\mathrm{eff}} \approx 1500\ \Omega\mathrm{m}$. Anomalous resistivity of this magnitude can account for a significant fraction of the measured DC electric field.

In order to estimate the electromagnetic part of anomalous resistivity ν_{em} one has to correlate the current δj and magnetic field fluctuations δB. Although the STAFF (Spatio-Temporal Analysis of Field Fluctuations) experiment onboard Cluster directly measures magnetic field fluctuations δB up to high frequencies, Cluster cannot measure the electric current fluctuations $\delta \mathbf{j}$ with the necessary temporal and spatial resolution. Instead, one can estimate the order of magnitude of the relevant components of current fluctuations $\delta \mathbf{j}$ putting an upper limit on the possible $\nu_{\mathrm{em,eff}}$. Assuming that the fluctuations in magnetic field propagate with the typical speed of 200 km/s in the frequency range $20 \rightarrow 100$ Hz, a typical amplitude $\delta B = 1$ nT results in $\delta j \approx 1\ \mathrm{mA/m}^2$. Such values reveal an average of $\langle [\delta \mathbf{j} \times \delta \mathbf{B}]_y \rangle$ of the order of $5 \times 10^{-9}\ \mathrm{V/m}^4$ and $\nu_{\mathrm{em,eff}} \approx 70$ Hz. Although one cannot test whether the direction of magnetic portion of the effective collision rate is consistent with current flow, the order-of-magnitude estimate allows the conclusion that $\nu_{\mathrm{em,eff}}$ can be of the same order as $\nu_{\mathrm{es,eff}}$ (Panov *et al.*, 2006).

These Cluster results lead us beyond the predictions of the quasi-linear LHD wave theory. To understand the reason for the strong collisionless scattering kinetic plasma simulations have to be carried out. Most appropriate for this sake are Vlasov-code simulations which are practically noiseless and allow a fine resolution of the effects of wave-particle resonances. With the availability of modern parallel computer architectures attempts were started to investigate the nonlinear properties of the LHD instability by multidimensional Vlasov-codes, at least for reduced mass ratios M_i/m_e (Wiegelmann and Büchner, 2001). Using a 2D3V version of this code we calculated the collisionless resistivity by a lower hybrid drift instability in thin current sheets (Silin *et al.*, 2005). We found that the resonant wave-particle interaction creates pronounced deviations from the initial drifting particle distribution functions, which can no longer be properly described by the quasilinear theory. Using the results of our Vlasov-code simulations, we estimated the effective collision rate and anomalous resistivity for both particle species due to electrostatic and electromagnetic LHD wave oscillations. The effective collision frequency due to electron scattering was always approximately two orders of magnitude larger than that due to ions, at least for the reduced particle mass ratios M_i/m_e used in the simulations. The absolute values of the effective collision rate are of the order of lower-hybrid frequency $\nu_{\text{eff}} \sim F_{LH} = \Omega_{LH}/2\pi$. For higher particle mass ratios ν_{eff}/F_{LH} slightly increases. The electromagnetic contribution to the electron current dissipation $\nu_{\text{em},e}$ is comparable to the electrostatic counterpart $\nu_{\text{es},e}$ and is usually enhanced closer to the current sheet center (Büchner and Daughton, 2006). Thus, the electromagnetic term is important for the correct estimate of effective collision rate especially at the current sheet center, where resistivity is most important for reconnection. Unfortunately, multidimensional Vlasov code simulations are still at the limits of modern computer facilities. Their extension to higher, more realistic mass ratios might modify the results obtained so far for $M_i/m_e \leq 100$.

2.2. Ion-Acoustic Waves

Theories of the ion-acoustic instability usually assume $T_e \gg T_i$ where the linear theory predicts wave growth starting after the current carrying drift velocity $v_{d,e} j/en_e$ exceeds the ion sound speed $c_s = \sqrt{\kappa T_e/M_i}$. In this limit the quasi-linear theory predicts a saturation of the wave growth revealing the following prediction of an effective collision rate (Galeev and Sagdeev, 1984)

$$\nu_{\text{eff}}|_{IA} = \omega_{pe} \frac{\epsilon_o (\delta E)^2}{2n\kappa T_e} \approx 0.01 \omega_{pi} \frac{v_d}{c_s} \frac{T_e}{T_i} \tag{8}$$

where $\omega_{pi} = \sqrt{ne^2/\epsilon_o M_i}$ is the ion-plasma frequency. For the parameters characterizing the solar atmosphere the quasi-linear prediction (8) would suffice for fast reconnection observed in the solar atmosphere, however (Somov and Titov, 1985). However, for realistic space plasma parameters like, e.g., isothermal conditions instead of the $T_e \gg T_i$, assumed in the quasi-linear approach, the threshold of the ion-acoustic instability would be much higher (Gary, 1993) and the spectrum narrower (Büchner and Elkina, 2006a), so that the non-linear properties of the turbulence as well as its resulting collisionless dissipation have to investigated by kinetic computer simulations.

Due to the sensitivity of the Landau damping to fine resonance effects in the phase space, the usually used kinetic particle-in-cell (PIC) simulation approach (Dum, 1978) did not help much to solve the problem. Hence efficient methods of solving the Vlasov-equation (1) directly had to be developed. Recently, a sensational simulation result was published by Watt *et al.* (2002) who utilized a 1D1V (one spatial, one velocity space dimension) Vlasov-Amperé code developed by Horne and Freeman (2001). Indeed, for an electrostatic perturbation the IA instability in strongly magnetized plasmas can be considered in one spatial (x) and one velocity space dimension ($v_x \rightarrow v$). In this case the IA instability causes just electrostatic oscillations in the x-direction ($E_x \rightarrow \langle E \rangle + E$). In this case the Vlasov-Maxwell system of field equations can be reduced to a set of two one-dimensional Vlasov equations and a one-dimensional Ampère equation for the perturbed current ($j_x \rightarrow \langle j \rangle + \delta j$), from which the average current $j_{\rm ave} = \langle j \rangle$ is deduced. The average current balances the average magnetic field ($\nabla \times \langle \mathbf{B} \rangle = \langle j \rangle = j_{\rm ave}$) such that $\partial / \partial t \langle E \rangle = 0$ (Horne and Freeman, 2001) and $\partial E / \partial t = -\epsilon_0^{-1}(J + J_{\rm ave})$.

Watt *et al.* (2002) simulated the ion-acoustic collisionless resistivity problem for a typical isothermal $T_e \approx T_i$ space plasma, although for artificially low mass ratios only. Their simulations confirmed the theoretical prediction that similar ion and electron temperatures require higher electron drift velocities $u_{de} > v_{te} \gg c_s$. But most interestingly, for the effective collision rate they obtained a three to five orders of magnitude (!) larger value than the quasi-linear estimate (8). Notice that these calculation were restricted to artificially low mass ratios of $M_i/m_e = 25$. If they were confirmed for realistic mass ratios they would solve the collisionless resistivity problem in astrophysical reconnection for good. Hellinger *et al.* (2004) carried out 1D1V Vlasov-code simulations with a realistic mass ratio $M_i/m_e = 1800$, but again with periodic boundary conditions so a Fourier-code could be used. They could not verify at all the results of Watt *et al.* (2002) for the realistic mass ratio, finding just a small enhancement of the collisionless resistivity above the quasi-linear prediction. In order to clarify this point we developed a conservative Vlasov code (Elkina and Büchner, 2006) and revisited the collisionless resistivity caused by an ion acoustic instability (Büchner and Elkina, 2006b). In contrast to the above mentioned previous approaches we applied an external voltage to the simulation box and waited until waves and turbulence spontaneously arose. The initial acceleration of electrons (and, weaker, ions) in the external electric field continued until their relative drift reaches a critical speed $v_{\rm crit}$. Then the plasma became unstable and ion-acoustic waves were generated.

The left panel of Figure 1 depicts the evolution of the spatially averaged electron distribution function obtained for a mass ratio $M_i/m_e = 100$. One can see that after about $t\omega_{pe} = 400$, when the averaged electron velocity (drift) has reached $v_{\rm crit} = 2.3 v_{te}$ until $t\omega_{pe} = 500$, the growing waves scatter the electrons to lower velocities which reduces the drift of the current carrying electrons. After that the strongly nonlinear electron scattering drastically changes their distribution function away from the Maxwellian, assumed in the quasilinear theory.

The right panel of Figure 1 depicts the simulated effective collision frequency ν and the weak turbulence prediction (8) for the simulated electric field fluctuations for a run with $v_{di} = 2.4\, v_{te}$ and an applied external electric field as in the first example, but for the real mass ratio $M_i/m_e = 1836$ and $T_i = 4$ eV and $T_e = 10$ eV, carried out in a box with $L_x = c/\omega_{pe} \approx 460\lambda_D$, $v_e^{\rm max} = 12 v_{te}$, $v_i^{\rm max} = 12 v_{ti}$ and with a numerical grid resolution of $N_x \times N_v = 256 \times 256$, stretched near the resonance

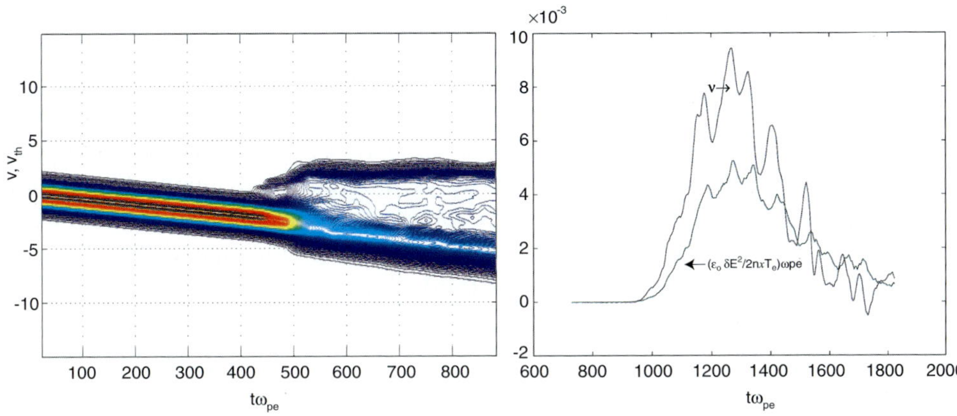

Figure 1. Left panel: Evolution of the spatially averaged electron distribution function (mass ratio 100). Right panel: Simulated effective collision frequency ν and its quasi-linear estimate according to Equation (8), mass ratio 1836, from Büchner and Elkina (2006a).

velocity (Elkina and Büchner, 2006). As one can see, the obtained level of anomalous resistivity does not exceed much the predictions of the weak turbulence theory. Despite of this, Cluster found much larger waves amplitudes, up to 50 mV/m, e.g., in the Earth's magnetotail between the dense plasma sheet and the lobe (Cattell *et al.*, 2005). As similar fluctuation field strengths were found in the auroral region, in the high-latitude cusp (Cattell *et al.*, 1999) and in the magnetopause current layer (Cattell *et al.*, 2002), they were interpreted as signatures of electron holes (see also Cattell *et al.*, 1999; Matsumoto *et al.*, 1999; Franz *et al.*, 1999). Electron holes are regions of electron density depletion with a bipolar variation of the electric field. A low mass-ratio two- and three-dimensional PIC-code simulation of reconnection revealed electron holes (Drake *et al.*, 2003) supporting the idea that strong nonlinear effects may enhance the resistivity by the formation of phase space holes. Unfortunately, these results were obtained by driving the turbulence after intense secondary electron beams were formed by reconnection, which already took place. Nevertheless, obviously a strong driving of the system can create essentially nonlinear structures leading far beyond the limits of applicability of the quasi-linear theory. Such nonlinear effects and their consequences for scattering and collisionless dissipation in realistic setups still have to be simulated, which will be possible in the future by means of advanced Vlasov-codes (Büchner and Elkina, 2006a).

3. The Structure of Reconnection in Three Dimensions

The second question, left open by Harry Petschek's theory, is about the nature of reconnection in three dimensions (3D). In fact, so far even the geometry of reconnection in three dimensions is not agreed upon. While the topology of the solar wind interaction with planetary magnetic fields is relatively simple, solar observations by SOHO, TRACE and RHESSI indicate that the solar coronal magnetic field is very complex. Unfortunately, while in two dimensions reconnection with only in-plan magnetic field components can uniquely be defined by a plasma flowing through magnetic separatrices, crossing each other in a magnetic X-type null point (Vasyliunas, 1975), the situation becomes more complicated already when adding a constant magnetic field in the direction perpendicular to the plane. In three dimensions, when changes in the third dimensions are allowed, however, the most general definition of reconnection as a process changing the magnetic connectivity (Axford, 1984) does not constrain the configuration at all. Multiple ways to reconnect the magnetic field are possible. For example, one can start with generalized X-points which become in 3D nulls of different types (see, e.g., Priest and Forbes, 2000). For reconnection at 3D nulls magnetostatic (e.g. Priest and Titov, 1996), kinematic (Lau and Finn, 1990) and kinetic approaches (Büchner, 1999) have been developed. In the presence of 3D nulls reconnection can take place through separatrices (surfaces through null points) and separators (lines of intersection of separatrices, see, e.g., Longcope and Cowley, 1996). Unfortunately, solar observations revealed that, despite of the overwhelming complexity of the coronal magnetic field, only a small number of magnetic nulls can be found in the corona (Schrijver and Title, 2002). This is even more true in the case of magnetospheres which are formed by the interaction of planetary magnetic fields with the solar wind (Greene, 1988). A generalized reconnection concept was developed by Hesse and Schindler (1988) who emphasized parallel electric fields as the main property of reconnection in 3D, independent of whether null points exist or not.

Simulating the coronal plasma dynamics for magnetic field configurations derived from photospheric observations we found that both null-point reconnection (Büchner *et al.*, 2004b) as well as well as finite-B reconnection without nulls can occur in the solar corona (Büchner *et al.*, 2004a; Büchner and Nikutowski, 2005). The same is true, perhaps, also for planetary magnetospheres and other magnetic field topologies. Let us demonstrate the two types of reconnection – with and without nulls – by simulations of two different situations in which EUV bright points (BPs) were observed. For the simulations we used our newly developed solar MHD code (see, e.g., Büchner *et al.*, 2005) to which we added the latest results about the value of "anomalous" resistivity for solar conditions as discussed in Section 2. Also, we started with the photospheric magnetic field observed below the BPs. Our code starts with a stratified initial equilibrium of the solar chromospheric and coronal plasma which includes the transition region, where the temperature strongly increases and the plasma density drops. In our model the lower solar atmosphere reproduces itself by an ionization – recombination equilibrium with the neutral gas component via collisions. The initial 3D magnetic field configuration is obtained by an MHD-compatible force-free extrapolation of the observed line-of-sight

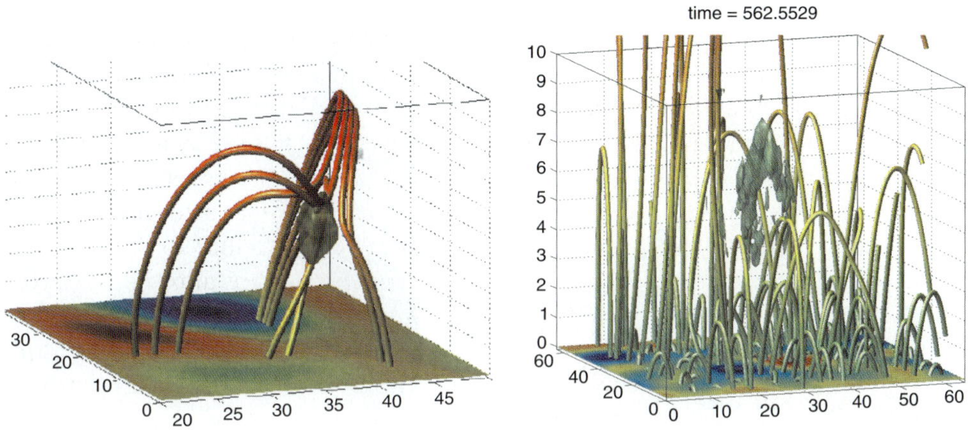

Figure 2. Left panel: Magnetic field lines, indicating a strong change of the magnetic connectivity across a magnetic null, surrounded by a nonideal plasma region of enhanced resistivity from Büchner *et al.* (2004b). Right panel: Site of enhanced parallel electric fields indicating 3D finite-B reconnection without a null from Büchner *et al.* (2004a).

component of the photospheric magnetic field (for details, see Otto *et al.*, 2006). The configuration is energized by the footpoint motion of the photospheric magnetic fluxes. In the simulation the observed footpoint motion was introduced as a boundary condition. The magnetic field around an EUV BP, observed by SOHO EIT (Madjarska *et al.*, 2002), contained a region of very weak magnetic fields, a magnetic null (Büchner *et al.*, 2004b). Figure 2, left panel, depicts the resulting magnetic field configuration around the BP location. It shows that the magnetic connectivity strongly changes through the null in the transition region. There a region of nonideal, resistive plasma is formed which can be seen as an isosurface of constant resistivity built up around the null. For comparison the right panel of Figure 2 depicts a situation of finite-B-reconnection. It was obtained by the simulation of another EUV-BP, observed by TRACE (Brown *et al.*, 2001). Again, starting with the information contained in the corresponding SOHO-MDI magnetogram, a rotation of one of the main polarities below the BP, we obtained a region of finite resistivity, through which reconnection occurs. The reconnection region is depicted by the parallel electric field, building up after the velocity of the current carriers has exceeded a critical speed (Büchner *et al.*, 2004a). Notice that there is no magnetic null at the site of parallel electric field and reconnection.

But how is finite-B-reconnection related to the geometry and topology of the three-dimensional magnetic field? Starting from the general definition of reconnection relating it to a connectivity change of the magnetic field (Axford, 1984) one should look for locations of as strong as possible changes of the magnetic connectivity for only small shifts of the footpoints of magnetic fluxes in the photosphere. But how strong does the connectivity change has to be? How can one quantify the

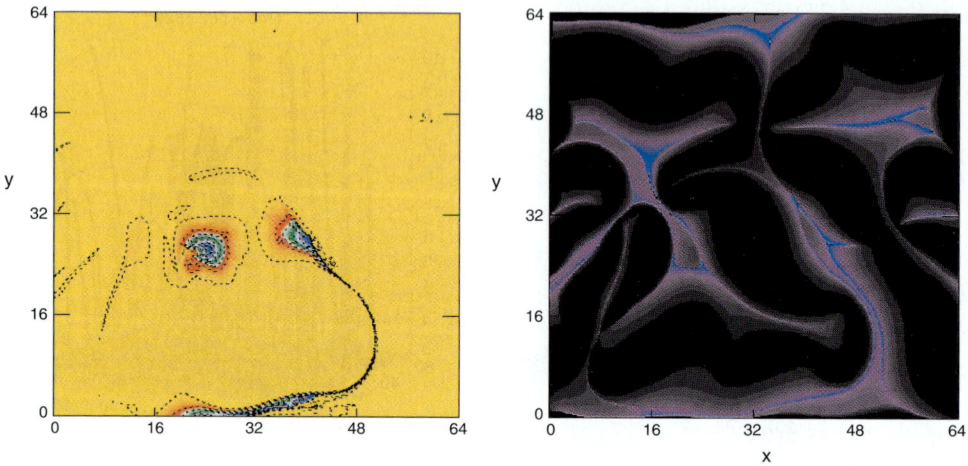

Figure 3. Left panel: Potential drop along the magnetic field lines, calculated for of the parallel electric field, mapped down to the photosphere. Right panel: Differential flux tube volume at $t = 0$ in the same area from Otto *et al.* (2006).

strength of connectivity change if there is no jump in quality as through magnetic null regions and separatrices? One suggestion is to generalize the separatrix concept to regions of strong connectivity change through so called quasi-separatrix layers (QSL) (Demoulin *et al.*, 1996). A quantitative measure enabling to find quasi-separatrix layers is the "squashing factor" Q, introduced by Titov *et al.* (2002). The squashing factor Q becomes maximum for so called hyperbolic flux tubes (HFT) (Titov *et al.*, 2003). Another way to quantify the strength of connectivity change might be the differential flux tube volume

$$DFTV_B = \int_B \frac{ds}{|\mathbf{B}|} \tag{9}$$

In order to demonstrate the special geometry of the magnetic field configuration, in which we found finite-B reconnection in the solar corona, we integrated the parallel (3D-reconnection) electric fields for the case, shown in the right panel of Figure 2, along the magnetic field lines. The left panel of Figure 3 shows the parallel potentials related to the photospheric footpoints of the magnetic field lines. For comparison one can see in the right panel of Figure 3 the $DFTV_B$ for the same photospheric area as shown in the left panel of the Figure. As one can see, enhanced differential flux-tube volumes $DFTV_B$ are indicative for the sites of reconnection. For a more detailed discussion of the relevance of V_B we refer to Büchner (2005). Anyway, whichever criterion one uses, it seems that the nature of three-dimensional reconnection is not sufficiently well understood, yet, in order to proceed to the development of a general theory of the onset of 3D reconnection, the rate of 3D reconnection and particle acceleration in the reconnection electric field.

4. Summary

At least two questions were left open after Harry Petschek's 1964 reconnection theory, the microscopic origin of collisionless dissipation and the nature of reconnection in three dimensions. For both problems Cluster and SOHO provided valuable and complementary information: Cluster about the microphysics of dissipation and SOHO about the geometry of reconnection. It appeared, however that these observations and existing theories alone were not able to solve the two issues which are naturally non-linear and complex, numerical simulation methods have to be applied.

In particular, collisionless dissipation and resistivity related Cluster observations of electric field and plasma fluctuations indicate that in thin current sheets gradient-driven LHD waves might cause more resistivity than predicted by the weak turbulence theory. Other observations indicate the possibility of a collapse of strong turbulence into localized hole structures in which the electron density is depleted. The search for the reason of enhanced LHD resistivity and of the birth and death of electron holes as well as the associated intense electric fields led beyond the quasi-linear approach and needs kinetic computer simulation.

Newly developed Vlasov-codes have been applied to better understand the dissipation necessary for reconnection in space, in the solar corona and in magnetospheres. Concerning multi-dimensional gradient driven and current instabilities so far only artificially low mass-ratio Vlasov code simulations are available, typical mass ratios are $M_i/m_e \approx 100$. Theses simulations indicate that in electron-scale thin current sheets a strongly nonlinear particle scattering in LHD waves might enhance the collisionless dissipation. As far as current driven instabilities are concerned, one-dimensional realistic mass-ratio Vlasov code simulations have shown that even strongly driven $T_i \approx T_e$ space plasmas provide only small amounts of collisionless resistivity, much less than the one, predicted by the quasi-linear theory, as long as periodic boundary conditions are used to close the system. In the more realistic case of open boundaries, allowing a permanent energy supply by external current drivers, phase space holes might enhance the resistivity to the necessary value as preliminary low mass-ratio PIC-code calculations indicate (Drake *et al.*, 2003) and realistic mass ratio simulations confirmed (Büchner and Elkina, 2006a).

Concerning the second question, left after Harry Petschek's first theory of fast magnetic reconnection, the nature of reconnection in 3D, still fundamental questions have to be solved before proceeding. Among the open questions is the role of magnetic nulls in 3D and the mechanism that ignites reconnection, i.e. which lets a magnetic configurations suddenly disrupt. Only after solving these questions one can find out whether there is a generic reconnection rate in 3D similar to the Petschek rate in 2D that can simply be related to the basic macroscopic features of the system and if yes, what is it? Another, follow-up and still unsolved question is about the ways particles are accelerated by reconnection.

In all the envisioned research directions a further development and application of numerical simulation methods is mandatory, though of different kind, kinetic in combination with fluid approaches are needed. Together with observations and non-linear theories such numerical simulation will help to solve the still open questions in magnetic reconnection. Since reconnection is a major commonality of solar and magnetospheric plasmas, future remote sensing of the sun by SDO, STEREO and SOLAR-B as well as further *in situ* observations of current sheets and their consequences for reconnection in magnetospheres by the MMS mission will have to be addressed.

Acknowledgements

I dedicate this paper to the late Harry Petschek who untimely passed away in March 2005. I was lucky enough to have a chance of joyful and enlightening discussions with Harry which in particular encouraged me to work in this field. I also thank N. Elkina, D. Longcope, B. Nikutowski, A. Otto, E. Panov, E. Priest, I. Silin, V. Titov and A. Vaivads with whom I enjoyed discussions and collaboration on different topics discussed in this paper. I further thank Steve J. Schwartz and an anonymous referee for carefully reading the manuscript and for their useful comments which helped to improve the paper.

References

André, M., Vaivads, A., Buchert, S. C., Fazakerley, A. N., and Lahiff, A.: 2004, *Geophys. Res. Lett.* **31**, L03803.

Axford, W. I.: 1984, in E. W. H. Jr. (ed.), *Reconnection in Space and Laboratory Plasma*, Vol. 30 of *Geophysical Monograph Series*, AGU, Washington D. C., p. 4.

Bale, S. D., Mozer, F. S., and Phan, T.: 2002, *Geophys. Res. Lett.* **29**, 2180.

Biskamp, D.: 1986, *Phys. Fluids* **29**, 1520.

Brown, D. S., Parnell, C. E., Deluca, E., Golub, L., and McMullen, R. A.: 2001, *Solar Phys.* **201**, 305.

Büchner, J.: 1999, *Astrophys. Space Sci.* **264**(1–4), 25.

Büchner, J.: 2005, in D. Danesy, P. K. U., A. D. Groof, and J. Andries (eds.), *Proc. of the 11th European Solar Physics Meeting The Dynamic Sun: Challenges for Theory and Observations, 11–16 September 2005, Leuven, Belgium*. 2200 AG Noordwijk, The Netherlands.

Büchner, J., and Daughton, W.: 2006, in E. Priest and J. Birn (eds.), *Magnetic Reconnection*, Cambridge, UK: Cambridge University Press, pp. 145.

Büchner, J., and Elkina, N.: 2006a, *Phys. Plasmas* **13**, 082304, doi: 10.1063/1.2209611, 2006.

Büchner, J., and Elkina, N.: 2006b, *Space Sci. Rev.* **121**(1–4), 237, doi: 10.1007/s11214-006-6542-6, 2006.

Büchner, J., and Nikutowski, B.: 2005, in D. Innes and D. Danesy (eds.), *Chromospheric and Coronal Magnetic Fields, Proc. CCMAG Lindau*, Vol. ESA-SP 596, Noordwijk: ESA, pp. 1.

Büchner, J., Nikutowski, B., and Otto, A.: 2004a, in J. Ireland and R. W. Walsh (eds.), *Proc. of the SOHO15 Coronal Heating Workshop, 6–9 September 2004*. Noordwijk, pp. 23.

Büchner, J., Nikutowski, B., and Otto, A.: 2004b, in A. V. Stepanov, E. E. Benevolenskaya, and A. G. Kosovichev (eds.), *Multi-Wavelength Investigations of Solar Activity*, Vol. 223 of *Proceedings IAU Symposium*. Cambridge, UK, pp. 353.

Büchner, J., Nikutowski, B., and Otto, A.: 2005, in D. Gallagher, J. Horwitz, J. Perez, R. Preece, and J. Quenby (eds.), *Particle Acceleration in Astrophysical Plasmas: Geospace and Beyond, AGU monograph*, Vol. 156, Washington DC.: AGU, pp. 1.

Cattell, C., Crumley, J., Dombeck, J., Wygant, J., and Moser, F. S.: 2002, *Geophys. Res. Lett.* **29**(5), 1065.

Cattell, C., Dombeck, J., Wygant, J., Drake, J. F., Swisdak, M., Goldstein, M. L., *et al.*: 2005, *J. Geophys. Res.* **110**, A01211.

Cattell, C., Dombeck, J., Wygant, J. R., and Hudson, M. K.: 1999, *Geophys. Res. Lett.* **26**, 425.

Coroniti, F. V.: 1985, *Space Sci. Rev.* **42**, 399.

Davidson, R. C., and Gladd, N. T.: 1975, *Phys. Fluids* **18**(10), 1327.

Demoulin, P., Henoux, J. C., Priest, E. R., and Mandrini, C. H.: 1996, *Astron. Astroph.* **308**, 643.

Drake, J. F., Swisdak, M., Catell, C., Shay, M. A., Rogers, B. N., and Zeiler, A.: 2003, *Science* **299**(5608), 873.

Dum, C.: 1978, *Phys. Fluids* **21**, 945.

Dungey, J. W.: 1953, *Phil. Mag.* **44**, 725.

Dupree, T. H.: 1970, *Phys. Rev. Lett.* **25**(12), 789.

Elkina, N. V., and Büchner, J.: 2006, *J. Comp. Phys.* **213**, 862.

Franz, J., Kintner, P., and Pickett, J.: 1999, *Geophys. Res. Lett.* **25**(8), 1277.

Galeev, A. A., and Sagdeev, R. Z.: 1984, *Handbook of Plasma Physics*, Vol. II, Amsterdam: North Holland Physics Publishing.

Gary, S.: 1993, *Theory of Space Plasma Microinstabilities*, Cambridge, UK: Cambridge University Press.

Giovanelli, R.: 1946, *Nature* **158**, 81.

Greene, J. M.: 1988, *J. Geophys. Res.* **93**, 8583.

Hellinger, P., Trávnícek, P., and Menietti, J.: 2004, *Geophys. Res. Lett.* **31**, L10806.

Hesse, M., and Schindler, K.: 1988, *J. Geophys. Res.* **693**, 5559.

Horne, R., and Freeman, M.: 2001, *J. Comp. Phys.* **171**, 182.

Lau, Y.-T., and Finn, J.: 1990, *ApJ* **350**, 672.

Longcope, D. W., and Cowley, S. C.: 1996, *Phys. Plasmas* **3**(8), 2885.

Madjarska, M. S., Doyle, J. G., Teriaca,, L., and Banerjee, D.: 2002, *Astron. Astrophys.* **398**, 775.

Matsumoto, H., Frank, L. A., Omura, Y., and Kojima, H.: 1999, *Geophys. Res. Lett.* **126**, 421.

Otto, A., Büchner, J., and Nikutowski, B.: 2006, *Astron. Astrophys.*, in press.

Panov, E., Büchner, J., Fränz, M., Korth, A., Khotyaintsev, Y., Nikutowski, B., *et al.*: 2006, *Adv. Space Res.*, in press.

Parker, E. N.: 1957, *J. Geophys. Res.* **62**, 509.

Petschek, H.: 1964, in W. Ness (ed.), *AAS/NASA Symposium on the Physics of Solar Flares*, Vol. SP-50, Washington, D.C., p. 425.

Priest, E. R., and Forbes, T. G.: 2000, *Magnetic Reconnection: MHD Theory and Applications*, New York: Cambridge Univ. Press.

Priest, E. R., and Titov, V. S.: 1996, *Phil. Trans. Roy. Soc.* **354**, 2951.

Schrijver, C., and Title, A.: 2002, *Sol. Phys.* **207**.

Silin, I., Büchner, J., and Vaivads, A.: 2005, *Physics of Plasmas* **12**(6).

Somov, B., and Titov, V. S.: 1985, *Solar Phys.* **102**, 79.

Spitzer, L.: 1956, *Physics of Fully Ionized Gases*, New York: Interscience.
Sweet, P. A.: 1958, in B. Lehnert (ed.), *IAU Symp. 6: Electromagnetic Phenomena in Cosmical Physics*. New York, p. 123.
Titov, V. S., Hornig, G., and Demoulin, P.: 2002, *J. Geophys. Res.* **107**, SSH 3, 1.
Titov, V. S., G. K., and Neukirch, T.: 2003, *Astroph. J.* **582**, 1172.
Treumann, R.: 2001, *Earth Planet. Sci.* **53**, 433.
Vasyliunas, V.: 1975, *Rev. Geophys. Space Phys.* **13**, 303.
Vlasov, A.: 1938, *J. Exp. Theoret. Phys.* **8**, 291 (in Russ.).
Watt, C., Horne, R., and Freeman, M.: 2002, *Geophys. Res. Lett.* **29**(1), 1002.
Wiegelmann, T., and Büchner, J.: 2001, *Nonlin. Proc. Geoph.* (8), 127.

THE LOCALIZATION OF PARTICLE ACCELERATION SITES IN SOLAR FLARES AND CMES

MARKUS J. ASCHWANDEN
Solar and Astrophysics Lab., Lockheed Martin ATC, 3251 Hanover St., Palo Alto, CA 94304, USA
(E-mail: aschwanden@lmsal.com)

(Received 21 July 2005; Accepted in final form 7 September 2005)

Abstract. We review the particular aspect of determining particle acceleration sites in solar flares and coronal mass ejections (CMEs). Depending on the magnetic field configuration at the particle acceleration site, distinctly different radiation signatures are produced: (1) If charged particles are accelerated along compact closed magnetic field lines, they precipitate to the solar chromosphere and produce hard X-rays, gamma rays, soft X-rays, and EUV emission; (2) if they are injected into large-scale closed magnetic field structures, they remain temporarily confined (or trapped) and produce gyrosynchrotron emission in radio and bremsstrahlung in soft X-rays; (3) if they are accelerated along open field lines they produce beam-driven plasma emission with a metric starting frequency; and (4) if they are accelerated in a propagating CME shock, they can escape into interplanetary space and produce beam-driven plasma emission with a decametric starting frequency. The latter two groups of accelerated particles can be geo-effective if suitably connected to the solar west side. Particle acceleration sites can often be localized by modeling the magnetic topology from images in different wavelengths and by measuring the particle velocity dispersion from time-of-flight delays.

Keywords: solar flares, coronal mass ejections, particle acceleration

1. Introduction

The knowledge of the location of particle accelerators in the solar corona is important in many respects: (1) to probe the electromagnetic fields and physical conditions in the acceleration region; (2) to predict the (adiabatic) trajectories of accelerated particles; (3) to predict the targets and energy losses of the accelerated particles; (4) to model and diagnose their radiation signatures in many wavelengths; and (5) to understand their origin and chemical composition when detected *in-situ*. In the context of solar flares and coronal mass ejections (CMEs), multi-wavelength images may constrain the magnetic topology at the acceleration site and along the particle trajectories. A powerful tool to constrain the location of particle acceleration sites in solar flares and CMEs is the method of time-of-flight measurements, which can be applied to solar hard X-rays as well as to *in-situ* detections of high-energy particles near 1 AU. We organize this review by considering various magnetic field configurations in the particle acceleration sites, which also can be used as diagnostic for localizing the acceleration sites and particle propagation paths.

Space Science Reviews (2006) 124: 361–372
DOI: 10.1007/s11214-006-9095-9

2. Particle Acceleration in Primary Flare Loops

It is generally believed that solar flares and CMEs are triggered by a magnetic instability in the lower corona, which is often accompanied with the eruption of an unstable filament and involves a magnetic reconnection process (for a summary see, e.g., Aschwanden, 2004, Sections 10–12). The various magnetic topologies can be classified into 2D and 3D configurations, of which the most common ones are the bipolar, tripolar, and quadrupolar geometry (Figure 1). These configurations involve a magnetic topology change between open and closed magnetic field lines, which in all three cases produce a compact post-reconnection field line, which we call the *"primary flare loop"*. Let us consider first the fate of accelerated particles in those primary flare loops, which consist of compact closed magnetic field lines.

The geometric reconstruction of the pre-reconnection field lines can be accomplished for each (new) post-reconnection field line (marked with double linestyle in Figure 1) by switching the magnetic polarities to obtain the alternative (old) connectivity (marked with dashed linestyle in Figure 1). This geometric reconstruction implies that each post-reconnection field line relaxes from a cusp-shaped field line (during reconnection) into a circular post-reconnection field line. The shortening from the cusp-shaped to the circular field line yields the free magnetic energy ΔW (per cross-sectionional area dA) that can be dissipated,

$$\frac{\Delta W}{dA} \approx \int_{\mathrm{cusp}} \frac{B^2(s')}{8\pi} ds' - \int_{\mathrm{relaxed}} \frac{B^2(s)}{8\pi} ds \approx \frac{\langle B \rangle^2}{8\pi}(s' - s) > 0. \tag{1}$$

The cusp area that is outlined during the shrinkage of the field line demarcates the primary acceleration site (marked with a hatched area in the third row in Figure 1). The altitude h_{acc} of the acceleration site in the cusp region has also been confirmed from electron time-of-flight (TOF) measurements between the coronal cusp and chromospheric footpoints, which was found to scale with the loop height $h_{\mathrm{loop}} = L_{\mathrm{halfloop}} \times (2/\pi)$ by

$$\frac{h_{\mathrm{acc}}}{h_{\mathrm{loop}}} \approx \frac{L_{\mathrm{TOF}}}{L_{\mathrm{halfloop}}} = 1.43 \pm 0.30, \tag{2}$$

based on energy-dependent hard X-ray time delays (Aschwanden *et al.*, 1996). The relaxation process implies that the accelerated particles are confined to the relaxing compact (primary) field line and have no escape possibility. Therefore, the accelerated particles have no other choice than to precipitate to the two footpoints of the primary postflare loop, either directly (within a half loop transit time) or after several mirror reflections as a consequence of wave-induced or collision-induced pitch angle scattering. The precipitating electrons are then stopped in the chromosphere and produce collisional bremsstrahlung in hard X-rays. The accelerated

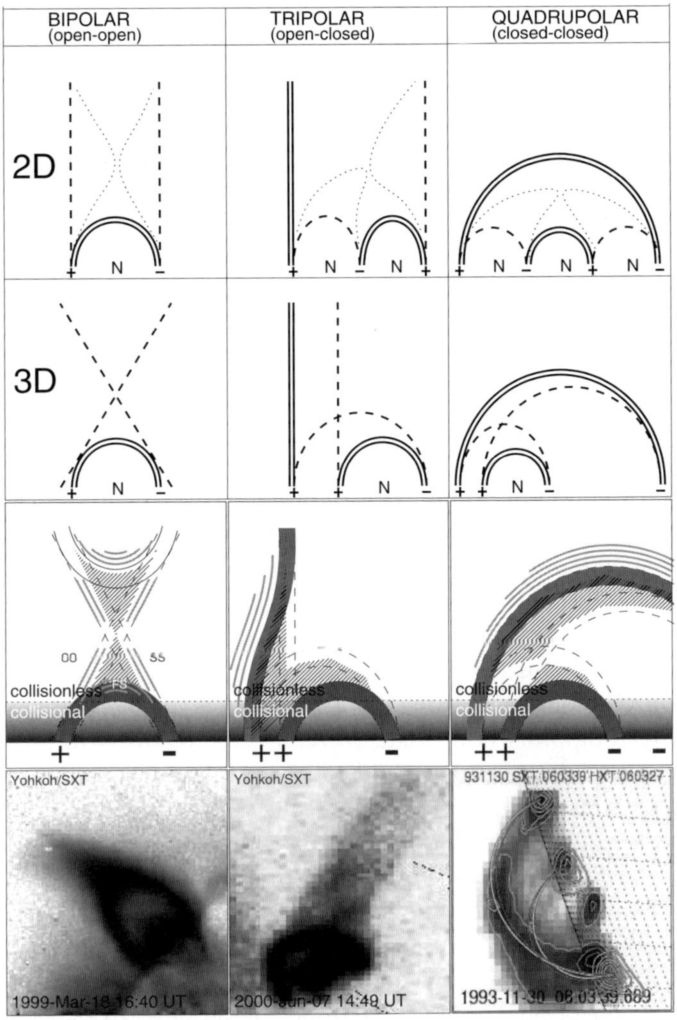

Figure 1. The topology of magnetic reconnection regions is classified into three combinations: bipolar or open-open (left column), tripolar or open-closed (middle column), and quadrupolar or closed-closed field line reconnection (right column). The 2D versions are shown in the top row, with the pre-reconnection field lines marked with dashed lines, during reconnection with dotted lines, and post-reconnection field lines with double solid linestyle. The 3D versions are indicated in the second row, where the pre-reconnection field lines are not coplanar, but located behind each other. The third row indicates the acceleration regions (hatched), the relative densities (greyscale), and upward/sideward directed shocks (grey lines). The bottom row shows flare observations from Yohkoh/SXT that correspond to the three different reconnection topologies (adapted from Aschwanden, 2004).

ions precipitate also but produce nuclear de-excitation lines in gamma-rays in the chromosphere. The relativistic electrons produce also gyrosynchrotron emission in radio wavelengths, which is generally delayed by seconds to minutes depending on the trapping time, which increases with (relativistic) energy.

None of these energized particles accelerated in the relaxing primary flare loop can escape into the upper corona and interplanetary space, but their sharply defined timing can often be used to determine the time at which energized particles are injected into secondary magnetic field lines, concomitantly accelerated in the same reconnection process, possibly injected onto an escape route into interplanetary space. The particles accelerated in the primary flare loop are therefore an independent population and magnetically decoupled from *solar energetic particles (SEPs)* that are detected in the heliosphere or near Earth. So, the two populations of high-energy particles have a different origin and are not expected to have the same energy or spectra.

3. Particle Acceleration in Secondary Flare Loops

The quadrupolar configuration (Figure 1, right-hand side) involves magnetic reconnection between two closed magnetic field lines, which relaxes into an alternative quadrupolar connection with switched polarities. This basic (3D) quadrupolar reconnection process has also been dubbed as a flaring process between *"two interacting loops"* and was physically interpreted as a magnetic flux transfer between two current-carrying loops (Melrose, 1997). Observational evidence for this type of reconnection was obtained from the detailed analysis of magnetic polarities at the flare loop footpoints (Hanaoka, 1996, 1997; Nishio *et al.*, 1997) and the 3D reconstruction of the magnetic field topology (Aschwanden *et al.*, 1999). In most of the cases observed by Hanaoka (1996, 1997) and Nishio *et al.* (1997), the secondary large-scale flare loop showed a brightening in soft X-rays and microwaves (see the four cases in Figure 2), simultaneously with the bright hard X-ray and soft X-ray emission in the primary (compact) flare loop, which strongly suggests a simultaneous injection of accelerated particles into the secondary (large-scale) loop. The injection of energized particles is likely to be accomplished from the common reconnection X-point, where acceleration is likely to occur in the upward-located cusp that is outlined by the relaxation of the secondary flare loop. The secondary flare loop generally is substantially bigger and thus the injected heated plasma and the number of accelerated particles is spread over a larger volume, causing fainter bremsstrahlung emission in soft X-rays and gyrosynchrotron emission in microwaves than from the (compact) primary flare loop. Also, hard X-ray emission has not been detected (or only at a very weak level) at the remote footpoint of the secondary flare loop, probably because the propagating electrons become mirrored or lose gradually their energy on the way to the remote footpoint. However, the relativistic speed of accelerated electrons in these secondary flare loops has been measured from time-of-flight measurements between the injection point and the remote footpoint (e.g., Hanaoka, 1999; Lang and Willson, 1989).

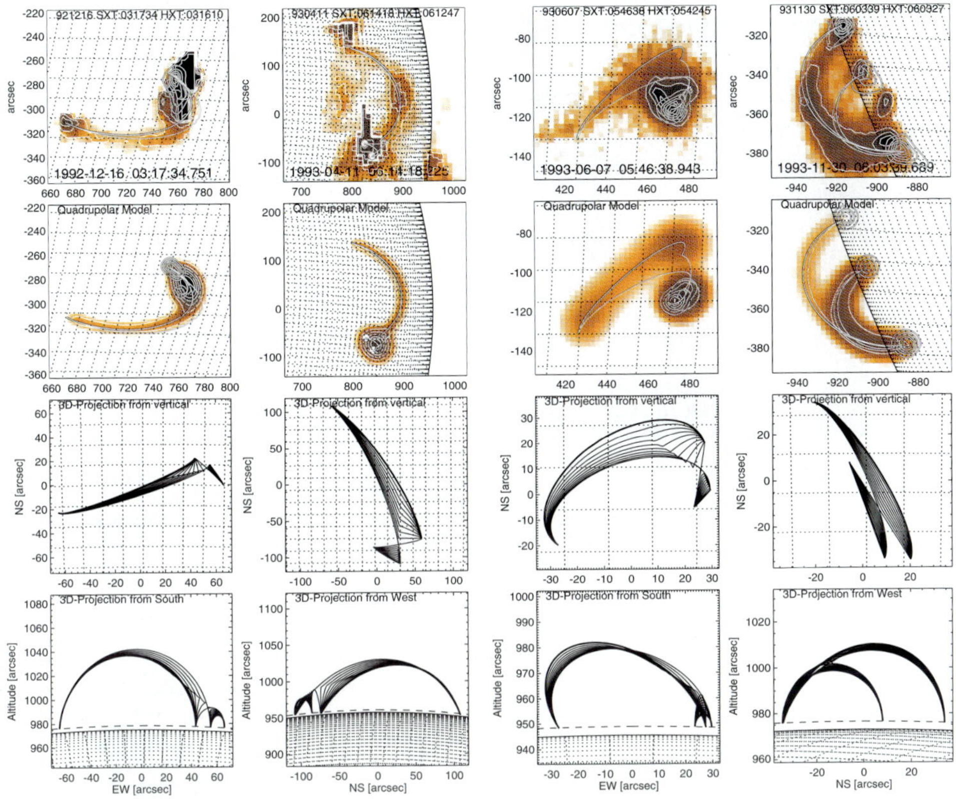

Figure 2. Reconstruction of reconnection geometry with a 3D quadrupolar model: *Top row:* soft X-ray image (logarithmic greyscale and thin contours) from Yohkoh/SXT and hard X-ray image (thick contours) from Yohkoh/HXT. The thin circular segments represent the pre-reconnection field lines, and the thick circular segments show the corresponding post-reconnection field lines, which coincide with the flare loops. *Second row:* simulated SXR and HXR maps constrained by the 3D quadrupolar model (shown in Figure 1), represented by identical greyscales and contour levels just like the original data (in the top row). *Third row:* the geometric solution of the 3D quadrupolar model is rotated so that the vertical z-axis coincides with the line-of-sight. Ten field lines are interpolated between the pre-reconnection and post-reconnection state, visualizing the relaxation process of field lines after reconnection. *Bottom row:* the same 3D model is rotated so that either the x-axis (view from west) or the y-axis (view from south) coincides with the line-of-sight. The spacing of the heliographic grid is 1° in all frames, corresponding to 12,150 km (Aschwanden *et al.*, 1999).

Other examples of particle acceleration in large-scale, closed magnetic field geometries comes from radio type-U bursts, which are likely to represent secondary flare loops as described above. The U-type morphology indicates the propagation of electron beams along closed magnetic field lines. A case of a U-burst that has also been simultaneously imaged with the *Very Large Array (VLA)* is shown in Figure 3. The turnover frequency of the type U-burst in the dynamic spectra corresponds to

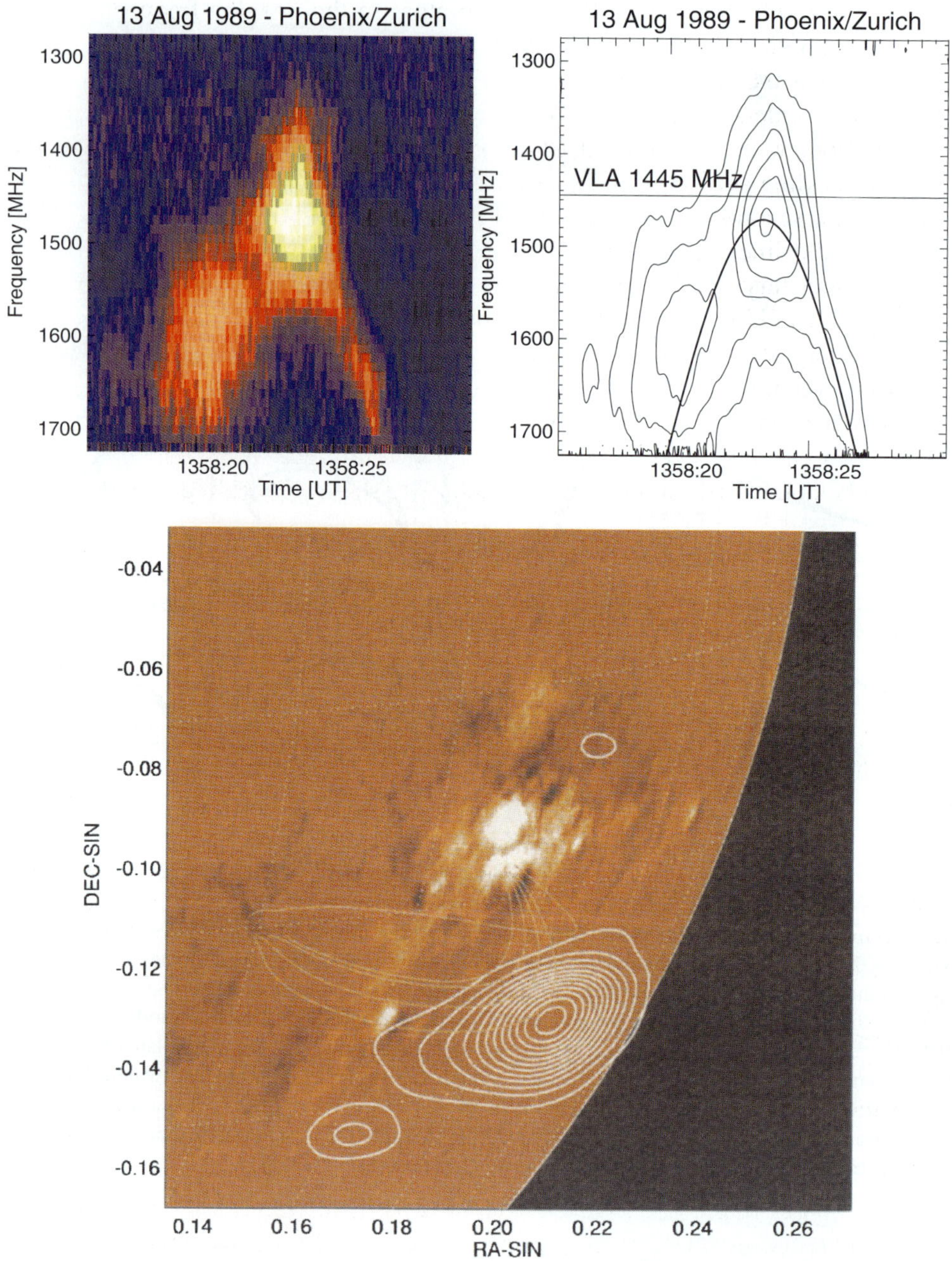

Figure 3. Top: Radio observations of a type U-burst on 1989-Aug-13 with the ETH Zurich spectrometer. The dynamic spectrum (frequency versus time) shows a duration of ≈ 7 s and a turnover frequency of $\nu \approx 1.5$ GHz for the U-burst. *Bottom:* The turnover of the U-burst was simultaneously imaged with the VLA at a frequency of $\nu = 1.445$ GHz. A potential magnetic field extrapolation at the same location based on a KPNO/NSO magnetogram confirms the presence of closed magnetic field lines (thin curves) at the location of the U-burst (Aschwanden *et al.*, 1992).

$\nu \approx 1.5$ GHz, which accidentally was also close to the imaging frequency of the VLA. A magnetic field extrapolation at the location of the U-burst turnover location confirmed that the radio source was confined at the top of closed magnetic field lines (Aschwanden *et al.*, 1992). The U-burst might well be part of a quadrupolar configuration, with the compact primary loop located near the footpoint of the (western) leading sunspot.

Since the secondary loop in a quadrupolar configuration represents a closed magnetic field line, no accelerated particle can escape into the upper corona or into interplanetary space, and thus they are magnetically decoupled from SEP particles detected in the heliosphere or near Earth. Actually, since both the primary and secondary flare loops are closed in quadrupolar configurations, we do not expect any common acceleration site for solar flare particles and SEPs in such quadrupolar events. This might also explain why a number of large flares seem to have fully confined magnetic topologies, which produce strong gamma-rays but completely lack radio emission (e.g., Simnett and Benz, 1986; Rieger *et al.*, 1999).

4. Particle Acceleration in Open Magnetic Fields

Energetic particles accelerated in solar flares can only escape into interplanetary space if some open magnetic fields are involved in the magnetic reconnection process. Both the bipolar and tripolar case (Figure 1, left and middle column) involve open pre-reconnection field lines. In addition, the tripolar case (Figure 1, middle column) transforms again to an open post-reconnection field line. While the three reconnection topologies shown in Figure 1 encompass only basic components between two reconnecting field lines, a real flare may be considered as a temporal sequence of many such basic reconnection processes, which can occur in arbitrary sequences, and thus most flares may involve a time phase where open field lines are involved. For instance, a flare may start with reconnection of highly-sheared low-lying field lines above the neutral line that are all closed (i.e., quadrupolar) initially, while it may later evolve into reconnection of less-sheared high-lying field lines that could involve open field lines at some particular locations (i.e., tripolar). Such a flare would initially confine the accelerated particles, but inject them into interplanetary space in a later phase. A temporary opening of the magnetic field configuration is, for instance, envisioned in the so-called *"magnetic break-out"* flare model (Antiochos *et al.*, 1999).

On the other hand, open field regions that connect directly to the interplanetary space exist not only in coronal holes, but also to a substantial fraction in active regions (Figure 4). Schrijver and DeRosa (2003) found from potential-field extrapolations of the global magnetic field over the entire solar surface that the *interplanetary magnetic field (IMF)* originates typically in a dozen disjoint regions, around the solar cycle maximum. While active regions are often ignored as a source for the interplanetary magnetic field, Schrijver and DeRosa (2003) found that the fraction of the IMF that connects directly to magnetic plages of active regions in-

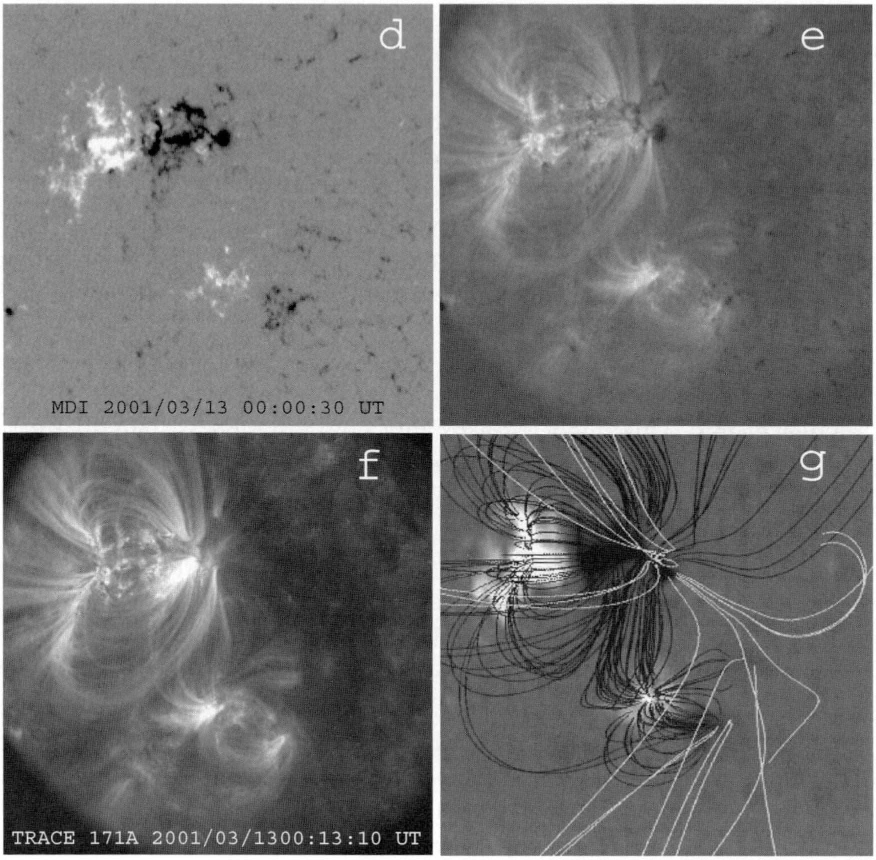

Figure 4. Top left: Magnetogram recorded with the *Michelson Doppler Imager (MDI)* onboard the *Solar and Heliospheric Observatory (SoHO)*; *Top right: MDI* magnetogram overlayed on an EUV 171 Å image from the *Transition Region and Coronal Explorer (TRACE)*; *Bottom left:* TRACE 171 Å image of 2001-Mar-13, 00:13 UT; *Bottom right:* Potential field extrapolation using a source-surface model. Closed field lines of active regions are indicated with black color, the open field lines that connect to interplanetary space with white. The spatial scale of a panel is a half solar radius (350,000 km), (courtesy of Schrijver and DeRosa, 2003).

creases from $\lesssim$10% at cycle minimum up to 30–50% at cycle maximum, with even direct connections between sunspots and the heliosphere (Figure 4). Additional support for the magnetic connectivity comes also also from the establishment of a connection between the interplanetary field and active region fields (Neugebauer *et al.*, 2002; Liewer *et al.*, 2004).

The most common diagnostic of particle acceleration along open magnetic field lines are radio type III bursts, which are detected in the majority of flares. For flare-associated acceleration at coronal heights we expect decimetric starting frequencies (corresponding to electron densities of $n_e \gtrsim 10^9$ cm^{-3}) or metric starting frequencies ($n_e \gtrsim 10^8$ cm^{-3}), which both may have extensions to interplanetary type III

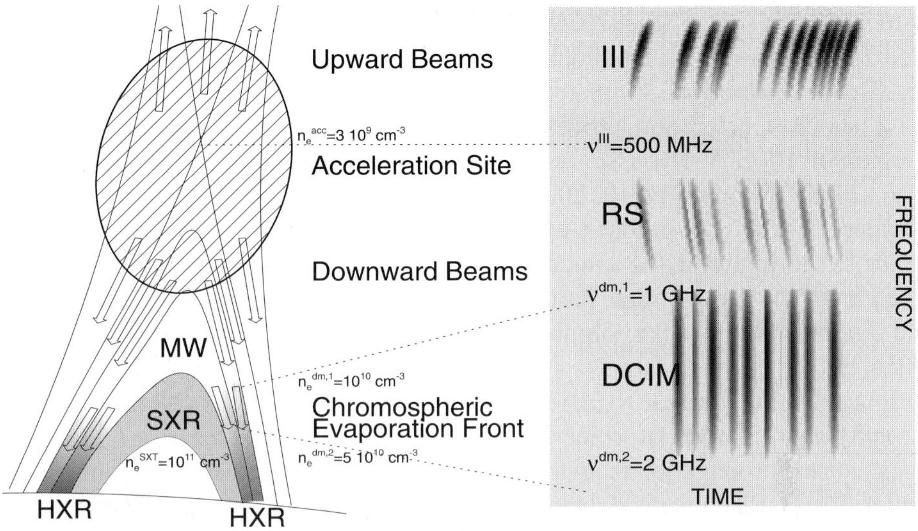

Figure 5. A diagram of a flare model is shown, inferred from the magnetic topology constraint of simultaneously detected upward and downward electron beams, in radio and hard X-rays. The diagram on the right illustrates a dynamic radio spectrum with radio bursts indicated in the frequency-time plane. The acceleration site is located in a low-density region (in the cusp) with a density of $n_e^{\mathrm{acc}} \approx 10^9$ cm^{-3} from where electron beams are accelerated in upward (type III) and downward (RS bursts) directions. Downward-precipitating electron beams produce pulses of chromospheric thick-target bremsstrahlung emission, possibly intercepting chromospheric upflows. Those loops that have already been filled with heated chromospheric plasma brighten up in soft X-rays and have higher densities of $n_e^{\mathrm{SXR}} \approx 10^{11}$ cm^{-3} than the acceleration region. There is a filling delay of soft X-ray loops, during which the magnetic reconnection point rises higher, widening the hard X-ray emitting footpoints (Aschwanden and Benz, 1997).

bursts (but not always). Inversely, however, almost all interplanetary type III bursts are found to be rooted in coronal type III bursts (e.g., Poquérusse *et al.*, 1996). So, the localization of acceleration regions along open magnetic field lines can be constrained from the electron density that corresponds to the (fundamental) plasma frequency of the starting frequency of type III bursts, which is often found around $n_e \approx 10^9$ cm^{-3} (Aschwanden and Benz, 1997), and thus most likely corresponds to the upward-directed cusp of reconnection X-points in coronal flare sites (Figure 5).

5. Particle Acceleration in CMEs

Besides the flare-associated acceleration sites described above, there are also CME-associated acceleration sites, produced in the shocks associated with super-Alfvénic CME fronts that propagate through interplanetary space. CMEs have speeds from

20 to 2500 $\mathrm{km\,s^{-1}}$. The fast CMEs with speeds in excess of the ambient solar wind ($\gtrsim$800 $\mathrm{km\,s^{-1}}$) drive shocks ahead and are capable of accelerating particles. Recent numerical simulations of realistic CME shocks show that the diffusive shock acceleration process can accelerate solar energetic protons up to energies of 10 GeV (Roussev *et al.*, 2004; Manchester *et al.*, 2005; Kocharov *et al.*, 2005; Tsurutani *et al.*, 2003). The acceleration efficiency, however, varies considerably along a circular CME shock wave, depending on the angle between the shock normal and the incident magnetic field. Quasi-perpendicular shocks, which are found at the outer flanks of CMEs, accelerate particles at a higher rate than quasi-parallel shocks, supposedly occurring near the CME front (Giacalone, 2005a, b).

Solar energetic particle (SEP) events detected at Earth are subdivided into *impulsive* and *gradual events*, of which the latter type clearly originates from CME-driven shocks during propagation from the corona through the heliosphere (Reames, 1999; Kahler, 2001). SEP events detected near Earth indicate magnetic connectivity between the CME site and the Earth, a necessary condition for *geo-effective events*. CME shocks accelerate at least two different seed populations, flare suprathermals as well as solar-wind particles. Distinctive differences in the elemental composition suggest that quasi-parallel shocks (in the CME front) generally draw their seeds from solar-wind suprathermals, while quasi-perpendicular shocks (in the CME flanks) preferentially accelerate flare supra-thermals, because they require a higher initial speed for effective injection (Tylka *et al.*, 2005).

The acceleration site of CME-accelerated particles can also be determined with time-of-flight measurements of their velocity dispersion. Recent measurements with the *Wind* spacecraft have shown two classes of accelerated protons: one class that traveled essentially scatter-free over a path length of $\approx$1.1–1.3 AU, while the other shows a path length of $\approx$2.0 AU. The first class seems to be accelerated in the range of $\approx 1-10\ R_{\odot}$, while the second class seems to be accelerated successively later (Krucker *et al.*, 1999; Krucker and Lin, 2000).

Since the plasma in interplanetary space is collisionless, superthermal and high-energy particles can propagate through interplanetary space and form particle beams that are responsible for *interplanetary radio type III bursts*. There is also a particular type of interplanetary type III-like bursts, called *"shock-associated (SA) events"*, believed to be electron beams that are produced by collisionless shocks associated with passing CMEs and propagate in the antisunward direction from the (type II-emitting) shock (Cane *et al.*, 1981). Radio type II emission is the most common signature of shocks propagating in the corona and interplanetary space, and thus is used as a robust diagnostic of propagating CMEs. Enhanced radio emission and SEP production is observed when a fast CME passes a slow CME, which is interpreted as a consequence of shock strengthening (Gopalswamy *et al.*, 2001, 2002). Occasionally, there occur also interplanetary radio type IV-like bursts, i.e., synchrotron emission caused by energetic electrons confined in a magnetic trap

created behind an interplanetary shock wave (Bastian *et al.*, 2001). The spatial size of interplanetary radio bursts can be very large, since the extent of the radio source grows with distance from the Sun. Nevertheless, radio imaging at multiple frequencies aids to reconstruct the CME-associated acceleration sites and particle kinematics.

6. Final Remarks

The acceleration sites of electrons in flares have been identified in the immediat neighborhood of magnetic reconnection sites, based on electron time-of-flight measurements and coronal hard X-ray emission, but we have much less diagnostics on the acceleration sites of ions. The acceleration sites of energetic particles in CMEs are less known for several reasons: (1) we are lacking accurate magnetic field models in CME environments, (2) the method of velocity dispersion applied to distances of 1 AU has uncertainties in the order of solar radii, and (3) *in-situ* detection of particles is mostly carried out at distances of 1 AU. The STEREO mission will constrain particle acceleration sites in CMEs significantly better.

Acknowledgements

Part of the work was supported by NASA contract NAS5-98033 (RHESSI mission) through University of California, Berkeley (subcontract SA2241-26308PG) and the NASA contract NAS5-38099 (TRACE mission). Workshop participation was supported by the *International Space Science Institute (ISSI)*, Bern, Switzerland.

References

Antiochos, S. K., DeVore, C. R., and Klimchuk, J. A.: 1999, *ApJ* **510**, 485.

Aschwanden, M. J., Bastian, T. S., Benz, A. O., and Brosius, J. W.: 1992, *ApJ* **391**, 380.

Aschwanden, M. J., Wills, M. J., Hudson, H. S., Kosugi, T., and Schwartz, R. A.: 1996, *ApJ* **468**, 398.

Aschwanden, M. J., and Benz, A. O.: 1997, *ApJ* **480**, 825.

Aschwanden, M. J., Kosugi, T., Hanaoka, Y., Nishio, M., and Melrose, D. B.: 1999, *ApJ* **526**, 1026.

Aschwanden, M. J.: 2002, *Space Sci. Rev.* **101**(1–2), 1.

Aschwanden, M. J.: 2004, *Physics of the Solar Corona – An Introduction*, Praxis Publishing Ltd., Chichester UK, and Springer, New York, ISBN 3-540-22321-5, 842 p.

Aschwanden, M. J.: 2005, in Gopalswamy, N., Mewaldt, R., and Torsti, J. (eds.), AGU Monograph "Solar Eruptions and Energetic Particles", Vol. 165, p. 189.

Bastian, T. S., Pick, M., Kerdraon, A., Maia, D., and Vourlidas, A.: 2001, *ApJ* **558**, L65.

Cane, H. V., Stone, R. G., Fainberg, J., Stewart, R. T., Steinberg, J. L., and Hoang, S.: 1981, *JGR* **8/12**, 1285.

Giacalone, J.: 2005a, *ApJ* **624**, 765.

Giacalone, J.: 2005b, *ApJ* **628**, L73.
Gopalswamy, N., Yashiro, S., Kaiser, M. L., Howard, R. A., and Bougeret, J.: 2001, *ApJ* **548**, L91.
Gopalswamy, N., Yashiro, S., Michalek, G., Kaiser, M. L., Howard, R. A., Reames, D. V., *et al.*: 2002, *ApJ* **572**, L103.
Hanaoka, Y.: 1996, *Solar Phys.* **165**, 275.
Hanaoka, Y.: 1997, *Solar Phys.* **173**, 319.
Hanaoka, Y.: 1999, *Publ. Astron. Soc. Japan* **51**, 483.
Kahler, S. W.: 2001, *JGR* **106/A10**, 20947.
Kocharov, L., Lytova, M., Vainio, R., Laitinen, T., and Torsti, J.: 2005, *ApJ* **620**, 1052.
Krucker, S., Larson, D. E., Lin, R. P., and Thompson, B. J.: 1999, *ApJ* **519**, 864.
Krucker, S., and Lin, R. P.: 2000, *ApJ* **542**, L61.
Lang, K. R., and Willson, R. F.: 1989, *ApJ* **344**, L77.
Liewer, P. C., Neugebauer, M., and Zurbuchen, T.: 2004, *Solar Phys.* **223**, 209.
Manchester, W. B., Gombosi, T. I., De Zeeuw, D. L., Sokolov, I. V., Roussev, I. I., Powell, K. G., *et al.*: 2005, *ApJ* **622**, 1225.
Melrose, D. B.: 1997, *ApJ* **486**, 521.
Neugebauer, M., Liewer, P. C., Smith, E. J., Skoug, R. M., and Zurbuchen, T. H.: 2002, *JGR* (Space Physics) **107**, Issue A12, pp. SSH 13-1, CiteID 1488, DOI 10.1029/2001JA000306.
Nishio, M., Yaji, K., Kosugi, T., Nakajima, H., and Sakurai, T.: 1997, *ApJ* **489**, 976.
Poquérusse, M., Hoang, S., Bougeret, J. L., and Moncuquet, M.: 1996, in Winterhalter, D., Gosling, J. T., Habbal, S. R., Kurth, W. S., and Neugebauer, M. (eds.), *Solar Wind Eight*, Internat. Solar Wind Conference, held in Dana Point, California, June 1995, AIP Press, New York, American Institute of Physics Conference Proceedings AIP CP-382, pp. 62–65.
Reames, D. V.: 1999, *Space Sci. Rev.* **90**, 413.
Rieger, E., Treumann, R. A., and Karlicky, M.: 1999, *Solar Phys.* **187**, 59.
Roussev, I. I., Sokolov, I. V., Forbes, T. G., Gombosi, T. I., Lee, M. A., and Sakai, J. I.: 2004, *ApJ* **605**, L73.
Schrijver, C. J., and DeRosa, M. L.: 2003, *Solar Phys.* **212**, 165.
Simnett, G. M., and Benz, A. O.: 1986, *Astron. Astrophys.* **165**, 227.
Tsurutani, B. T., Wu, S. T., Zhang, T. X., and Dryer, M.: 2003, *Astron. Astrophys.* **412**, 293.
Tylka, A. J., Cohen, C. M. S., Dietrich, W. F., Lee, M. A., Maclennan, C. G., Mewaldt, R. A., *et al.*: 2005, *ApJ* **625**, 474.